Geomechanical Aspects of Abandoned Room and Pillar Mines and Remediation Measures

Geomechanical Aspects of Abandoned Room and Pillar Mines and Remediation Measures addresses the occurrence, prediction, and control of subsidence issues encountered in room and pillar mining, including creep, degradation, and seismic aspects. The book distinguishes itself by:

- Presenting a unified treatise on room and pillar mines.
- Using clear definitions and explanations of fundamental concepts.
- Considering long-term aspects (creep and degradation), seismic effects, remedial measures, contact types, and remote sensing monitoring techniques.
- Including physical and numerical modeling.
- Demonstrating an in-depth view of procedures and solution techniques.

This book provides a clear understanding of active and/or abandoned room and pillar mines. It is aimed at academics and engineers with a mining and/or civil engineering background, but it also serves as a reference for undergraduate and graduate students.

ISRM Book Series

International Society for Rock Mechanics and Rock Engineering

ISRM

The ISRM Book Series comprises significant scientific publications in rock mechanics and related disciplines. The aim of the Series is to promote the scientific output of ISRM. The Series is published on a regular basis, according to the emergence of noteworthy scientific developments. Each volume in the Series is prepared independently and focuses on a topical theme. Contributions are rigorously peer-reviewed by experts in the field, ensuring strong scientific standards, and are expected to be leading in rock mechanics and related disciplines. The Series may include edited volumes, textbooks and reference works.

Rock Reinforcement and Rock Support
Ömer Aydan

Continuum and Computational Mechanics for Geomechanical Engineers
Ömer Aydan

Storage of LPG in Large Rock Caverns
Kenji Aoki

Geomechanical Aspects of Abandoned Room and Pillar Mines and Remediation Measures
Ömer Aydan

For more information about this series, please visit: www.routledge.com/ISRM-Book-Series/book-series/ISRM

Geomechanical Aspects of Abandoned Room and Pillar Mines and Remediation Measures

Ömer Aydan

CRC Press
Taylor & Francis Group
Boca Raton London New York Leiden

CRC Press is an imprint of the
Taylor & Francis Group, an **informa** business
A BALKEMA BOOK

Designed cover image: An abandoned lignite mine in Mitake, Gifu Prefecture, Japan. Courtesy of Filling Technology Association, Nagoya, Japan.

First published 2025
by CRC Press/Balkema
4 Park Square, Milton Park, Abingdon, Oxon, OX14 4RN

and by CRC Press/Balkema
2385 NW Executive Center Drive, Suite 320, Boca Raton FL 33431

CRC Press/Balkema is an imprint of the Taylor & Francis Group, an informa business

British Library Cataloguing-in-Publication Data
A catalogue record for this book is available from the British Library

ISBN: 978-1-032-85073-3 (hbk)
ISBN: 978-1-032-85234-8 (pbk)
ISBN: 978-1-003-51721-4 (ebk)

DOI: 10.1201/9781003517214

Typeset in Times New Roman
by codeMantra

Contents

Author Biography

Dr. Ömer Aydan was born in 1955. He studied Mining Engineering at the İstanbul Technical University, Turkey (B.Sc., 1979), and Rock Mechanics and Excavation Engineering at the University of Newcastle upon Tyne, UK (M.Sc., 1982), and finally, he received his Ph.D. in Geotechnical Engineering from Nagoya University, Japan, in 1989. He worked at Nagoya University as a research associate (1987–1991) and at the Department of Marine Civil Engineering at Tokai University, first as Assistant Professor (1991–1993), then as Associate Professor (1993–2001), and finally as Professor (2001–2010). He then became a Professor at the Institute of Oceanic Research and Development at Tokai University, and he moved to the Department of Civil Engineering at the University of Ryukyus (Nishihara, Okinawa, Japan) as a Professor of Material Science, retiring at the end of March 2020 as Emeritus Professor. He has been an adjunct staff member of the University of the Ryukyus Disaster Prevention Research Center for Island Regions since 2014 and served as the director of the Center until his retirement. He established the natural disaster simulation facility. He has played an active role in numerous ISRM, JSCE, JGS, SRI, and Rock Mechanics National Group of Japan committees and has organized several national and international symposia and conferences. He organized the 2019 Rock Dynamics Summit in Okinawa from May 7 to May 11, 2019, as a world-class conference in Rock Dynamics. He was also made an Honorary Professor in Earth Science by Pamukkale University in 2008. He served as the Vice President at Large of the International Society for Rock Mechanics and Rock Engineering for the term of 2019–2023. He was appointed as the associate editor of the *Journal of Rock Mechanics and Rock Engineering* in April 2020. He is the author and/or one of the authors of more than 900 scientific papers, the author of seven international books on rock mechanics, and co-author of five more international books.

Preface

Shallow lignite seams have commonly been exploited using the room and pillar technique in the past. However, the design of these mines is mostly based on the short-term mechanical properties of lignite and the surrounding rock mass. The dominant failure mode in shallow room and pillar mines is sinkholes, which occur due to the failure of roof layers. In contrast, deep room and pillar mines primarily experience large ground subsidence as a result of pillar failures. Failures in room and pillar mines, whether in the short or long term, are attributed to overburden load, creep, and degradation of the surrounding rock masses. Therefore, understanding the long-term characteristics and degradation of surrounding rocks is necessary in the design of room and pillar mines. In addition to long-term characteristics and degradation properties, seismic events are also critically important during exploitation and after abandonment.

Collapsed chimneys in the roof, which may occur partially or fully, are often observed at mine junctions. Periodic caving, collapses, and subsidence in areas above abandoned room and pillar mines present significant concerns for municipalities and residents in urbanized areas. Furthermore, in areas with high seismicity, the risk of collapses due to earthquakes exacerbates these concerns. The author has been involved with the stability problems of some abandoned room and pillar mines in the Tokai Region of Japan since 2000. He has investigated their long-term performance and responses, especially regarding the effects of earthquakes on the abandoned room and pillar mines in Tokai Region, and also he. He specifically participated in reconnaissance efforts in the Yamoto area following the 2002 M6 Miyagi Hokubu Earthquake and in the Tohoku region from Ibaraki to Iwate after the M9 Great East Japan Earthquake. His investigations in the Tohoku region include lignite and coal mines as well as underground stone (tuff) quarries.

This book has evolved from specific observational, experimental, analytical, numerical, and monitoring studies conducted by the author in the Tokai Region of Japan since 2000, as well as reconnaissance visits to sinkholes in Mitake, Nagakute, and Tsu City. These sinkholes were caused by long-term creep and degradation of surrounding rock, as well as by the 2003 Miyagi Hokubu Earthquake and the 2011 Great East Japan Earthquake in the Tohoku Region (Iwate and Miyagi Prefectures). In addition, the author was a member of the Mine Closure and Post-Mining Management Commission of the International Society for Rock Mechanics and Rock Engineering (ISRM) from 2004 to 2008, which motivated him to write this book.

This book consists of nine chapters, with eight chapters dedicated to major aspects of abandoned room and pillar mines and remedial measures. **Chapter 1** addresses the social, scientific, and engineering necessities concerning the topic. **Chapter 2** briefly describes the fundamentals of the room and pillar mining method and discusses observations of sinkhole issues and the associated damage caused by abandoned lignite mines, particularly in the Mitake region of Gifu

Prefecture and Aichi Prefecture of Japan, as well as in other areas worldwide due to short- and long-term overburden, creep, and degradation. The duration of long-term events is generally greater than half a century and typically occurs in areas where the room and pillar technique has been exploited, and urbanization has taken place. Although examples of damage due to earthquakes are quite rare, the damage and sinkholes caused by abandoned lignite and coal mines during the 2003 Miyagi Hokubu and 2011 Great East Japan earthquakes in the Tohoku region are explained. In addition, some sinkhole issues related to abandoned room and pillar mines and antique stone quarries that have occurred worldwide are summarized.

Chapter 3 is devoted to laboratory and in-situ experimental studies on the short- and long-term physico-mechanical characteristics of surrounding rocks, which are generally soft rocks. The experimental results, including the testing methods that are crucial for assessing the stability of abandoned room and pillar mines, are described. Furthermore, several in-situ characterization techniques are presented to evaluate the excavation state of abandoned lignite mines, particularly. Specific examples of such investigations in abandoned lignite mines in Mitake town of Gifu Prefecture and the Nagakute district of Aichi Prefecture are presented. In addition, the applications of modern rock mass classification techniques, such as RMR, Q-value, and RMQR, for in-situ characterization of surrounding rock, including lignite seams, are specifically discussed.

Model tests are essential for physically and visually understanding the governing mechanisms of stability and failure modes under static and dynamic conditions. **Chapter 4** describes some static and dynamic physical model tests utilizing breakable materials and photo-elasticity techniques, and explains and illustrates the fundamental governing mechanisms of instability issues and sinkhole formation. In addition, major geological faults and underground excavations beneath existing abandoned lignite room and pillar mines may pose significant concerns for structures to be built on the surface. Such situations are also simulated by physical model tests, and the results are described.

As a remedial measure for abandoned lignite room and pillar mines, the backfilling technique is commonly utilized. **Chapter 5** focuses on physical model tests using lignite pillars, with experimental results described to provide insight into the effects of backfilling in abandoned room and pillar mines using various backfilling materials, such as granular and cohesive materials, under both static and dynamic conditions. Furthermore, some non-destructive tests are implemented to gain additional information about the state and effectiveness of the backfilling in room and pillar mines for geophysical explorations.

Empirical, analytical, and numerical methods are used to assess the stability of room and pillar mines under static and dynamic conditions. **Chapter 6** presents a comprehensive discussion on the available empirical, analytical, and numerical methods. While empirical methods are based on rock mass classifications, analytical methods involve beam theory and arching theory for the roof, with the tributary area concept being used to assess the stability of pillars. These analytical methods consider not only the short-term properties of surrounding rocks but also the effects of creep, saturation, degradation, and seismic activity. Numerical methods primarily involve analyses using a simplified multi-layer shearing method, along with 2D and 3D elasto-plastic finite element modeling techniques under static and dynamic conditions. In addition, numerical simulations of the vibration characteristics of ground above abandoned lignite room and pillar mines are described. The discrete finite element method is utilized to simulate fracturing phenomena, alongside continuum-type finite element analyses. Furthermore, studies evaluating the effect of backfilling in abandoned lignite mines through geophysical explorations, as well as various 2D and 3D simulations of these mines beneath critical structures and large tunnels, are presented.

The stability analyses of active and abandoned room and pillar mines always require information on the initial in-situ stress state. **Chapter 7** briefly presents several techniques for measuring and inferring in-situ stresses in mines, with specific applications to various abandoned lignite and stone room and pillar quarries.

Remediation measures for areas above abandoned lignite room and pillar mines are necessary when such areas become urbanized or when infrastructure, such as roadways or railways, is to be constructed. Backfill techniques or immersion may be used as remediation measures, and the depth of backfilling is another important consideration when implementing these methods. **Chapter 8** focuses on the implementation of remediation techniques, describing detailed procedures for backfilling, analytical and numerical methods for simulating the movement and spreading of backfill materials in abandoned lignite mines, and presenting several specific actual implementations.

Monitoring techniques are necessary for observing ground settlement over abandoned room and pillar mines. Furthermore, monitoring is necessary if remedial measures are implemented, including groundwater quality assessment. Monitoring of abandoned room and pillar mines can be conducted using contact-type monitoring as well as remote sensing techniques (such as laser technology and SAR and its derivatives). Furthermore, contact-type measurements such as displacement monitoring, acoustic emission, electric potential response, and various environmental parameters may be implemented. **Chapter 9** addresses various aspects of monitoring abandoned lignite room and pillar mines, detailing these techniques and presenting actual implementations. In addition, some monitoring techniques for evaluating the effectiveness of remediation measures are described, along with specific examples.

This book is intended for readers interested in the room and pillar mining method and its short-term and long-term stability aspects. Furthermore, the effects of earthquakes on such mines are discussed, and actual case history data, as well as physical tests, analytical studies, and numerical studies, are presented. This book also covers the rock mechanics aspects of lignite and the surrounding rock mass and their characterization. Although lignite and surrounding rocks are generally considered soft rocks, the fundamentals can be applied to room and pillar mines consisting of harder rocks.

This book is quite unique in combining the design and stability issues of active and/or abandoned lignite room and pillar mines, considering short- and long-term factors, as well as ground motions resulting from earthquakes and remedial measures, all within a single entity. This book addresses the occurrence, prediction, and control of subsidence issues associated with room and pillar mining, including aspects of creep, degradation, and seismic effects. Some specific merits of this book are as follows:

1 A unified treatise on room and pillar mines.
2 Clear definitions and explanations of fundamental concepts.
3 Consideration of long-term aspects (creep and degradation), seismic effects, remedial measures, contact type, and remote sensing monitoring techniques.
4 Inclusion of physical and numerical modeling.
5 In-depth view of procedures and solution techniques.

This book is intended for readers with a mining and/or civil engineering background and can serve as a reference for undergraduate and graduate students. It aims to provide a clear understanding of active and/or abandoned room and pillar mines, benefiting academicians, engineers, and students alike.

Acknowledgments

The author sincerely acknowledges Prof. Reşat Ulusay and Prof. Xia-Ting Feng, former Presidents of ISRM, as well as Prof. Seokwon Jeon, the current President of ISRM, for inviting the author to contribute to the prestigious ISRM Book Series on *Geomechanical Aspects of Abandoned Room and Pillar Mines and Remediation Measures*. The content of this book is an outcome of over four decades of learning and teaching at Istanbul Technical University in Türkiye, Newcastle upon Tyne University in the United Kingdom, Nagoya University, Tokai University, and the University of the Ryukyus in Japan. The author would like to sincerely thank Emeritus Prof. Dr. Toshikazu Kawamoto of Nagoya University, Prof. Dr. Hisataka Tano of Nihon University, Prof. Dr. Takashi Ito, and Prof. Dr. Naohiko Tokashiki of the University of the Ryukyus for their encouragement, collaboration, and guidance over these years. Emeritus Prof. Dr. T. Kawamoto emphasized the importance of this topic, which inspired the author to study the geomechanical aspects of abandoned mines and to develop experimental, analytical, and numerical methods, as well as monitoring tools. The author also sincerely acknowledges the great help and guidance provided by Prof. H. Tano, particularly regarding field monitoring. The invitation from Prof. Nielen van der Merwe, the former ISRM President, and Dr. Christophe Didier of INERIS to join the mine closure commission, along with the support of the commission members, is gratefully acknowledged. These international interactions also motivated the author to contemplate this book. The author appreciates Prof. Ismet Canbulat from the University of the South Wales for reviewing the book during its draft stage. The author would like to sincerely acknowledge Mr. K. Sugiura, Dr. A. Sakamoto, and Mr. K. Wada of the Nagoya Branch of Tobishima Corporation, Japan, who are directly involved in the remediation of abandoned lignite mines and other abandoned underground cavities, particularly in the Tokai Region of Japan. Their financial support for the author's projects related to abandoned mines, as well as their assistance during field studies and in-situ monitoring in Mitake Town, has been invaluable. Prof. Dr. Melih Geniş of Bülent Ecevit University, Zonguldak, Türkiye, is gratefully acknowledged for his significant assistance in performing some of the static and dynamic numerical analyses presented in this book. The author would also like to thank his former students at Tokai University and the University of the Ryukyus for their help during experiments, in-situ monitoring, and analyses. In particular, special thanks go to his former students, Mr. Mitsuo Daido and Dr. Yoshimi Ohta of Tokai University (currently with the Nuclear Regulation Authority of Japan), and Mr. S. Kodate of Chuden Engineering Consultants (Hiroshima), who was formerly an undergraduate and graduate student at the University of the Ryukyus.

Finally, I want to thank my wife, Reiko; my daughter, Ay; my son, Turan Miray; and my late parents for their continuous help and understanding, without which this book could not have been completed.

Chapter 1

Introduction

The room and pillar mining method is one of the most commonly used techniques due to its lower exploitation costs compared to other mining techniques, and it has been employed since ancient times. Therefore, there are many abandoned mines in various parts of Japan and worldwide. Figure 1.1 shows some examples of abandoned room and pillar mines in Japan and other countries.

As seen in Figure 1.2, sinkholes and subsidence occur from time to time in areas where abandoned room and pillar mines exist (e.g., Brady and Brown, 1999; Beck and Herring, 2001; Ghabraie, 2016). The extraction of lignite and building stones using the room and pillar mining technique was quite extensive in various parts of Japan until the 1960s. Most of these mines are now abandoned, and some of these areas have since become urbanized. The Tokai region in Central Japan is a well-known example of such a case. The author was involved with the

Figure 1.1 Some examples of abandoned room and pillar mines in Japan and other countries (a) examples from Bazda and Kuşini (Türkiye) and Qurna (Egypt); (b) examples from Japan: Oya, Asuwayama, Muraiwa, Mitake.

DOI: 10.1201/9781003517214-1

Figure 1.2 Some examples of sinkholes occurred in recent years.

stability of abandoned mines beneath urbanized residential areas in the Tokai region, which presents significant concerns for municipalities as well as for people living in those areas.

If the seismicity of areas above the abandoned mines is high, the risk of collapses, sinkholes, and subsidence due to earthquakes may exacerbate these concerns. The 2003 Miyagi Hokubu and the 2011 Great East Japan earthquakes clearly indicated that such areas may experience roof collapses and subsidence, as well as the ejection of underground waters from the mines (Aydan and Kawamoto, 2004; Aydan and Tano, 2012a,b). Therefore, there is growing concern regarding the safety of areas above these abandoned mines in relation to the anticipated Nankai–Tonankai–Tokai earthquake. Unfortunately, Japan's current seismic risk assessment does not consider the effects of abandoned mines on ground motions in these areas.

In 2004, the International Society for Rock Mechanics and Rock Engineering established a special commission called the Mine Closure and Post-Mining Management Commission to address issues related to the closure of various types of mines worldwide and to create a state-of-the-art report on the topic (Didier et al., 2008, 2009). The author was invited to be a member of this commission in 2004, and the contributions made by the author to the activities and report of this commission from 2004 to 2008 constitute a significant part of this book. Furthermore, it is also intended that the content of this book fulfills some of the goals of the International Society of Rock Mechanics and Rock Engineering (ISRM) related to the establishment of the Mine Closure and Post-Mining Management Commission.

There are almost no studies on the dynamic characteristics of ground and the amplification of ground motions induced by earthquakes above abandoned mines exploited by room and pillar and long-wall mining techniques, except for those conducted by the authors. In this book, the dynamic characteristics of ground and the amplification of ground motions above abandoned mines are explained through experiments, micro-tremor measurements on sites, and numerical studies, and their implications are discussed. The effect of plastic zone formation in relation to the stability of abandoned mines under dynamic conditions is described. Additionally, the effects of certain parameters associated with extraction and some topographic effects on ground motions in areas with abandoned mines are explained.

When new structures are to be built over or below existing natural or man-made cavities, the potential effect of the collapse of underground openings on the ground surface is crucial for the design of new structures (Aydan and Tokashiki, 2012; Geniş and Aydan, 2013). The construction of new structures over or beneath areas with man-made underground openings or natural cavities has been increasing in recent years. For example, Shizuoka Airport was constructed over a tunnel of the Tokaido Shinkansen Railway Line. Similarly, New Ishigaki Airport was built over an area with karstic cavities, which are home to precious bat species that must be protected. There are also several examples of highways and railways built over lava tubes in the vicinity of Mt. Fuji in Japan, Cheju Island in Korea, and Hawaii. When such underground openings or natural cavities are known prior to construction, some countermeasures can be undertaken. However, the most critical situation occurs if their existence is unknown and collapse happens after the construction. Collapses of underground openings may occur due to disturbances from construction or natural causes, such as long-term degradation of the rock mass and/or seismic shaking.

The author has been involved with cavity filling for some time in the Tokai region. The cavity filling technique involves several fundamental steps, such as exploring the dimensions of cavities, establishing filling plant facilities, backfilling, and monitoring. Non-hazardous industrial by-products are also utilized as filling materials. Particularly, the effect of the quality of backfilling in abandoned mines beneath urbanized areas is very important under both static and dynamic conditions. The effect of the quality of backfilling on ground subsidence associated with abandoned lignite mines under static and dynamic conditions is experimentally investigated, and its implications for urbanized areas above abandoned lignite mines are discussed.

The dominant failure mode of shallow room and pillar mines is sinkholes due to the failure of roof layers, while the failure of deep room and pillar mines primarily results in large ground subsidence. Besides seismic loads, the failures of room and pillar mines in the short or long term are due to overburden load, creep, and degradation. The collapse and chimneying of the roof, which may occur partially or fully, are often observed at mine junctions. The effects of creep and degradation processes are analytically presented, and several practical applications are described.

Photo-elasticity has been used in various engineering fields since its invention by Brewster in 1815 to investigate stress concentrations. This is the only method that allows for the visual observation of stresses, and it has served well in various fields of engineering. However, this technique has received less attention nowadays due to advances in numerical techniques and computational technology. Nevertheless, the computational results from numerical techniques may still require validation by this method, particularly in fracturing and dynamic problems. Furthermore, validating computational codes for simulating the behavior of discontinuum is especially necessary. The outcomes of physical tests and numerical simulations are presented. Additionally, the author discusses the implications of these outcomes in engineering and their renewed potential for both educational and research purposes, considering advances in digital recording, processing of data, and LED technology.

The book consists of nine chapters. **Chapter 1** addresses the background of social, scientific, and engineering necessities. **Chapter 2** briefly explains the room and pillar mining method, along with observations on sinkhole issues and associated damage caused by abandoned lignite mines on surface effects and structures. Specifically, the damage and sinkholes from the abandoned lignite and coal mines caused by the 2003 Miyagi Hokubu and 2011 Great East Japan earthquakes in the Tohoku region are explained.

Chapter 3 focuses on laboratory and in-situ experimental studies of the short- and long-term physico-mechanical characteristics of surrounding rocks. It also describes in-situ characterization techniques to assess and evaluate the excavation state of abandoned lignite mines.

Chapter 4 explains some static and dynamic physical model tests utilizing breakable material and photo-elasticity techniques, as well as the fundamental governing mechanisms of instability issues and sinkhole formation. In addition, the effects of major geologic faults and underground excavations beneath existing abandoned lignite room and pillar mines are simulated through physical model tests, and the results are described.

Chapter 5 is concerned with physical model tests using pillars of various rocks, exploring the effects of backfilling abandoned room and pillar mines with different materials, such as granular and cohesive materials, under both static and dynamic conditions.

Chapter 6 describes available empirical, analytical, and numerical methods. The analytical methods consider not only the short-term properties of surrounding rocks but also the effects of creep, saturation, degradation, and seismic events. Numerical methods mainly involve analyses using simplified multi-layer shearing methods and 2D and 3D elasto-plastic finite element modeling techniques. Numerical simulations on the vibration characteristics of ground above abandoned lignite room and pillar mines are described, with the discrete finite element method utilized to simulate fracturing phenomena. The effects of backfilling abandoned lignite mines through geophysical explorations, as well as various 2D and 3D simulations of abandoned lignite mines beneath critical structures and the excavation of large tunnels beneath abandoned lignite room and pillar mines, are presented.

Chapter 7 presents several techniques for measuring and inferring in-situ stresses in mines and describes specific applications to several abandoned lignite and stone room and pillar quarries.

Chapter 8 discusses the implementation of remediation techniques, including a detailed procedure for backfilling, as well as analytical and numerical methods for simulating the movement and spreading of backfill materials in abandoned lignite mines, along with several specific actual implementations.

Chapter 9 explains various techniques for monitoring abandoned lignite room and pillar mines and includes actual implementations. In addition, some monitoring techniques for evaluating the effectiveness of remediation measures are described, with specific examples provided.

This book uniquely combines the design and stability issues of active and/or abandoned lignite room and pillar mines, considering short- and long-term stability as well as ground motions resulting from earthquakes and remedial measures, all within a single entity. It is hoped that it will serve as a reference book for undergraduate and graduate students and provide a clear understanding of active and/or abandoned room and pillar mines.

Chapter 2

Abandoned Room and Pillar Mines and Issues

2.1 Room and Pillar Mining Techniques

The shallow lignite, coal, halite, and ulexite seams, as well as building stones, are commonly exploited using the room and pillar technique, as illustrated in Figure 2.1a, together with possible modes of failure. The dominant failure mode of shallow room and pillar mines appears as sinkholes due to the failure of roof layers, while the failure of deep room and pillar mines mainly

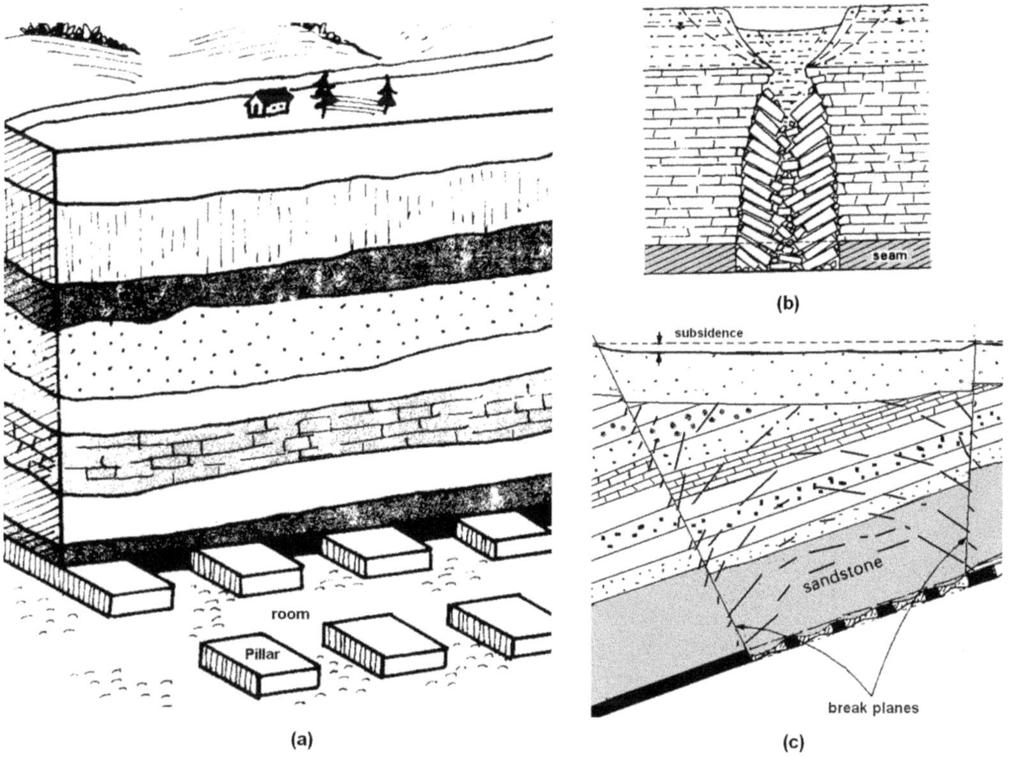

Figure 2.1 Illustration of the room and pillar method and likely instability modes. (a) Room and pillar method. (b) Sinkhole due to roof failure. (c) Settlement due to pillar failure. (Aydan et al., 2012, 2013).

DOI: 10.1201/9781003517214-2

results in large ground subsidence due to pillar failures (Figure 2.1c). In addition to seismic loads, failures of room and pillar mines in the short or long term may occur due to overburden load, creep, and degradation of the surrounding rock mass (Figure 2.2b). The chimney-like collapse of the roof, which may occur partially or fully, is often observed at mine junctions.

The long-wall technique was employed for exploiting the coal/lignite seams, as illustrated in Figure 2.2a. For example, the coal fields of Jobando and Shirakawa in Japan were excavated using the long-wall technique. As illustrated in Figure 2.2b, the overburden ground would subside into the excavated space following the extraction of coal/lignite seams. Depending on the backfilling state, the overall subsidence would be a fraction of the total height of extraction (Aydan 1994). For longwall mining operations in Japan, Hiramatsu et al. (1979) reported that the overall subsidence would be 70%–90% of the total extraction height due to the bulking of the broken overburden layers when no backfilling is employed. However, the broken overburden layers or goaf (gob) might further subside in the long term due to creep, degradation, and vibration resulting from earthquakes or other sources. It should also be noted that the geomechanical characteristics of the goaf, which is fractured and broken, would be different from the original unbroken ground.

The lignite seams in the Tohoku region of Japan are gently inclined, and their thickness changes between 0.9 m and 3 m depending on the location, with one or two seams of lignite extracted. The overburden depth of the lignite seam ranges between 2 and 30 m, and one example from the Komatsu lignite field in East Matsushima is shown in Figure 2.3 (Kawamoto et al., 2004; Nishida et al., 1984). The access to mines was through inclined and vertical shafts as also seen in Figure 2.3. The abandoned mines are submerged by underground water, with the groundwater level generally at a depth of 2–6 m below the ground surface.

2.2 Sinkhole and Settlement Occurrence

2.2.1 *Partial Instabilities*

The author and his group have been involved with the stability problems of several abandoned mines in the Tokai Region and other regions of Japan, investigating their long-term performance

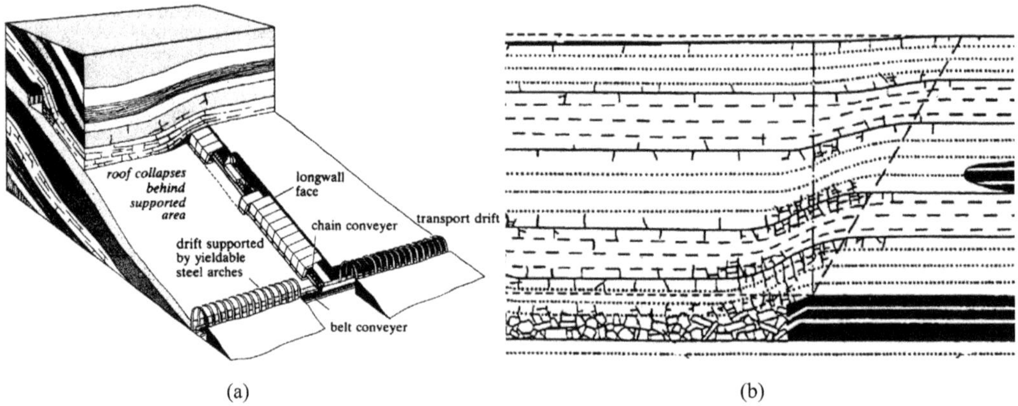

Figure 2.2 Illustration of the longwall mining method and ground subsidence. (a) Longwall mining method. (b) Settlement due to extraction of seam. (From Aydan et al., 2012, 2013.)

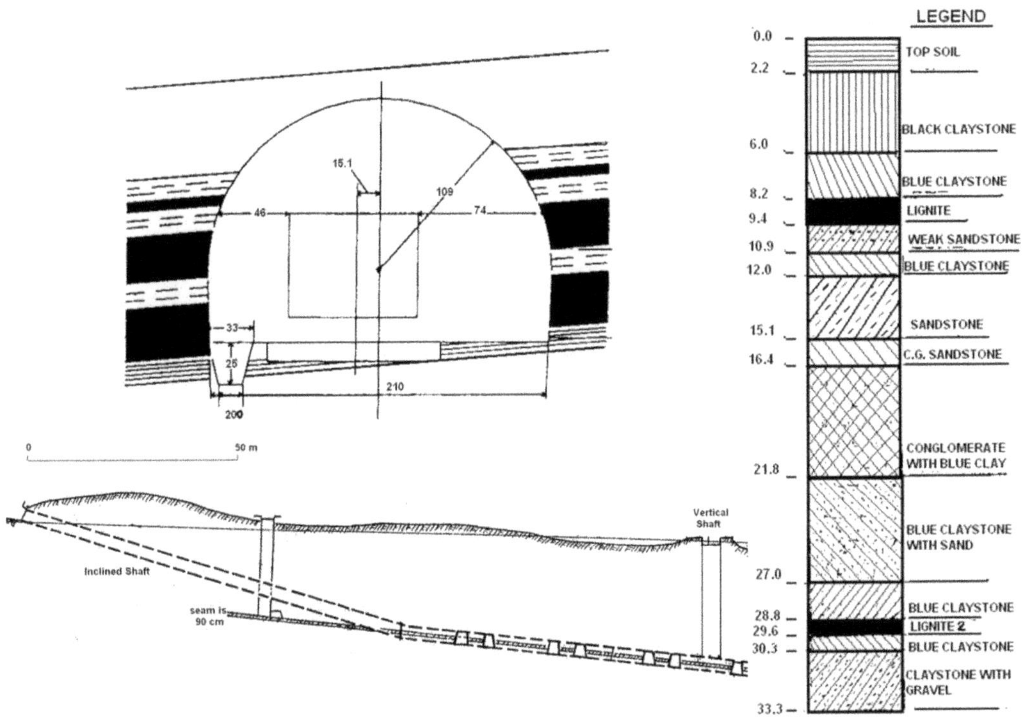

Figure 2.3 Dimensions of the main drift, longitudinal cross section and a geologic log at the Komatsu lignite field in East Matsushima. (Arranged and translated from the report by Kawamoto et al., 2004.)

and responses during earthquakes. Some of the findings described in this sub-section are based on observations in the abandoned lignite mines in the Tokai region.

The instability modes of the roof include bending failure, flexural toppling failure, and rockfall (Aydan et al., 2006; Aydan, 2018). Bending failure occurs due to sagging of the roof layers (Figure 2.4) and is observed when the span of the opening is large relative to the layer thickness. Flexural toppling failure of the roof happens when the roof is cut through a large discontinuity, such as a fault. Rockfall occurs when the roof layers are disturbed by cross-cutting discontinuities, allowing kinematically free blocks to fall.

The instability modes of pillars before their ultimate failure include splitting or slabbing failure, buckling of slabs, and block sliding (Figure 2.5). Splitting failure results from high stress concentrations on pillars, with the splitting cracks occurring almost parallel to the vertical axis of the pillars. This failure is observed in pillars that support relatively larger areas. Buckling failure is also observed when a rock slab created by a splitting crack extends from the top to the bottom. Block sliding failure is associated with existing discontinuities in the rock mass, which are generally associated with faults and their conjugates.

2.2.2 *Global Collapses: Sinkholes and Subsidence*

As a result of degradation, partial collapses of roof layers, reduced cross-sectional areas of supporting pillars, and even pillar failure can occur, although their effects may not be immediately

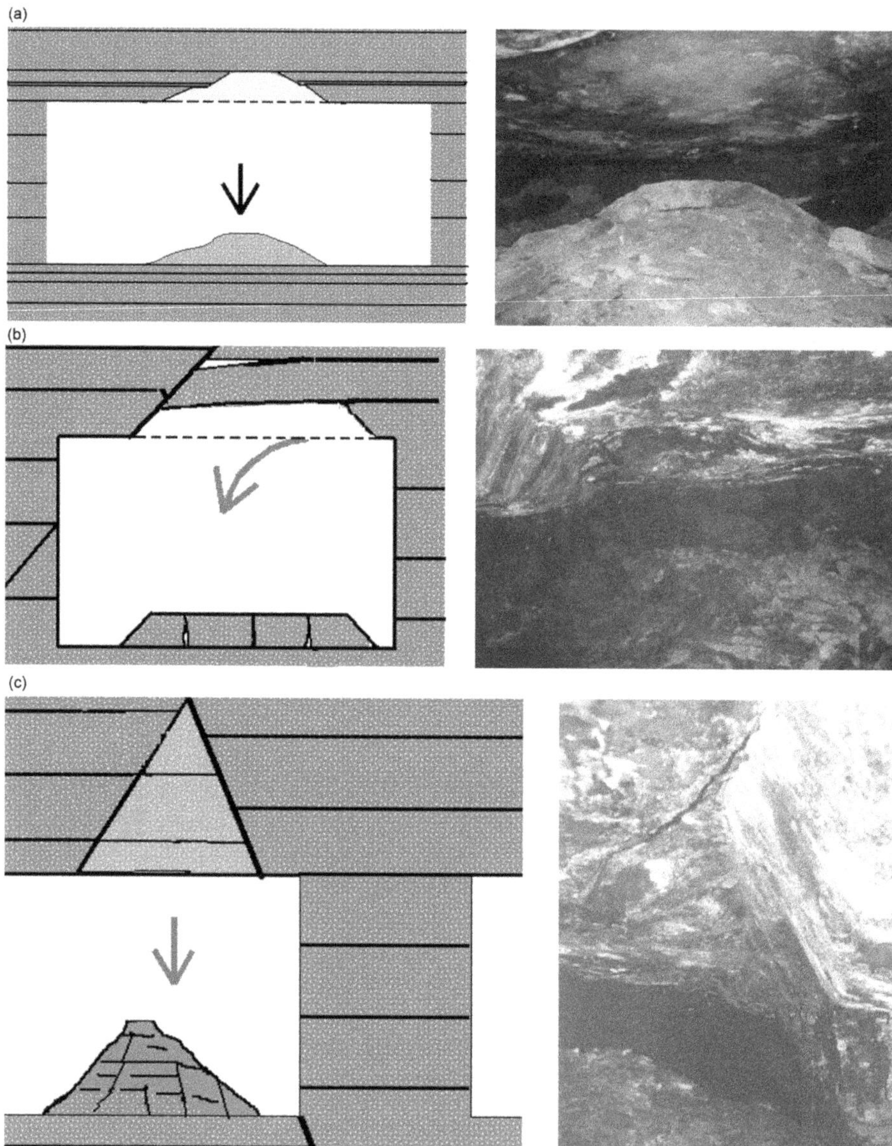

Figure 2.4 Instability modes of roofs in abandoned lignite mines. (a) Bending failure. (b) Flexural toppling failure. (c) Roof fall. (Based on Aydan et al., 2005; 2006.)

visible on the ground surface. There have been occurrences of sinkholes and significant ground subsidence from time to time in Mitake Town, as shown in Figure 2.6. The dimensions of these sinkholes and subsidence have been increasing in recent years due to the combined action of the degradation due to environmental variations, vibrations induced by traffic, construction as well as earthquakes and the creep characteristics of rock masses of pillars. With more than 50–60 years having passed since the abandonment of the mines, the long-term strength (creep strength) of the pillars is estimated to be only a fraction of their short-term strength.

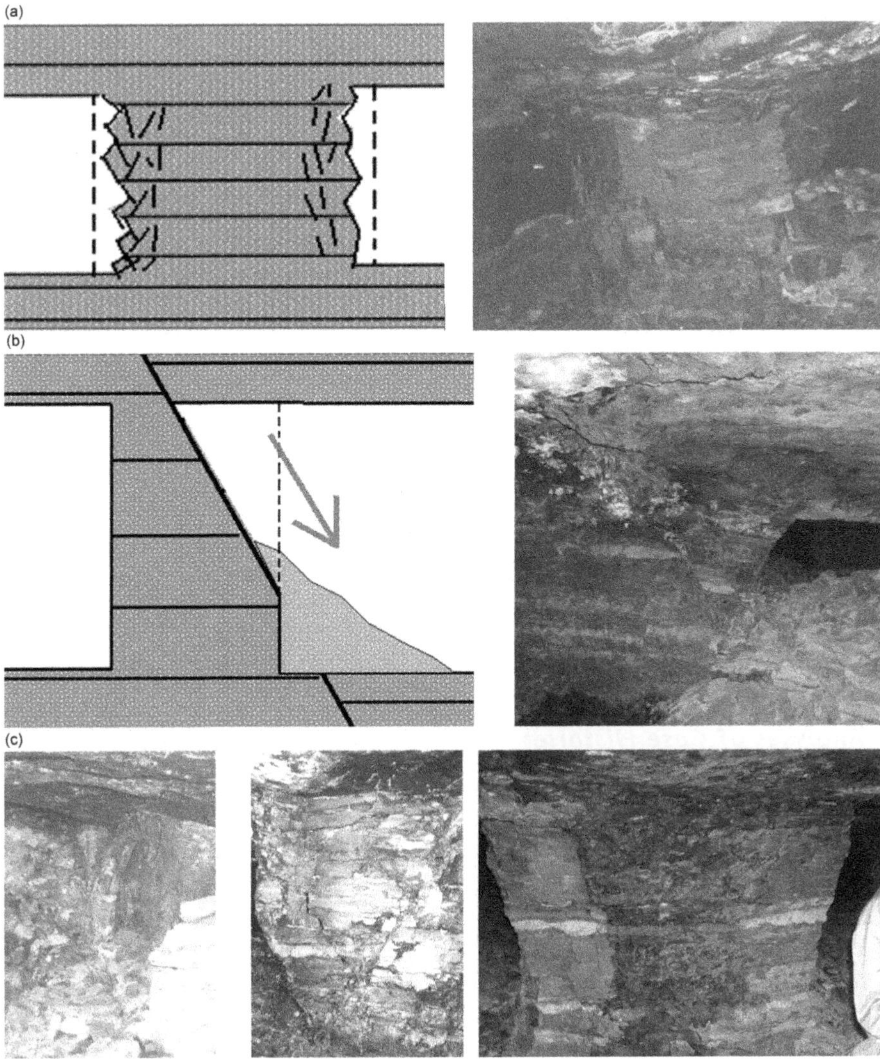

Figure 2.5 Instability modes of pillars. (a) Splitting or slabbing failure. (b) Sliding failure. (c) Buckling and other failure forms. (Based on Aydan et al., 2005a,b, 2006.)

Figure 2.6 Large sinkholes that occurred recently in Mitake Town. (Arranged from Aydan et al., 2015.)

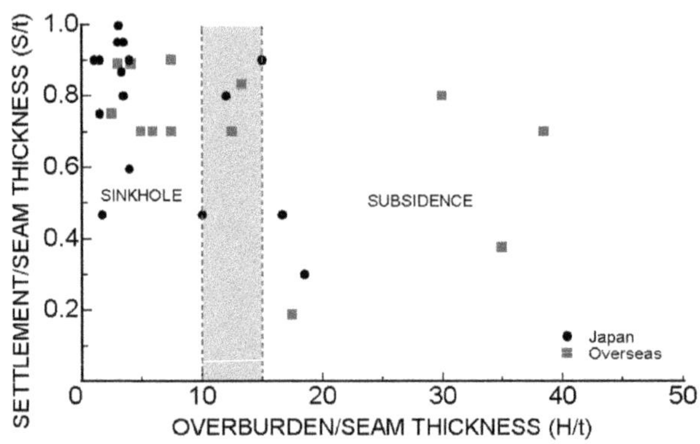

Figure 2.7 Normalized seam height versus normalized settlement. (From Aydan and Ito, 2015.)

When the shafts of abandoned mines and quarries are not deep, they may commonly be filled with gang material or debris, behaving similarly to non-cohesive granular material. When ground motions are strong enough to cause yielding of the filling material, the non-cohesive filling can flow into adjacent openings, resulting in subsidence of the filling material in the shafts. This phenomenon was observed in the shafts of abandoned quarries in Oya during the 2011 Great East Japan Earthquake and in the abandoned lignite mines of the Yamoto area (currently Higashi Matsushima City) during the 2003 Miyagi Hokubu Earthquake.

2.2.3 *Analyses of Case Histories of Sinkholes and Subsidence*

The author compiled various data on the formation of sinkholes and subsidence reported in the literature. Sinkholes are generally caused by the degradation of roof layers under different environmental conditions, groundwater flow, and vibration effects. On the other hand, subsidence involves the failure of rock pillars, with overburden depth being a critical parameter, including the degradation of rock materials. Figure 2.7 shows the case history data plotted in the space of overburden (H) and settlement (S) normalized by the seam extraction height (t). Although it is difficult to discuss every aspect here, sinkholes are observed when the overburden is less than 10–15 times the seam extraction height. This implies that the overburden would be 20–30 m for a seam extraction height of 2 m. However, subsidence may occur at any depth, provided that pillars fail. Subsidence may take a longer time following the termination of mining, with some examples of delayed failures occurring after about 40 to 130 years.

2.3 Effect of Earthquakes on Abandoned Room and Pillar Mines

In this subsection, earthquakes and associated sinkholes and subsidence are briefly explained.

2.3.1 *1978 Off-Miyagi Earthquake*

Nishida et al. (1984) reported that damage to abandoned lignite mines occurred in the Ohira and Esashi lignite fields in the Tohoku region during the 1978 Off-Miyagi Earthquake (M7.4), resulting in 212 instances of sinkholes and subsidence damage. They also mentioned that damage occurred in the Esashi lignite field, which was about 30 m deep, even though the estimated ground acceleration was only 50 gals.

2.3.2 *2003 Miyagi Hokubu Earthquake—Yamoto Town*

The damage surveys on abandoned lignite surfaces are restricted to surface damage. Although there is no doubt that some damage in the form of roof collapse and pillar failure may take place inside the abandoned mines, it is very difficult to observe this damage due to limited access to old workings. The local authorities (Yamoto Town) identified 28 locations where surface damage was observed. The damages were in the form of caving (sinkholes) and water discharge (Kawamoto et al., 2004; Sakamoto et al., 2006). Figure 2.8 shows some examples of the observed damage. Some of the water discharge was associated with sand and silty-soil boiling, as seen in Figure 2.8. Although some reports have associated this sand boiling phenomenon with the liquefaction of subsoil, it is quite difficult to contemplate such a phenomenon in ground that generally consists of rocky layers. Some soil sampling was conducted, and their grain size distributions were determined (Aydan, 2004). The grain size distributions of soil samples fall within the liquefiable soil limits. The grain size distributions of the overlaying weakly cemented sandstone layer above the lignite seam were the same as that of the ejected sand. It seems that the sand accumulated on the floor of the abandoned-lignite mine workings due to the collapse of the overlaying sandstone layer may be ejected to the ground surface during the sloshing of groundwater in the abandoned lignite mines as a result of ground shaking caused by the earthquake. Local residents indicated that some wells adjacent to damaged locations had very high water levels, implying that the abandoned-mine workings were fully submerged

Figure 2.8 Sinkholes, subsidence, and sandy water ejection in Yamoto Town. (From Aydan, 2004; Aydan and Kawamoto, 2004.)

with groundwater. Furthermore, it is reported that the water discharge from caving locations continued for a considerable period. This also implied that the motion of the groundwater might breach submerged old mining workings at higher elevations and flow toward the old mine workings at lower elevations, as such events generally took place at locations with lower elevations.

As seen in Figure 2.8, a peculiar liquefaction phenomenon was observed at a rice paddy over the abandoned-lignite mine workings, resembling a doughnut. Local authorities informed that the pillars at the abandoned lignite mines had a square shape with a side length of 1.5–2 m.

The sand and/or groundwater that spouted from the abandoned mines are often mistakenly interpreted as ground liquefaction by people with a background in soil mechanics. Although there is no doubt that the ejected sand has characteristics similar to liquefied soil, it is actually sand that accumulated at the bottom of mine adits and was ejected to the ground surface through vertical and inclined shafts or collapsed openings connected to the surface (Aydan, 2004).

2.3.3 Effect of Far-Field Earthquakes in an Abandoned Mine in Mitake Town

The far-field earthquakes designated as the 2004 September 5 Off Tokaido earthquakes (Mw 7.2; Mw 7.3) occurred, with epicenters approximately 236 km from Mitake Town. The maximum ground accelerations measured near Mitake Town were in the range of 22–25 gals. When the abandoned mine was visited on September 17, 2004, it was observed that blocks had fallen from the roof of the openings, as seen in Figure 2.9a. Rockfalls and splitting of a pillar shown in Figure 2.9b occurred following the 2007 Chuetsu earthquake (Mw 6.6), with the epicenter about 253 km away from Mitake Town. Despite the maximum ground acceleration being estimated at less than 5 gals, the rockfall and pillar splitting might have been induced by long-period ground motions.

2.3.4 The 2011 Great East Japan Earthquake

The Great East Japan Earthquake, with a moment magnitude of 9.0, took place at 14:46 (JST) on March 11, 2011 (JMA, 2011). This earthquake was a subduction plate-boundary earthquake, with a rupture area approximately 450 km long and 200 km wide (i.e., Aydan et al., 2011; Koketsu, 2012). The earthquake caused gigantic tsunami waves that destroyed many cities and towns along the shores of the Tohoku and Kanto regions of Japan. The casualties caused by this tsunami exceeded 20,000 people. The tsunami destroyed and heavily damaged buildings

(a) (b)

Figure 2.9 Partial collapses in abandoned mines of Mitake induced by recent earthquakes. (a) 2004 Tokaido-oki earthquake. (b) 2007 Chuetsu-oki earthquake. (From Aydan and Tano, 2012b.)

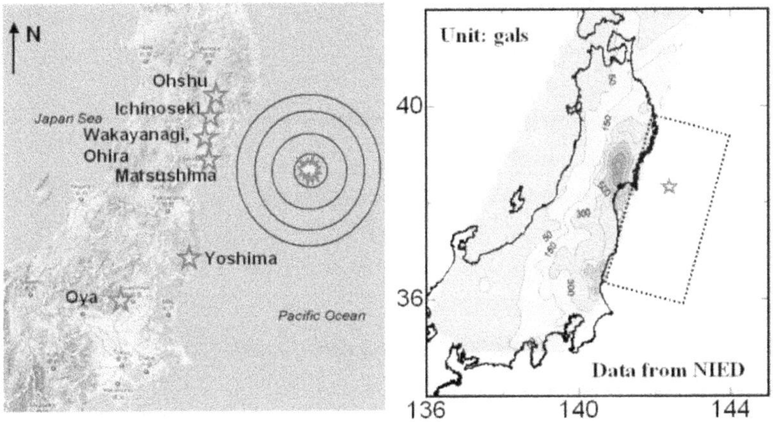

(a) (b)

Figure 2.10 Investigation locations and contours of maximum ground acceleration. (a) Investigation locations. (b) Maximum acceleration contours. (From Aydan and Tano, 2012a,b; Aydan et al., 2013.)

of various types, transportation facilities, and infrastructure. The authors visited Oya (Tochigi Prefecture) in the Kanto region and Iwaki (Fukushima Prefecture), Wakayanagi, Ohira (Miyagi Prefecture), and Ichinoseki (Iwate Prefecture) in the Tohoku region, making observations on the collapsed abandoned lignite and coal mines, as well as underground stone quarries and the associated damage to superstructures (Figure 2.10a).

The earthquake occurred along the subduction zone between the North American plate (NAM) and the Pacific plate (PAC), with an estimated inclination of the rupture plane of 14–16 degrees. The faulting was due to thrust faulting, with an estimated offset of 24 m, and the seabed was uplifted by 5 m while the land sank by about 0.8 m with a horizontal movement of 4.6 m (GSI, 2011). The rupture initiated in an area where an earthquake with a magnitude of 7.5 was anticipated and it propagated bilaterally.

Figure 2.10b shows the maximum ground acceleration contours, which are obtained using data from K-NET (2011) and KiK-net (2011). The highest ground acceleration, reaching a level of 3 g, was recorded at the Tsukidate strong motion station of K-NET. Although the ground conditions at this station are very good, it is still difficult to understand the cause of the high ground motions recorded there. Except for a top 1 m thick clay layer, this station is founded on sedimentary rocks, which are commonly observed in lignite mine areas. Figure 2.11 shows the acceleration record at the Tsukidate strong motion station of K-NET. The maximum ground acceleration at the Tsukidate strong motion station was 2,933 gals at the ground surface. However, the maximum ground acceleration measured at the strong motion station in Wakayanagi, operated by the JMA, was only 630 gals.

The earthquake caused sinkholes at five locations in Iwaki City in Fukushima Prefecture, 11 locations in Kurihara City, seven locations in Osaki, 11 locations in Higashi Matsushima, and three locations in Kurogawa in Miyagi Prefecture, as well as 11 locations in Ichinoseki City and 10 locations in Oshu City in Iwate Prefecture, according to the first preliminary report released by the Ministry of Economy, Trade, and Industry (METI) on March 30, 2011 (Aydan and Tano, 2012a). However, a press release on July 25, 2011, by METI revealed that the number of events was more than 316 (Aydan and Tano, 2012b,c). Figure 2.12 also shows the number of events at

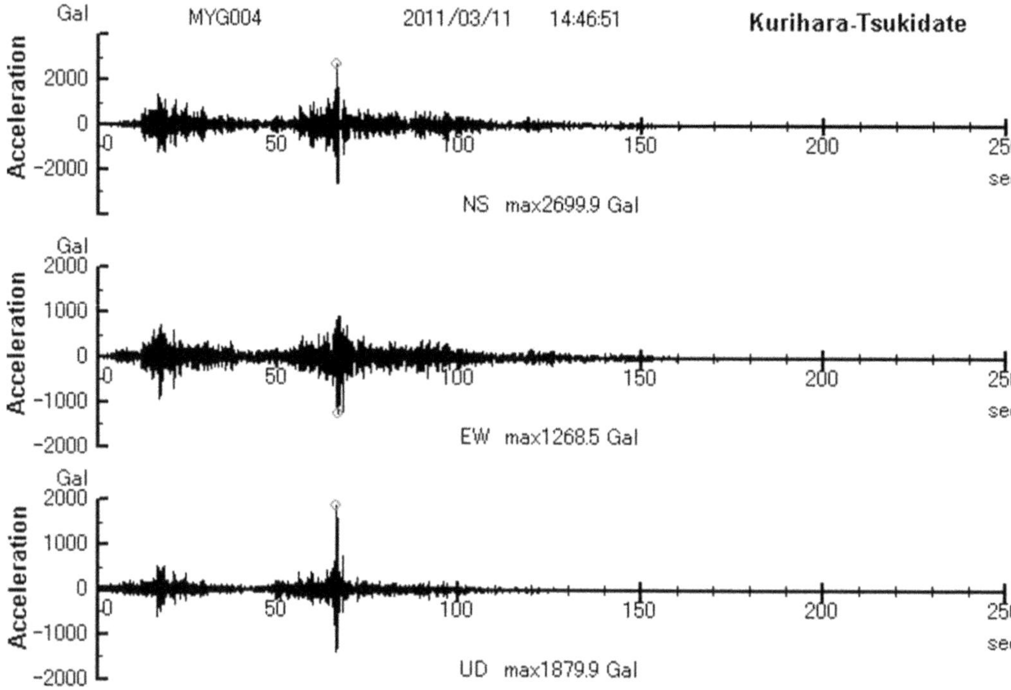

Figure 2.11 Acceleration records at Tsukidate station. (From Aydan and Tano, 2012a,b.)

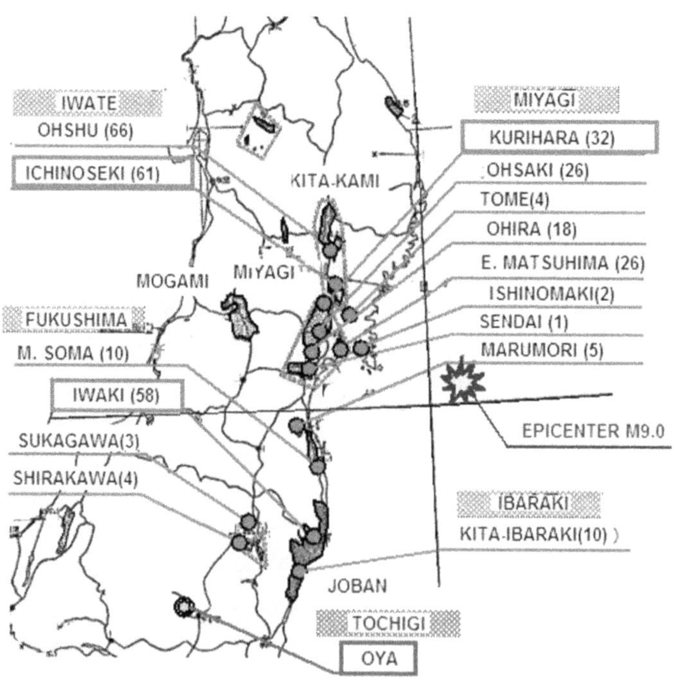

Figure 2.12 Coal and lignite fields in the Tohoku Region, along with sinkhole event numbers (Aydan & Tano, 2012a,b; Aydan et al., 2013).

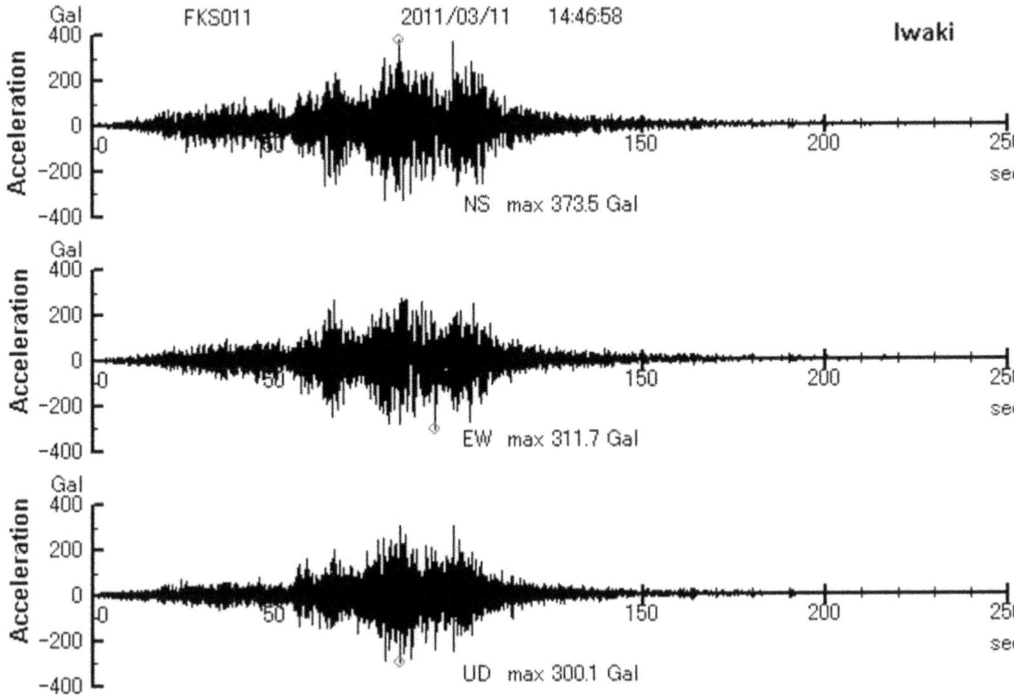

Figure 2.13 Acceleration records at Iwaki station. (From Aydan and Tano, 2012a,b.)

each major abandoned mine area. The drastic increase in the number of events may be related to the ministry's initial focus on the Fukushima nuclear power plant incident. As the effects of the Fukushima incident became gradually controlled, the ministry started to pay attention to the damage caused by the abandoned mines. It was reported that the number of sinkhole occurrences increased almost 20 times the annual number of sinkholes in Iwate and Miyagi prefectures. The direct observations of the author and his group on abandoned lignite and coalmines are presented in this sub-section.

The number of events related abandoned mines in Iwaki was 58. Figure 2.13 shows the acceleration record at Iwaki strong motion station of KiK-net. The maximum ground acceleration at Iwaki strong motions station of KiK-net was 481 gals at the ground surface. There were five sinkholes at the abandoned Yoshima mine of Jobando coal field, as shown in Figure 2.14. The collapses were associated with inclined shafts. The sloshing of underground water in the inclined shafts caused the failure of roof rock, resulting in sinkholes, as seen in Figures 2.15 and 2.16. Groundwater spouted from the sinkholes and ejected fragments of lignite, surrounding rock, and sandy material. Similar types of events were also observed in previous earthquakes, such as in the 2003 Miyagi Hokubu Earthquake (Aydan and Kawamoto, 2004).

In Wakayanagi Town of Kurihara City, five sinkholes occurred in a residential area, as seen in Figure 2.16. Fortunately, these sinkholes did not cause the collapse of any residential houses on the ground surface. However, the fracturing of the base concrete occurred, and one corner of the building was overhanging with a slight tilt. The overburden of the sinkholes ranged between 1 and 3 m at this locality (Figure 2.17).

In a large rice field near Ohira Village in Kurihara City, more than 10 sinkholes with a diameter ranging from 1 to 5 m were observed, as seen in Figure 2.18. Besides ground shaking, the

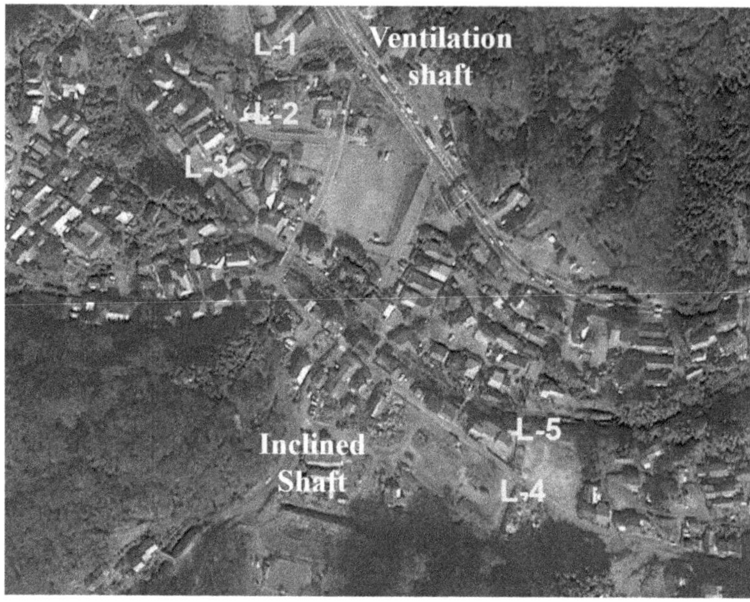

Figure 2.14 Locations of sinkhole at the abandoned Yoshima Coal mine. (From Aydan and Tano, 2012a,b.)

Figure 2.15 Views of sinkhole locations L1-L3. (From Aydan and Tano, 2012a,b.)

Figure 2.16 Views of sinkholes locations L4-L5. (From Aydan and Tano, 2012a,b.)

Figure 2.17 Some views of sinkholes in Wakayanagi Town. (From Aydan and Tano, 2012a,b.)

Figure 2.18 Views of sinkholes in a rice field near Ohira Village. (From Aydan and Tano, 2012a,b.)

sloshing phenomenon of groundwater in abandoned lignite mine areas contributed to the collapse of the roof layers. As a result of this phenomenon, sandy material, along with fragments of rocks and lignite, appeared on the ground surface. This ejected sandy material is sometimes misinterpreted as ground liquefaction by soil mechanics researchers and engineers. The overburden of the sinkholes ranged between 2 and 4 m at this locality.

In addition to the sinkholes in the rice field, some sinkholes were observed along the roadway and in the parking lot next to residential buildings, as seen in Figure 2.19. The overburden of the sinkholes, which had a diameter of 2–3 m, ranged between 1 and 2 m at this locality. The ground shaking caused the collapse, as there was no trace of groundwater spouting from the sinkhole. Two sinkholes in residential areas and one sinkhole in a rice field occurred in the Tsukidate Town of Kurihara City. Although there were no casualties in these events, they caused some anxiety among the people living in such areas.

The 2011 East Japan Mega Earthquake caused a number of sinkholes in both residential areas and fields in Ichinoseki. The sinkholes enlarged in size after each large aftershock. Figures 2.20 and 2.21 show the sinkholes beneath or in close proximity to houses in Hanaizumi Town of Ichinoseki City. Figure 2.22 shows a sinkhole along a roadway in Hanaizumi Town of Ichinoseki City. In addition to these, sinkholes were also observed in rice fields.

In addition to the sinkholes explained in this article, there were also some sinkhole formations in abandoned lignite mine areas in Oshu City (Iwate Prefecture) and Matsushima City (Miyagi

Figure 2.19 Views of sinkholes along a roadway and parking lot near Ohira Village. (From Aydan and Tano, 2012a,b.)

Figure 2.20 A sinkhole beneath a house in Hanaizumi Town of Ichinoseki. The size of the sinkhole enlarged after the April 7, 2011 aftershock. (From Aydan and Tano, 2012a,b.)

(a)

(b)

Figure 2.21 Sinkholes beneath a house in Hanaizumi Town of Ichinoseki. (a) A large sinkhole extending below the storage room and the main building. (b) Cracking of the concrete base and walls in the main building. (From Aydan and Tano, 2012a,b.)

Prefecture). Sinkholes were observed in both residential areas and fields. However, there were no casualties in any incident.

Damage to abandoned underground quarries occurred in Oya Town (Tochigi Prefecture). Oya tuff, or Oya stone, has been quarried in the Oya region of Utsunomiya, Japan. Over 200

Figure 2.22 Sinkhole beneath the roadway in Hanaizumi Town of Ichinoseki. (From Aydan and Tano, 2012a,b.)

underground quarries have been exploited for more than 100 years, and some of those are now beneath residential areas. Oya Town is about 295 km away from the earthquake epicenter.

Although ground motions were recorded in the vicinity of abandoned underground quarries in Oya, the instruments are mainly velocity-based and do not measure all three components of ground motion. Figure 2.23 shows the acceleration record at the Utsunomiya strong motion station of K-NET near Oya Town. The maximum ground acceleration at the Utsunomiya strong motion station of K-NET was 358 gals at the ground surface.

Total or partial collapse of semi-underground quarries occurred, as seen in Figures 2.24 and 2.25. The toppling of pillar (P1) caused the collapse of thin roof of the abandoned quarry, as seen in Figure 2.24. In some cases, partial collapses of the overhanging parts occurred, as seen in Figure 2.25.

The subsidence of backfilling material in three shafts of abandoned quarries was observed. In the example shown in Figure 2.26, the subsidence of backfilling material was more than 5 m, and the diameter of the shaft was about 7 m. The backfilling material was non-cohesive.

The backfilling material of the abandoned shaft settled by more than 3 m, as seen in Figure 2.27. There was a hut on top of this shaft, and both the hut and the car parked next to it fell into the settled shaft. These observations are very similar to those noted in Yamoto Town during the 2003 Miyagi Hokubu Earthquake and model experiments on shafts with backfilling described briefly in the previous section (Aydan, 2004; Aydan and Kawamoto, 2004; Kawamoto et al., 2004).

In addition to the possibility of caving in abandoned mines during earthquakes due to pillar failure and/or roof failure, groundwater in the mines may present additional effects on the immersed abandoned mines. These effects may be observed as sloshing, which could further weaken the surrounding rock mass and cause additional collapses.

The 2011 Great East Japan earthquake caused damage to structures and collapses of abandoned mines in Iwate and Miyagi prefectures, where mining activity was high about 40 years ago.

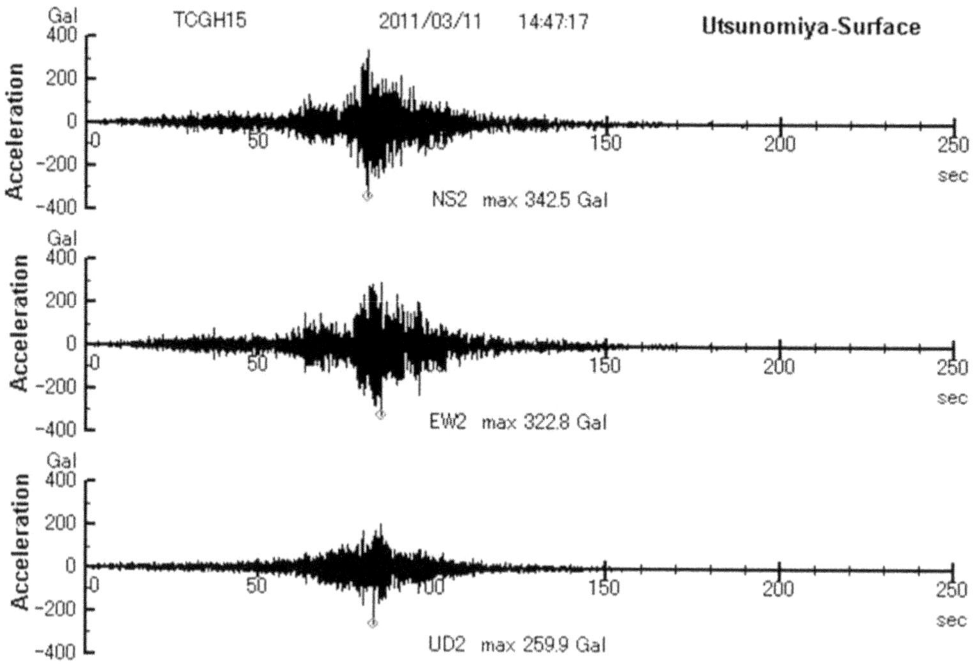

Figure 2.23 Acceleration records at Utsunomiya. (From Aydan and Tano, 2012a,b.)

Figure 2.24 Totally collapsed semi-underground opening (From Aydan and Tano, 2012a,b.)

Figure 2.25 Partially collapsed semi-underground opening (same as above). (From Aydan and Tano, 2012a,b.)

Figure 2.26 Subsidence of backfilling material of the shaft. (From Aydan and Tano, 2012a,b.)

Figure 2.27 Subsidence of backfilling material of the shaft and the fall of the hut and car into the shaft. (From Aydan and Tano, 2012a,b.)

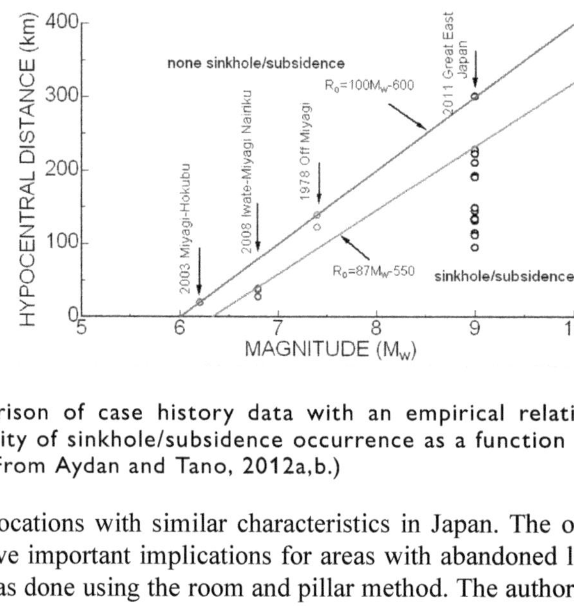

Figure 2.28 Comparison of case history data with an empirical relation for assessing the possibility of sinkhole/subsidence occurrence as a function of earthquake magnitude. (From Aydan and Tano, 2012a,b.)

There are several locations with similar characteristics in Japan. The observations from this earthquake may have important implications for areas with abandoned lignite and coal mines where extraction was done using the room and pillar method. The authors plotted the case history data described in the previous section along with data from other earthquakes that occurred in the Tohoku region, in a graph showing earthquake moment magnitude versus hypocentral distance of the locality where sinkholes or large subsidence occurred, as shown in Figure 2.28. The data was fitted to a linear function, with the coefficients displayed in the same figure. The line in the figure should be interpreted as a limiting line between surface damage and non-damage on the ground surface. This figure may serve as a guideline to assess the risk of sinkholes or large subsidence due to abandoned mines and quarries exploited using the room and pillar method, as a function of earthquake magnitude. However, there is still a great need to compile more data to confirm the validity of the proposed relation.

2.4 Effect of Urbanization over Abandoned Mines

The history of mining and underground quarries in many countries dates back several millennia (Figure 2.29). Quarries such as the Qurna underground quarry near Luxor in Egypt and the Bazda

(a) (b) (c)

Figure 2.29 Views of some abandoned underground limestone quarries. (a) Bazda (Türkiye). (b) Kuşini (Türkiye). (c) Qurna (Egypt).

Figure 2.30 Views of a partially inundated abandoned lignite mine in Mitake. (Courtesy of Tobishima Corporation, Japan.)

underground quarry in southeast Türkiye were excavated approximately 4,000–5,000 years ago (e.g., Aydan et al., 2014; Ağan et al., 2013). Over time, such excavations have mostly been forgotten, and urbanization has taken place. Furthermore, the increase in population in many countries has led to the inhabitation of areas above the abandoned underground mines and quarries. New town developments in Akaike, Nagakute, Nissin, and Kasugai in Aichi Prefecture, Japan, serve as examples. Additionally, the construction of new railways, roadways, and tunnels beneath presents further issues related to abandoned underground mines.

2.5 Effect of Groundwater in Abandoned Underground Mines

Groundwater in abandoned mines is generally considered to prevent the degradation of pillars and roof layers (Figure 2.30). Furthermore, the effective vertical stress acting on the pillars decreases due to groundwater, and pillar stress can be expressed as

$$\sigma'_p = \frac{1}{1-e}(\gamma_r H - \gamma_w H_w) \tag{2.1}$$

where $\gamma_r, \gamma_w, H, H_w$, and e are the unit weights of rock, water, overburden, water depth, and excavation ratio, respectively. If an abandoned mine is fully inundated and the water height is

(a) (b) (c)

Figure 2.31 Views of some abandoned room and pillar mines. (a) Iron mine in France. (b) Zonguldak. (c) Tuzköy rock salt mine.

equivalent to the overburden, the stress acting on the pillar would decrease by about 43%, which would act on the safer side. However, as discussed and presented in Chapter 3, the strength of the surrounding lignite layer decreases by about 60%–75% under fully saturated conditions. In other words, groundwater is likely to further decrease the overall safety of abandoned lignite mines.

2.6 Some Examples of Abandoned Room and Pillar Mines in Other Countries

In addition to many ancient underground quarries and mines, the exploitation of coal, lignite, rock salt, gypsum, potash, and building stones has been conducted using the room and pillar mining method. Such extensive room and pillar mines are mostly abandoned and can be found in the USA, UK, France, other European countries, Russia, South Africa, Latin America, and Australia. Numerous reports of sinkhole and/or subsidence issues arise from time to time (Figure 2.31). Groundwater further decreases the overall safety of abandoned lignite mines. The ISRM established a commission on mine closure to address issues related to mines developed using the room and pillar technique. The report produced by the mine closure commission of ISRM describes the state of the art on this issue, and interested readers are encouraged to refer to the reports produced by Didier et al. (2008, 2009).

Chapter 3

Assessment of Mechanical Properties of Surrounding Rocks and Their Degradation

This chapter focuses on the mechanical properties of lignite and the rocks surrounding abandoned lignite mines and their degradation. First, a brief summary of the geology of the Tokai region is presented. The geological age of soft rocks associated with lignite deposits varies from location to location. Figure 3.1 shows the distribution of lignite deposits in the Tokai region. While the age of the lignite field within the Nakamura unit of the Mizunami formation in the Mitake area belongs to the Miocene era, the lignite field belonging to the Yadagawa unit of the Seto formation in the Nagakute area is from the Pliocene. Therefore, the rocks in the Mitake area are older than those in the Nagakute area, and as a result, their physical and mechanical characteristics are expected to differ. There are four lignite seams in the Mitake area, of which the second

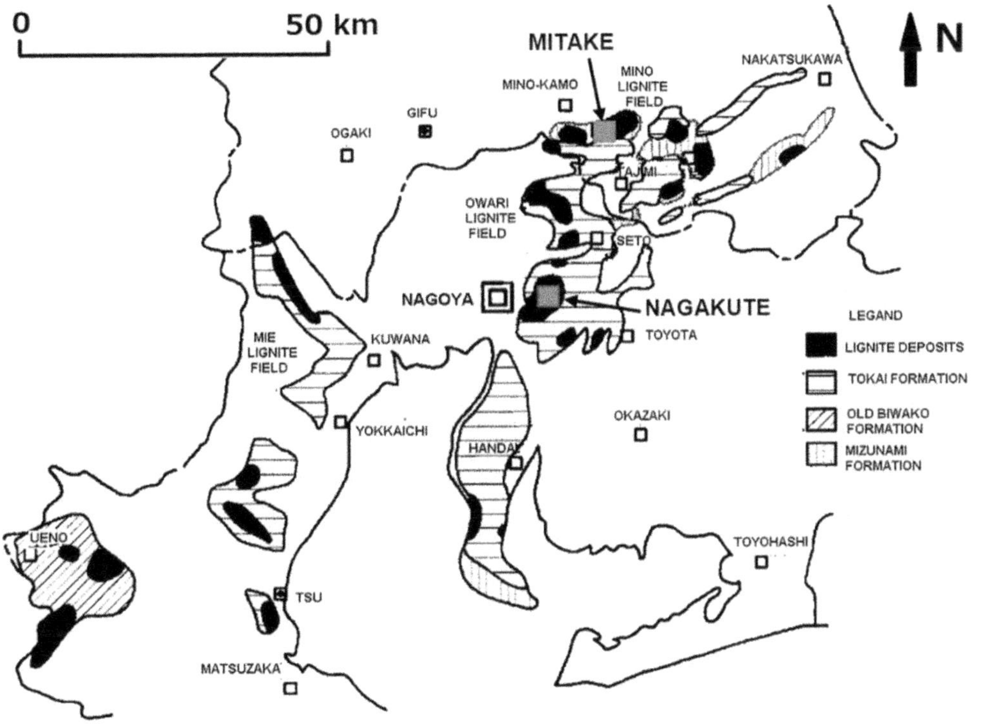

Figure 3.1 Distribution of lignite deposits in the Tokai region of Japan. (From Aydan et al., 2005a,b; modified and translated from Cavity Filling Association.)

DOI: 10.1201/9781003517214-3

and third seams were extracted. The thickness of the third seam is 1–3 m. Although the number of lignite seams in the Nagakute area is also four, the seams are much thinner (less than 1 m). Sedimentary rocks in the Mitake area are broadly classified as lignite, sandstone, siltstone, and mudstone, with conglomerate layers at the base of the formation above the pre-Tertiary rocks. Although the sedimentary rocks in the Nagakute area can be geologically classified in the same manner, the cementation is poor, and sometimes non-cemented sand layers are found beneath the lignite seams.

3.1 Laboratory Tests

3.1.1 *Short-Term Experiments*

Physical tests involve unit weight, water content, porosity, needle penetration index, and elastic p- and s-wave velocities, while mechanical tests include uniaxial compression and Brazilian tests (Aydan et al., 2005a), implemented according to the Suggested Methods of ISRM (Ulusay, 2014; Ulusay and Hudson, 2007). In addition, some creep and relaxation tests are also performed according to the ISRM Suggested Method on creep tests on rocks (Aydan et al., 2014b). Figure 3.2 shows views of some samples obtained from an abandoned lignite mine in Mitake Town.

The mechanical properties of soft sedimentary rocks are influenced by water content. As the compressive strength and elastic Young's modulus of soft rocks generally decrease with increasing water content, the effect of water content on the mechanical properties of the soft rocks was also investigated. During some experiments, electric potential (EP), electrical resistivity, magnetic field, and acoustic emissions (AEs) were measured in addition to conventional load and displacement parameters. The main purpose of such measurements was to establish some experimental bases for the real-time monitoring of multiple parameters for the stability assessment of abandoned mines in Mitake (Aydan et al., 2005a).

Unit weight, water content, porosity, needle penetration index, and elastic wave velocity were measured while mechanical tests, including uniaxial compression and Brazilian tests, were conducted using the devices shown in Figures 3.3 and 3.4. Tables 3.1 and 3.2 summarize

Figure 3.2 Views of samples from the Mitake abandoned lignite mines.

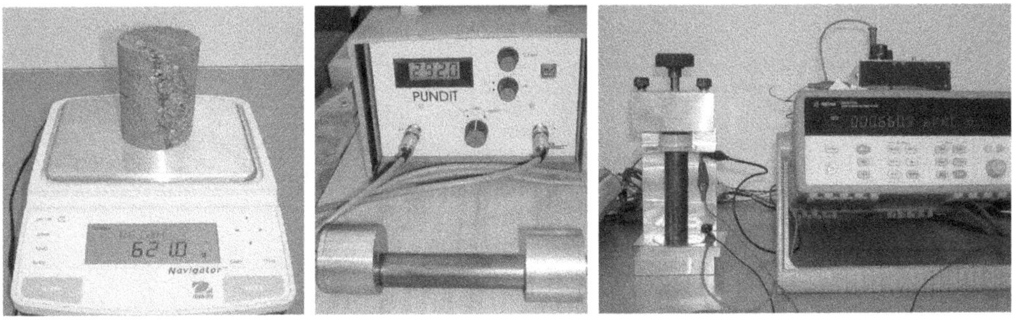

(a) (b) (c)

Figure 3.3 Views of physical property measurements. (a) Unit weight. (b) Wave velocity. (c) Electrical resistivity.

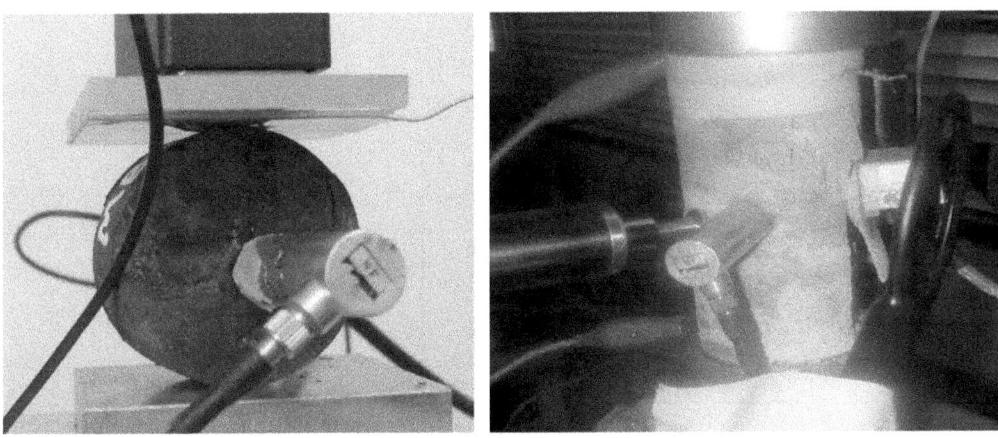

(a) (b)

Figure 3.4 Views of strength tests: (a) Brazilian tensile, (b) Uniaxial compression.

Table 3.1 Physical and mechanical properties of soft rocks from the Nagakute area

Rock Type	*Unit Weight (kN/m^3)*	*Water Content (%)*	*Compressive Strength (MPa)*	*Elastic Modulus (MPa)*	*NPI (N/mm)*
Lignite	7.1–10.2	30.2–35.3	2.3–3.2	86.8–167.9	16–28
Mudstone	15.6–16.5	7.3–17.1	0.3–2.2	15.2–77.8	2–22

Table 3.2 Physical and mechanical properties of soft rocks from Mitake area

Rock Type	*Unit Weight (kN/m^3)*	*Water Content (%)*	*Compressive Strength (MPa)*	*Elastic Modulus (MPa)*	*Vp (km/s)*	*NPI (N/mm)*
Lignite	10.6–14.1	12.6–27.2	4.0–6.0	1800–1860	1.6–2.3	22–33
Sandstone	16.8–19.2	15.4–19.8	2.3–7.0	328–	1.6–2.7	20–71(dry)
Mudstone	15.6–16.5	9.5–13.3	1.0–1.5	38–150	1.3–1.5	2–22

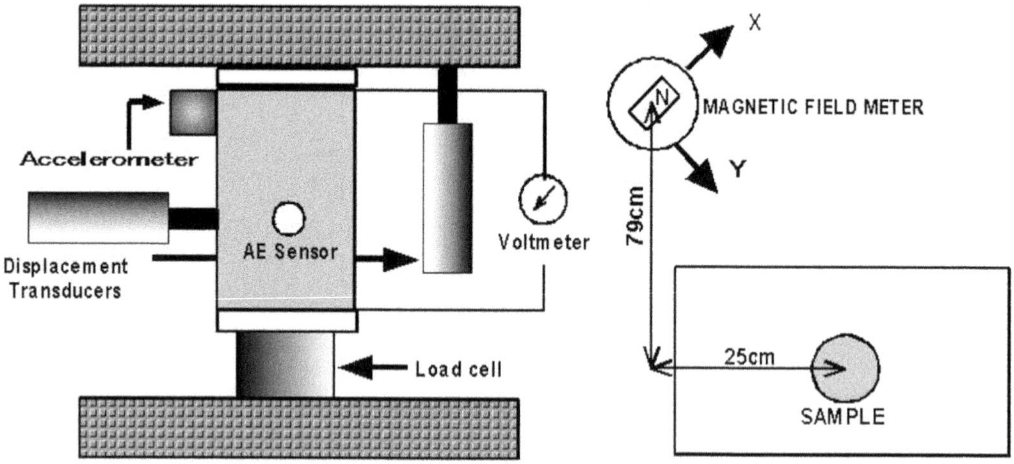

Figure 3.5 Illustration of the multi-parameter measurement system.

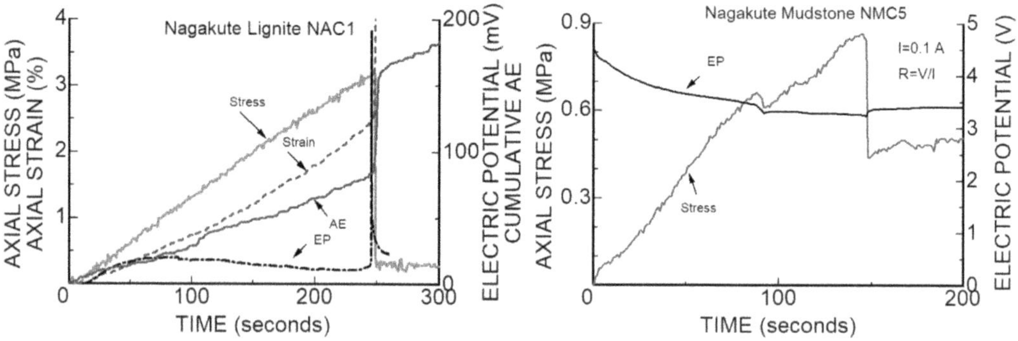

Figure 3.6 Multi-parameter response of rocks from the Nagakute area during uniaxial compression tests. (From Aydan et al., 2005a.)

the physical and short-term mechanical properties of rock samples from the Nagakute and Mitake areas.

The authors (Aydan et al., 2001, 2003, 2005a,d) developed a multi-parameter measurement system that includes EP, electrical resistivity, AEs, rock temperature, and magnetic field variations, which were measured in some of the tests reported herein. EP was measured directly as DC or AC using the AGILENT 34970A, while the Keithley 2400 Series electric current generator was used to measure electrical resistivity. Figure 3.5 illustrates the multi-parameter measurement systems. In addition, accelerations induced during the fracturing of rock samples were measured by attaching accelerometers directly to the samples. While most acceleration measurements were taken simultaneously with other parameters, stand-alone QV3-OAM-SYC accelerometers set to "Trigger Mode" were also utilized. Figures 3.6–3.10 show some of the experimental multi-parameter responses measured during experiments on samples from the Mitake and Nagakute areas.

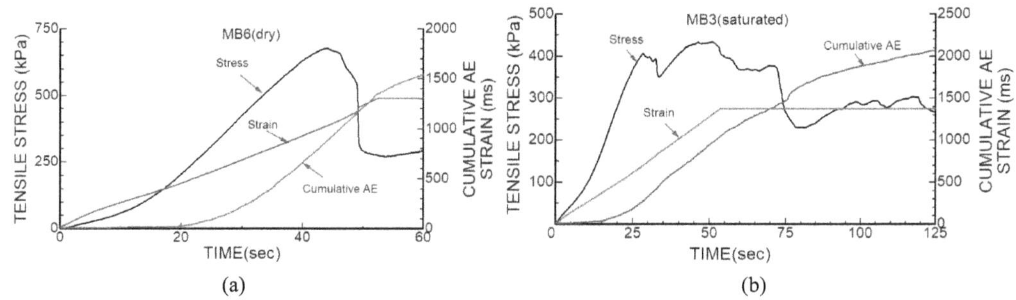

Figure 3.7 Responses of (a) dry and (b) saturated lignite samples during Brazilian tests. (From Aydan et al., 2005a.)

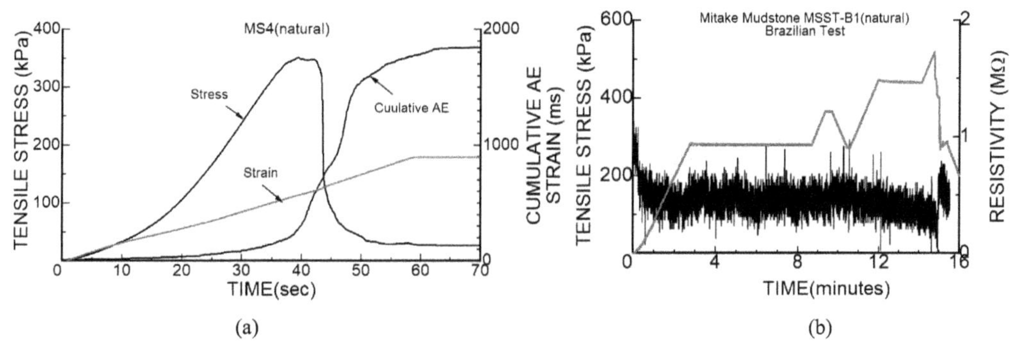

Figure 3.8 Responses of (a) sandstone and (b) mudstone samples with natural water content during Brazilian tests. (From Aydan et al., 2005a.)

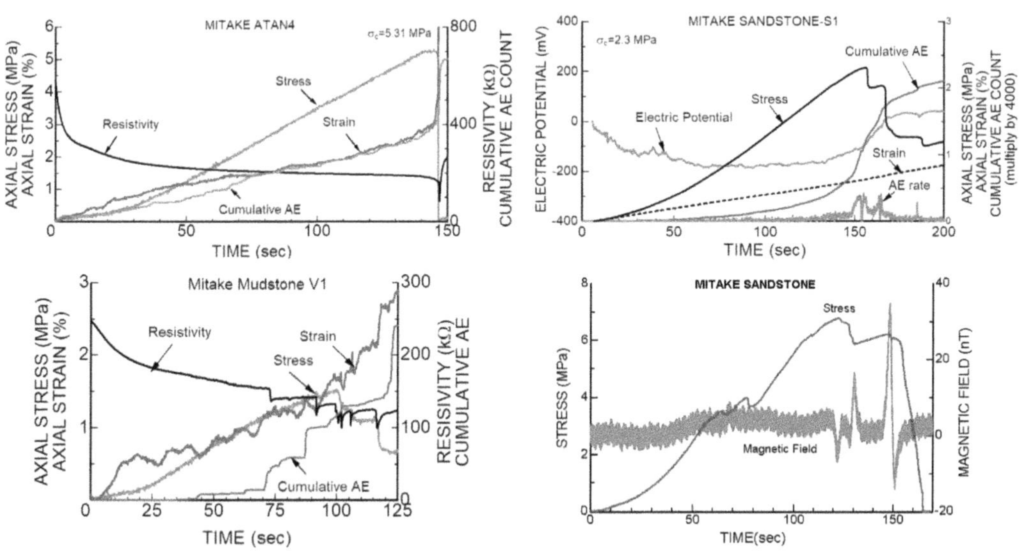

Figure 3.9 Multi-parameter responses of rocks from the Mitake area during uniaxial compression tests. (From Aydan et al., 2005a.)

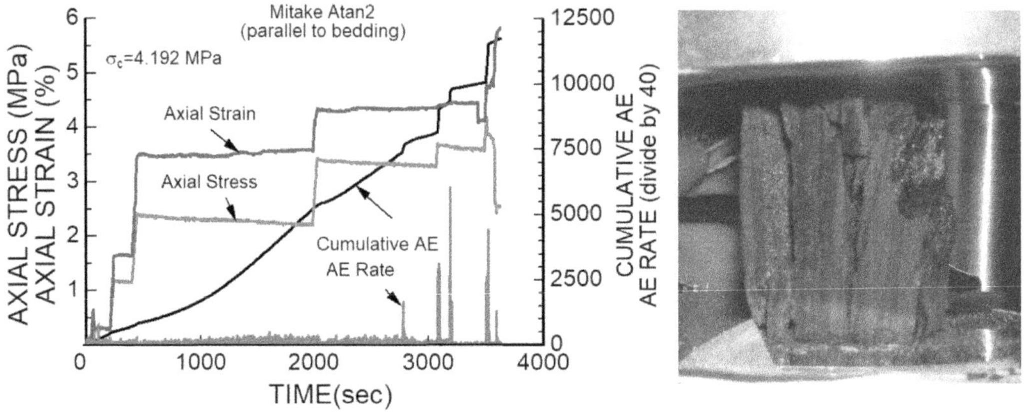

Figure 3.10 Multi-parameter responses of a lignite sample under step-like loading. (From Aydan et al., 2005a.)

Table 3.3 Parameters of yield functions for soft rocks from the Mitake area

Rock Type	*Mohr–Coulomb*				*Aydan*					
	Lower		*Upper*		*Lower*			*Upper*		
	c MPa	*ϕ (°)*	*c MPa*	*ϕ (°)*	*A (MPa)*	*b*	*σ_t (MPa)*	*A (MPa)*	*b*	*σ_t (MPa)*
Lignite	0.67	52.4	1.05	52.4	3.0	0.66	0.430	4.0	0.50	0.670
Sandstone	0.40	52.4	0.70	63.4	2.5	0.66	0.250	5.0	0.50	0.351
Mudstone	0.35	16.7	0.40	28.8	1.1	0.58	0.518	1.7	0.58	0.518

As discussed by Aydan et al. (2012b), it is also possible to determine the triaxial strength of rocks using the results of the Brazilian tensile strength and uniaxial compression experiments. In addition to the Mohr–Coulomb yield criterion, Aydan et al. (2012b) modified Aydan's criterion (Aydan, 1995) by relating its parameters to those of the Mohr–Coulomb criterion for isothermal conditions as follows (Aydan et al., 2014):

$$\tau = c_\infty - (c_\infty - c_o)e^{-\sigma_n/b} \tag{3.1}$$

where

$$b = \frac{\sigma_t}{\ln\left(\dfrac{c_\infty}{c_\infty - c_o}\right)}$$

Table 3.3 provides the parameters of the Mohr–Coulomb and Aydan's failure criteria determined from uniaxial compression and Brazilian tests, and Figures 3.11–3.13 compare the yield criteria for dry and saturated conditions.

Needle penetration index tests are also useful for inferring the mechanical properties of soft rocks (Aydan, 2012; Aydan et al., 2014c). Figures 3.14 and 3.15 show the results for Mitake and Nagakute soft rocks together with, alongside those of other soft rocks.

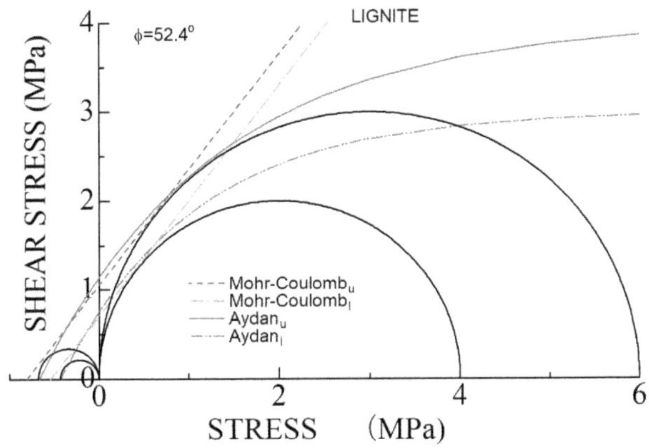

Figure 3.11 Yield criteria for lignite seam.

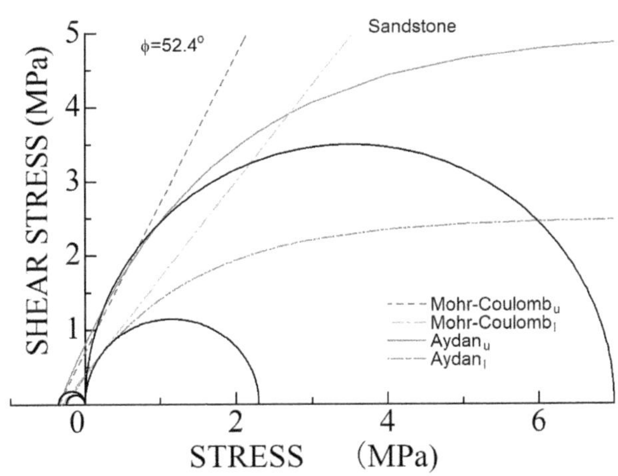

Figure 3.12 Yield criteria for sandstone.

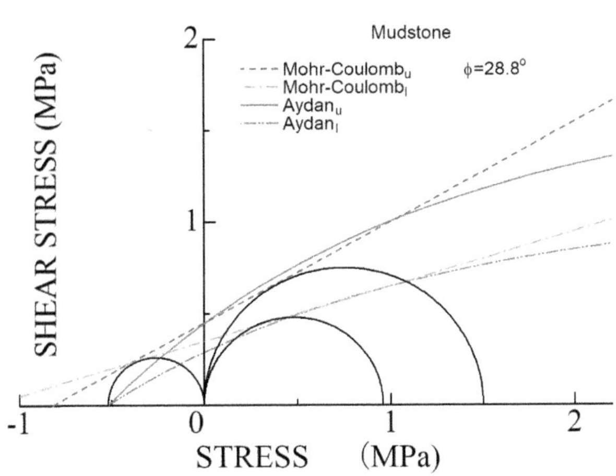

Figure 3.13 Yield criterion for mudstone.

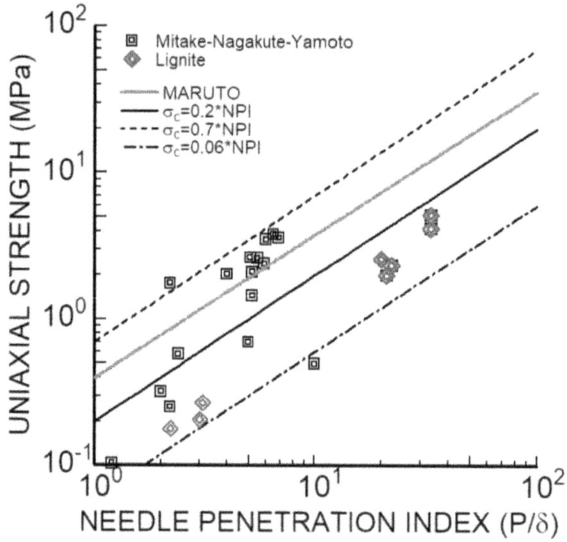

Figure 3.14 NPI vs compressive strength. (From Aydan, 2012; Aydan et al., 2014c.)

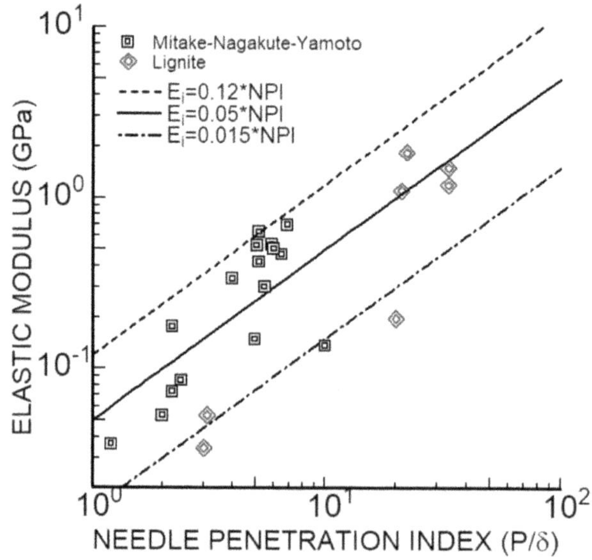

Figure 3.15 NPI vs elastic modulus. (From Aydan, 2012; Aydan et al., 2014c.)

3.1.2 Long-Term Experiments

The experiments were primarily conducted using a specially designed creep-loading device at Toyota National College of Technology (the device has since been moved to the University of the Ryukyus) and a conventional compression testing machine at the Marine Civil Engineering Department of Tokai University and the University of the Ryukyus (Aydan et al., 2014b, 2016a; Ito et al., 2016; Aydan, 2016). The creep testing device, shown in Figure 3.16, has a loading

Figure 3.16 A view of the device and a close-up view of the experimental setup. (From Ito et al., 2016.)

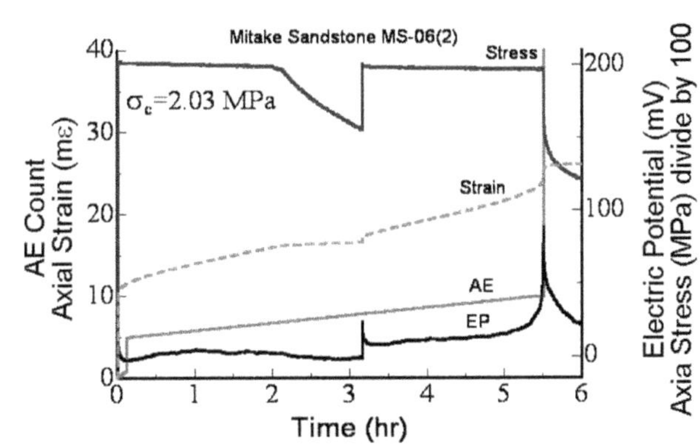

Figure 3.17 Creep test result on sample MS-06. (From Aydan et al., 2005a.)

capacity of 50 kN. It can impose loads in increments of 0.1 kN, and all loading operations are performed manually to avoid any electric disturbances during loading. The loads for the conventional compression testing machine are also applied manually.

The applied load and induced displacement were automatically measured and stored on the hard disk of a laptop computer through an electronic logger. The EP induced during the deformation of samples was measured using two electrodes attached to the top and bottom of the samples with a voltmeter and logger unit (AGILENT A8670A), with data simultaneously stored on the laptop's hard disk. The electrodes were isolated from the loading frame using isolators. The EP was measured either as DC or AC. One of the main parameters influencing the long-term stability of abandoned mines is the creep strength characteristic of rocks (e.g., Griggs, 1939; Singh, 1975; Aydan et al., 2005a, b, c, 2006; Aydan, 2016, 2018). A series of creep tests on samples from Mitake Town were carried out, measuring variations in EP and AEs in addition to the applied axial stress and induced axial strain. Figure 3.17 shows the responses measured in one of the creep tests.

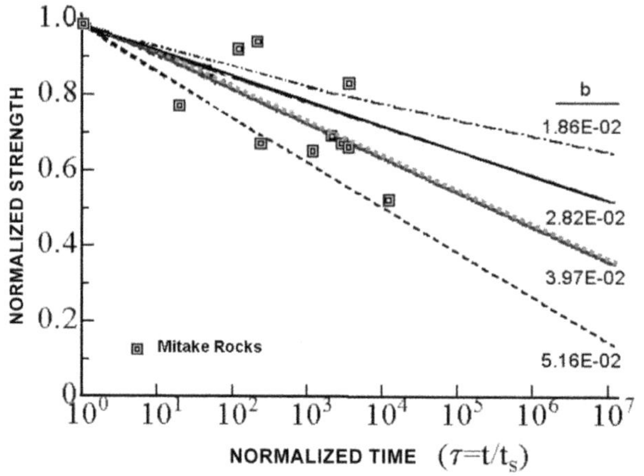

Figure 3.18 Comparison of normalized long-term strength of Mitake soft rocks with other rocks. (From Aydan et al., 2005a.)

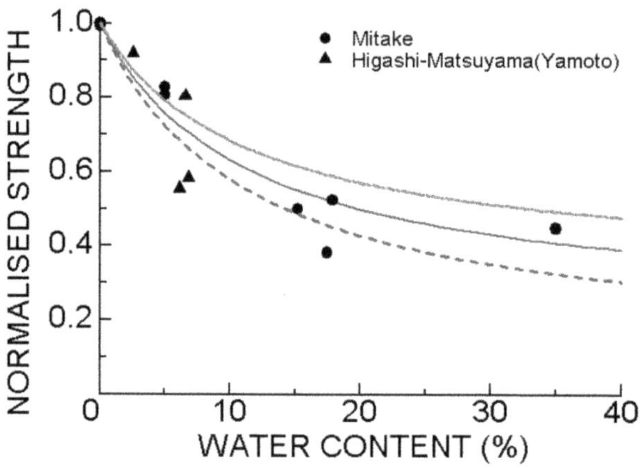

Figure 3.19 Reduction of uniaxial compressive strength as a function of water content. (From Aydan and Ito, 2015.)

Aydan and Nawrocki (1998) proposed the following function for the long-term strength of rocks normalized by their short-term strength.

$$\frac{\sigma_{\mathrm{cl}}}{\sigma_{\mathrm{cs}}} = 1 - b \ln\left(\frac{t}{t_s}\right) \tag{3.2}$$

The coefficient b in the above function is determined to be 0.0397 from the creep experiments shown in Figure 3.18.

When lignite mines are abandoned, they may become wholly or partially submerged in groundwater. Experiments have shown that the long-term strength of lignite and surrounding soft rocks drastically reduces upon submergence in groundwater, as shown in Figure 3.19.

Some sedimentary rocks are prone to degradation due to cycles of wetting and drying. This mainly occurs due to the suction of water by certain minerals, such as clay minerals. As a result, rocks disintegrate after each cycle of wetting and drying. Soft rocks in Mitake and Nagakute exhibit such behavior, as shown in Figure 3.20.

The same phenomenon is observed in abandoned mines. The degree of degradation varies depending on the environmental conditions within the mines. However, the degradation of rocks near the entrance of the mines is significantly higher than that of rocks located farther away. Such variations cause flaking of surrounding rocks from the roof and pillars of the lignite seam due to water content loss, resulting in volumetric shrinkage and a reduction in the support area of pillars and thinning of roof layers. Figure 3.21 shows some examples of degradation resulting from cycles of wetting and drying.

A series of tests were carried out to understand the mechanism of degradation during wetting and drying cycles. The tests involve the measurement of weight, shrinkage, EP or electric resistivity, and AEs of the sample. Weight is measured to obtain the water content over time, while AEs and electrical potential are measured to observe if cracking occurs during the drying

Figure 3.20 Some examples of disintegration of soft rocks of Mitake. (From Aydan et al., 2005a.)

Figure 3.21 Some examples of disintegration in the abandoned mine. (From Aydan et al., 2005a.)

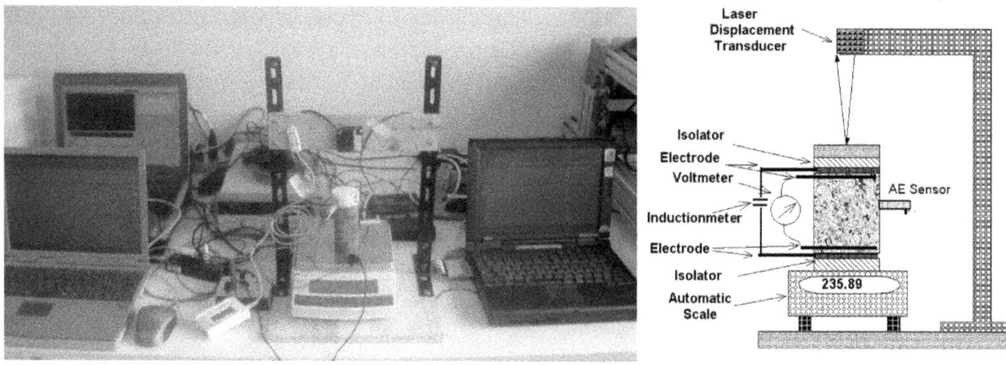

Figure 3.22 A view of the degradation test and its illustration. (From Aydan et al., 2005a.)

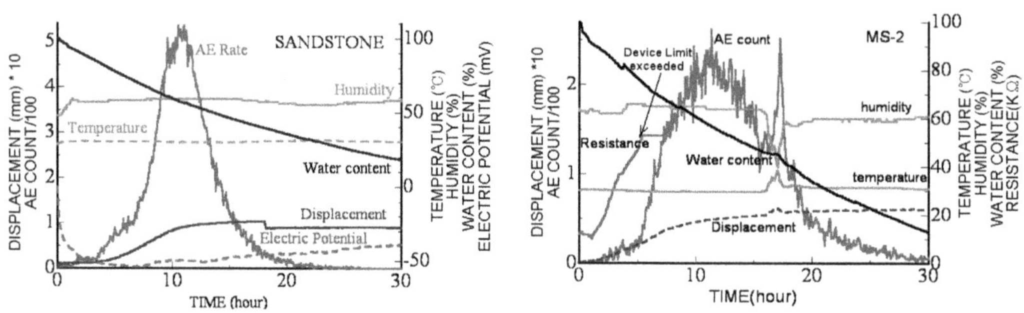

Figure 3.23 Degradation tests on the Mitake sandstone sample numbered MS-02. (From Aydan et al., 2005a.)

process. Electrical resistivity is measured to correlate water content with electrical resistivity, which could be utilized to infer water content in-situ as a non-destructive testing method. Furthermore, the temperature and humidity of the sample environment are measured simultaneously.

Figure 3.22 shows a view of a typical experimental setup. The experimental results are presented for a sandstone sample only, as shown in Figure 3.23. On the same sample, electrical potential and electrical resistivity tests were carried out. As noted from the figure, some cracking begins to occur after about 3 hours, and the cracking activity reaches its maximum after about 10–12 hours. This roughly corresponds to the peak of the drying rate. The sample shrinks, which implies that when the minerals lose water, they are volumetrically shrinking. As a result of this phenomenon, cracking would be induced in samples if their tensile strength is insufficient. It is also interesting to note that some samples were fractured or lost their cohesion during the saturation process (Figure 3.24a).

3.2 In-situ Investigations

3.2.1 Discontinuity Surveys

In this sub-section, the results of the surveys carried out on the discontinuities at one of the abandoned lignite mines in Mitake are explained. Discontinuities observed in the rock mass of the mine can be classified as bedding planes (BP), joints (J), and faults (F) (Figures 3.25 and 3.26).

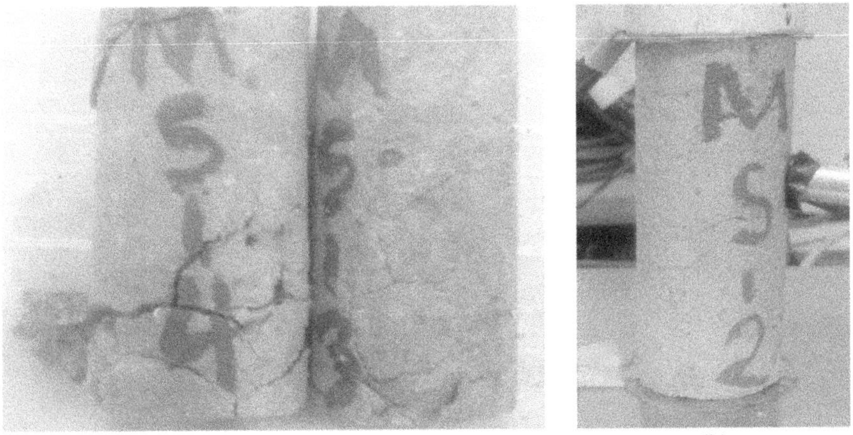

(a) (b)

Figure 3.24 Cracking of sandstone samples induced by submerging in water or drying. (a) Cracking upon submerging in water. (b) Drying-induced cracking. (From Aydan et al., 2005a.)

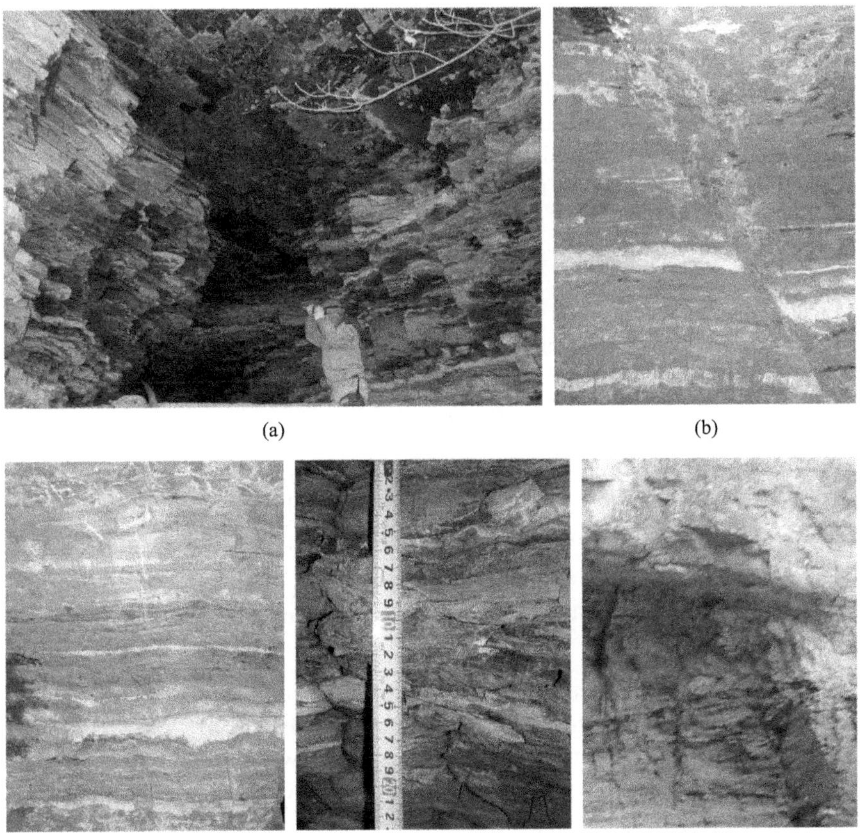

(a) (b) (c) (d)

Figure 3.25 Some views of discontinuities in an abandoned mine at Mitake Town. (a) A view of intercalated rock mass. (b) Normal fault seen in the mine. (c) Jointing in a pillar. (d) State of rock mass at the entrance.

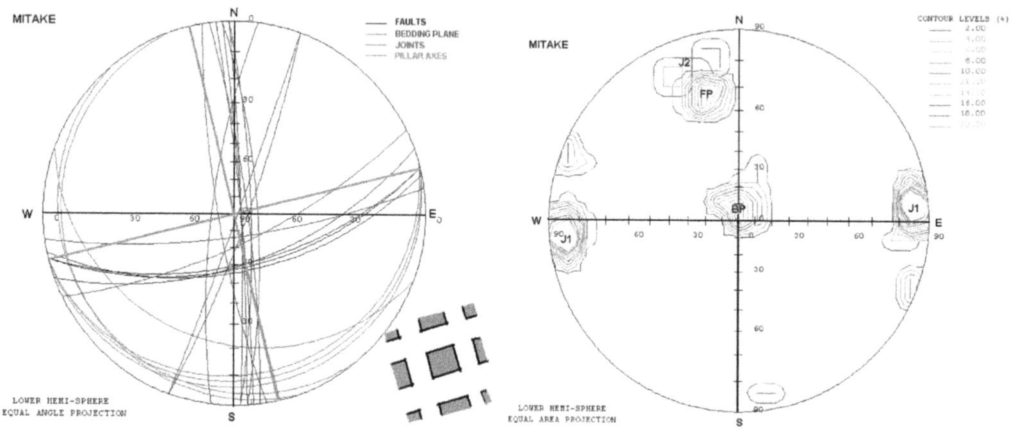

Figure 3.26 Stereo projection of discontinuities at the abandoned mine.

Rock layers generally dip towards the SE with a gentle inclination and have been folded. The folding axis is aligned in the direction of NE-SW. There are three lignite seams, and their thickness ranges between 1 and 6 m. The thickness of the lignite seam at the mine is about 2 m, consisting of intercalated mudstone and sandstone layers whose thickness ranges between 5 and 15 cm. Joints are generally perpendicular or sub-perpendicular to layering, and their spacing ranges between 10 and 60 cm. The joint interval becomes smaller within the lignite seam while it is widely spaced in mudstone and sandstone layers. Normal faults with a slight sinistral or dextral lateral sense are observed in the mine. The strike of major normal faults is aligned with the flow direction of the Kiso River, which passes nearby the mine.

3.2.2 *In-situ Index Tests*

Several in-situ index tests were performed at the abandoned mine. These include Schmidt hammer rebound tests, needle penetration tests, and elastic wave velocity measurements using McSEIS-III (Figures 3.27–3.29). The results from these index tests are summarized in Table 3.4.

In addition to McSEIS-III, a hand-held wave velocity measurement device called CardVibro 2000 was used to measure wave velocity parallel to the BPs. The p-wave velocity was measured at 867 m/s, which is almost half that of the intact lignite samples (Figure 3.30).

3.3 Characterization of Rock Mass through Rock Mass Classifications

The rock layers in the Mitake area generally dip toward the southeast with a gentle inclination of 5°–20° and have been folded. The folding axis is aligned in the direction of NE-SW. There are three lignite seams, with thicknesses ranging between 1 and 6 m. The thickness of the lignite seam at the mine is about 2 m and consists of intercalated mudstone and sandstone layers, with thicknesses ranging between 5 and 15 cm. Joints are generally perpendicular or sub-perpendicular to the layering, with spacing ranging between 10 and 60 cm. The joint interval becomes smaller within the lignite seam, while it is widely spaced in the mudstone and sandstone

Figure 3.27 NPI test.

layers. Normal faults with a slight sinistral or dextral lateral sense are observed in the mine. The strike of major normal faults is aligned along the NE-SW direction, parallel to the flow direction of the Kiso River, which passes nearby the mine.

Rock mass classifications are conducted according to Rock Mass Quality Rate (RMQR), Rock Mass Rate (RMR), and Q-system rock mass classification systems (Aydan et al., 2014a; Bieniawski, 1984; Barton et al., 1974). RMR values correspond to BASIC RMR values, indicating that no corrections are implemented. Rock masses are broadly classified as typical rock mass and fractured zone. Results are summarized in Tables 3.5–3.7.

3.4 Rock Mass Properties

Based on the rates of rock masses, some in-situ properties are inferred according to empirical relations. For this purpose, four approaches proposed by Aydan et al. (2014), Hoek-Brown (1980–1997), Kalamaras-Bieniawski (1995), and the wave velocity method (Ikeda, 1970) are

Figure 3.28 In-situ elastic wave velocity measurement.

Figure 3.29 Schmidt hammer rebound test.

Table 3.4 Results of in-situ index tests

Rock	*Rebound Number (R)*	*P-wave Velocity (km/s)*	*Needle Penetration Index (N/mm)*
Lignite	18–44	1.5	7.0–12.5
Sandstone	14–28		20.0–25.0
Mudstone	16–18	–	11.1–20.0

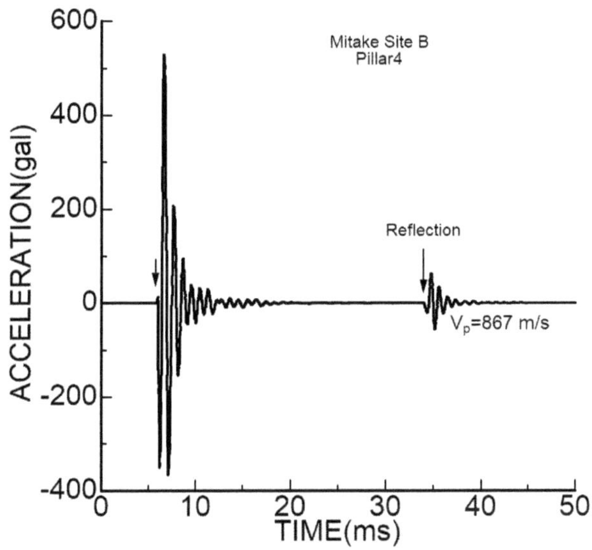

Figure 3.30 Measured wave response measured by the hand-held CardVibro2000.

Table 3.5 Rating of typical rock mass according to the RMQR system

Classification Item	*State*	*Rate*
Degradation degree (DD)	Slight to moderate	6–9
Discontinuity set number (DSN)	Three sets + random	8
Discontinuity spacing (DS/RQD)	10–60 cm	4–8
Discontinuity condition (DC)	Relatively smooth	13–15
Groundwater seepage condition (GSWC)	Damp to wet	5–7
Groundwater absorption condition (GWAC)	Moderately absorptive	3
Total RMQR		39–50

Table 3.6 Rating of fractured rock mass according to the RMQR system

Classification Item	*State*	*Rate*
Degradation degree (DD)	Slight to moderate	6
Discontinuity set number (DSN)	Four sets + random	4
Discontinuity spacing (DS/RQD)	Less than 7 cm	1–4
Discontinuity condition (DC)	Slicken-sided	7
Groundwater seepage condition (GSWC)	Wet to dripping	3–5
Groundwater absorption condition (GWAC)	Highly absorptive	2
Total RMQR		25–30

Table 3.7 Classification rates according to RMQR, RMR, and Q-systems

Rock Mass	*RMR*	*Q-value*	*RMQR*
Typical	37–53	2.08–2.50	39–50
Fractured	26–29	0.83–1.00	25–30

Table 3.8 Inferred normalized rock mass strength (σ_{cm}/σ_{ci})

Method	*Typical*	*Fractured*
Aydan et al. (2014a)	0.09–0.16	0.06
Hoek-Brown (1988–1997)	0.03–0.07	0.016
Kalamaras-Bieniawski (1995)	0.12–0.21	0.085
Elastic wave velocity $(V_{pm}/V_{pi})^2$	0.44–0.83	

Table 3.9 Inferred normalized elastic modulus (E_{cm}/E_{ci})

Method	*Typical*	*Fractured*
Aydan et al. (2014a)	0.09–0.16	0.06
Hoek (2007)	0.94–5.94	2.98–5.94
Elastic wave velocity $(V_{pm}/V_{pi})^2$	0.44–0.83	

used, and the results are shown in Table 3.8. The rock mass strength inferred from p-wave velocity is much higher than those inferred from other methods. However, the inferences by Aydan et al. (2014a) and Kalamaras-Bieniawski (1995) are similar to each other and are likely representative values for in-situ rock mass. Inferred normalized elastic moduli of rock masses using different methods are given in Table 3.9. As noted in previous applications of the approach proposed by Hoek (2007), the inferred elastic modulus of rock mass is several times that of intact rock. Therefore, the elastic modulus of rock mass inferred from the Aydan-Kawamoto or elastic wave velocity method would be reasonable.

It should be noted that Aydan et al. (2014a) provided relations for various properties of rock mass in terms of those of intact rock, in addition to uniaxial compressive strength and elastic modulus.

3.5 Inter-Relations of Various Properties of Rocks of Abandoned Lignite Mines

Interrelations among various properties of rocks are crucial for making quick assessments. Figure 3.31 presents some interrelations among several parameters, along with fitted empirical relations or available relations proposed by Aydan et al. (1992, 1993, 1996).

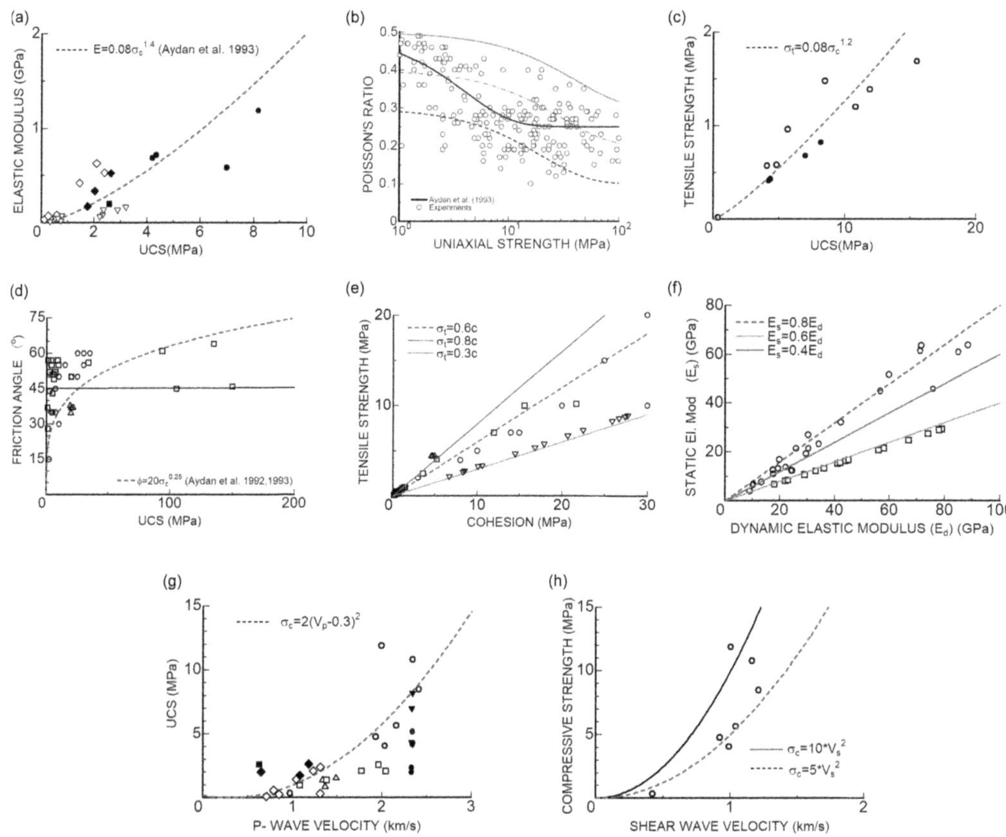

Figure 3.31 Inter-relations among various mechanical and physical parameters. (a) UCS vs elastic modulus; (b) UCS vs. Poisson's ratio; (c) UCS vs. Tensile strength; (d): UCS vs. Friction angle; (e) Cohesion vs. Tensile strength; (f) Dynamic elastic modulus vs. Static elastic modulus; (g) P-wave velocity vs UCS; (h) S-wave velocity vs. UCS. (Arranged from Aydan et al., 1992, 1993, 1996, 2014c.)

Chapter 4

Model Tests on Static and Dynamic Responses of Room and Pillar Mines

This chapter mainly focuses on model tests related to the mechanical stability, stress distributions, and seismic effects of earthquakes, faulting, and large-scale excavations on abandoned room and pillar mine models.

4.1 Laboratory Mechanical Tests on Pillars

Experiments on pillar samples prepared from Ryukyu limestone blocks were carried out. Short-term tests were performed using SHIMADZU loading test apparatuses with a capacity of 2,000 kN at the Universities of Ryukyus. The loading was applied to the samples either automatically with a pre-set loading rate of 0.1%/min or manually. Long-term tests were carried out at Toyota National College of Technology (NCT). The load on the samples was measured using load cells attached to the loading frame. Displacements in the direction of loading were measured using Tokyo-Sokki displacement transducers.

The temperatures of the samples were recorded with an ANRITSU AMS7000 Series Data Collector and five electronic thermistors attached directly to the samples, as shown in Figure 4.1. Three thermistors were attached to the sample, while the fourth thermistor was attached to the bottom loading platen. The remaining thermistor was fixed near the bottom platen to measure air temperature.

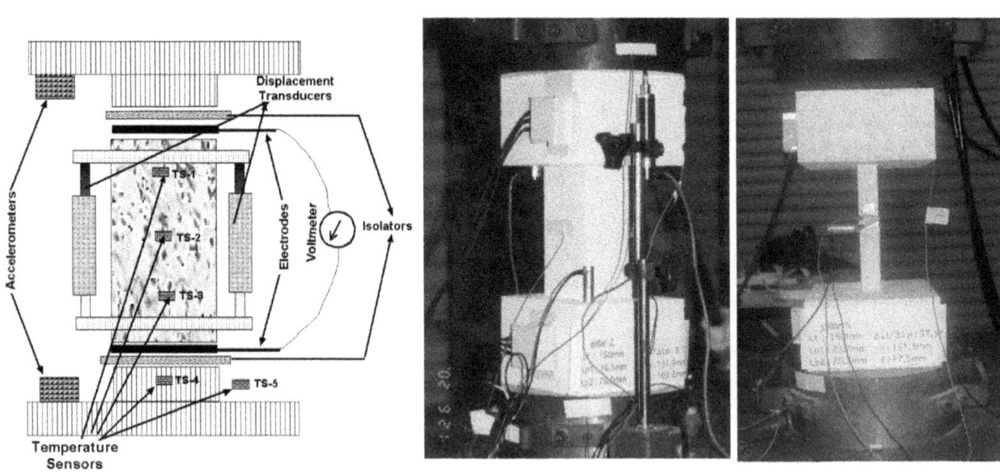

Figure 4.1 Experimental setup on pillars. (From Tokashiki et al., 2004, 2005.)

DOI: 10.1201/9781003517214-4

Acoustic emissions (AEs) were measured using a sensor attached to the bottom loading platen, as it was difficult to attach the sensors directly to the sample. The AE rate was counted for each time interval and plotted cumulatively. Two accelerometers were attached to the top and bottom ends of the sample. Each accelerometer measured accelerations in the direction of loading and two orthogonal directions to the loading direction. The device was set to trigger mode (Figures 4.1 and 4.2).

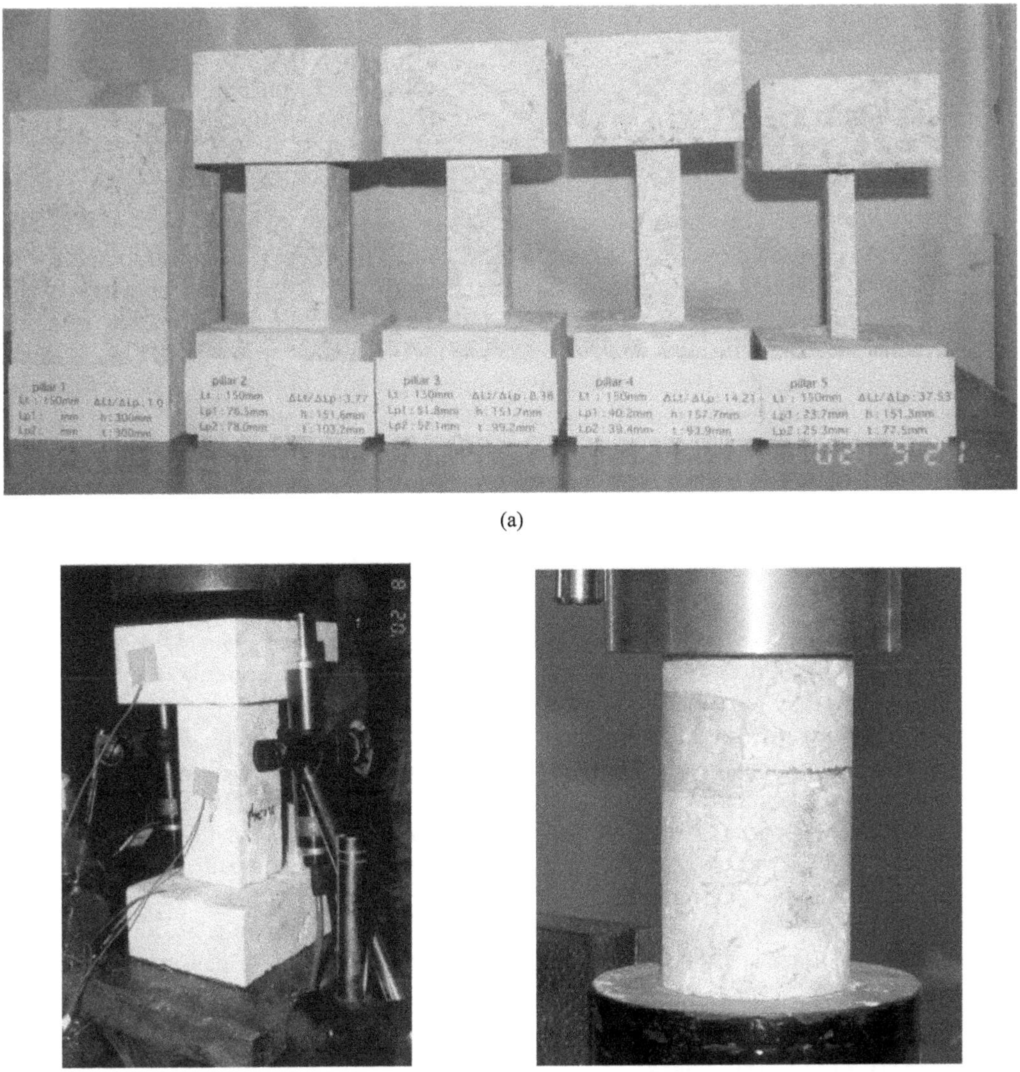

(a)

(b)

(c)

Figure 4.2 Views of samples at different institutes. (a) University of the Ryukyus. (b) Tokai University. (c) Toyota NCT. (From Tokashiki et al., 2004, 2005.)

Table 4.1 Properties of Ryukyu Limestone blocks

ρ (kN/m^3)	V_p (km/s)	E (GPa)	υ	σ_c (MPa)
19.6–23.4	4.5–6.3	8.1–27.8	0.15–0.3	20.0–42.0

Table 4.2 Dimensions of Ryukyu Limestone blocks at Tokai University

Test No	*Sample*			*Pillar*		
	Height (mm)	*Side1 (mm)*	*Side2 (mm)*	*Height (mm)*	*Side1 (mm)*	*Side2 (mm)*
TP1	100	40	40	100	40	40
TP2	180	100	80	100	40	40
TP3	180	100	80	100	40	40

The properties of Ryukyu limestone are provided in Table 4.1. Three model tests for each pillar model type were carried out, and the dimensions of the blocks are given in Table 4.2. Figure 4.3 shows the samples tested at Tokai University (TU), NCT, and the University of the Ryukyus (UR) along with instrumentation. The samples used in the long-term tests are cylindrical, with a height of 50 mm and a diameter of 25 mm (Table 4.3).

A total of eight samples were tested in short-term experiments. Figures 4.3 and 4.4 show the responses of axial stress–strain and AE in some of the experiments conducted at TU and the UR. As expected, both strength and deformability decrease as the pillar area decreases. AE responses are well correlated with loading strain and correspond well to variations in electric potential. Tables 4.4 and 4.5 summarize the measured peak uniaxial compressive strength and deformation modulus during the tests.

Figure 4.5 shows the measured acceleration during the failure of Pillar RP2. As noted in the figure, the acceleration at the topside (moving platen) is larger than that at the bottom side attached to the samples. This observation aligns with previous findings (Aydan, 2003).

Figure 4.6 compares the experimental data with the tributary area formula (the ratio of the total area supported by the pillar to its area, A_t/A_p) for computing the strength of pillars. With the exception of one pillar, the experimental results obtained at TU and the UR exhibit a tendency similar to the theoretical predictions based on the tributary area. Similar experiments were conducted using much softer model material, and the results confirmed the validity of the tributary area approach. Therefore, it can be concluded that the tributary area approach can be safely used for short-term stability assessments. Nevertheless, some considerations are necessary regarding the long-term aspects of the problem. To clarify this issue, long-term experiments are of great importance, and the utilization of creep strength may be quite useful for assessing the long-term stability of abandoned room and pillar mines. This issue is addressed in Chapter 6.

4.2 Photo-Elasticity Tests

The invention of the photo-elasticity technique is attributed to Brewster (1815), while Fresnel (1820) formulated the phase delay Δ related to the maximum shear stress acting in a stressed body

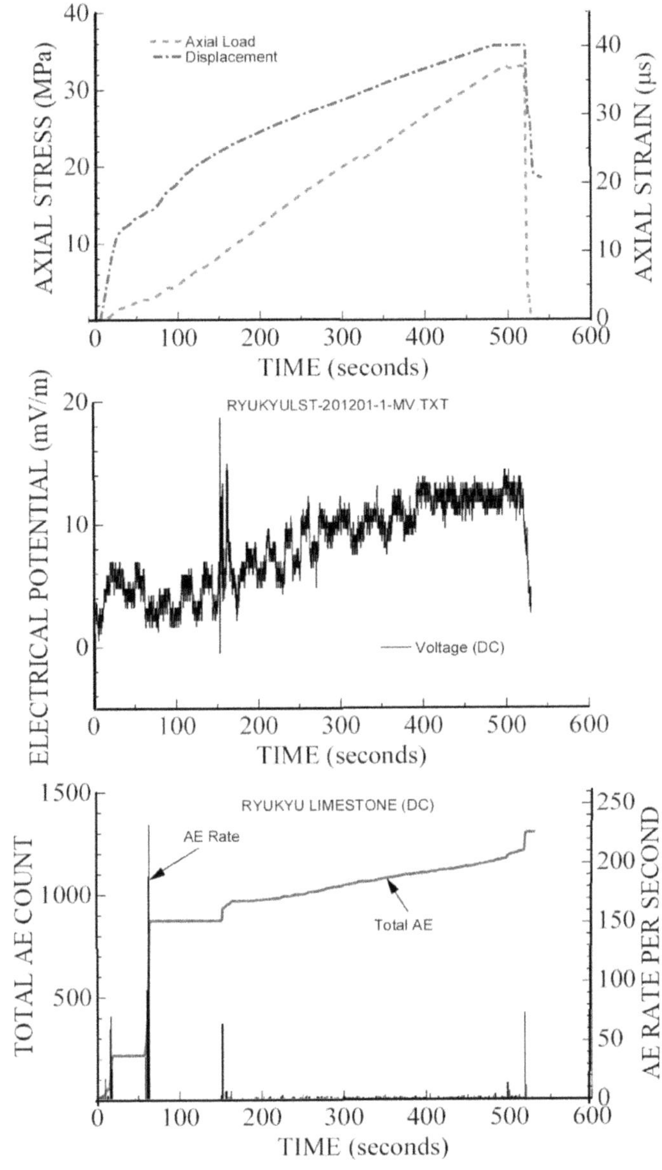

Figure 4.3 Multi-parameter responses of Pillar TP1. (From Tokashiki et al., 2004, 2005.)

with a given thickness *t*, which is associated with phase retardation (Figure 4.7). This relation is expressed in the following form:

$$\Delta = \frac{2\pi t}{\lambda} C\left(\sigma_1 - \sigma_3\right) \tag{4.1}$$

Table 4.3 Dimensions of Ryukyu Limestone blocks at University of the Ryukyus

Test No	*Sample*			*Pillar*		
	Height (mm)	*Side1 (mm)*	*Side2 (mm)*	*Height (mm)*	*Side1 (mm)*	*Side2 (mm)*
RP1	300	150	150	300	150	150
RP2	350	150	150	150	75	75
RP3	350	150	150	150	50	50
RP4	350	150	150	150	38	38
RP5	350	150	150	150	25	27

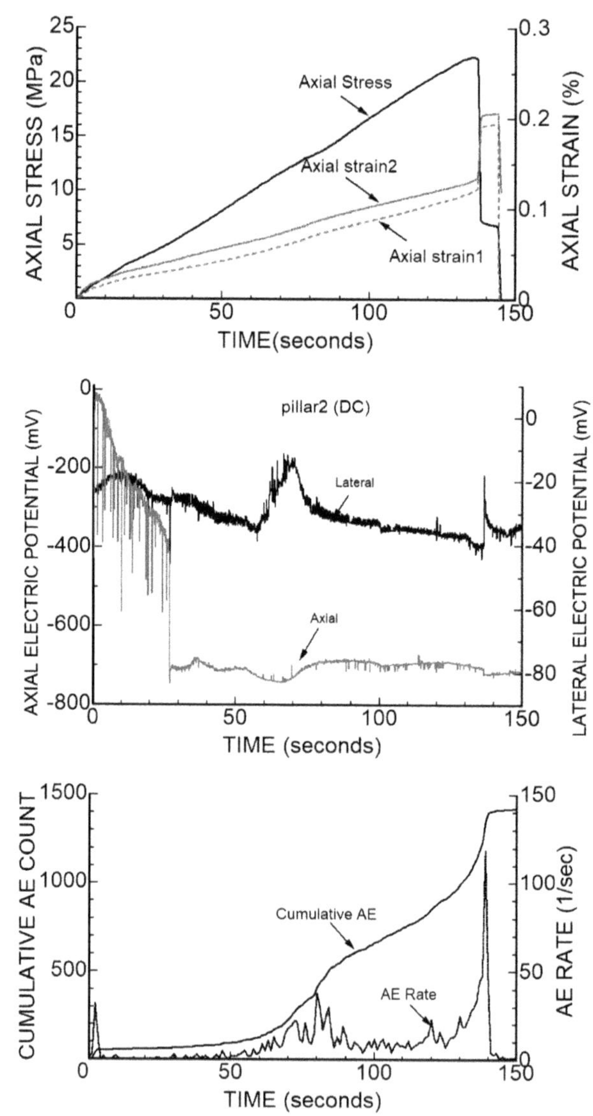

Figure 4.4 Multi-parameter responses of Pillar RP2. (From Tokashiki et al., 2004, 2005.)

Table 4.4 Properties of Ryukyu Limestone pillars tested at Tokai University

Test No	A_t / A_p	*E (GPa)*	σ_{max} *(MPa)*
TP1	1.00	12.4	34.6
TP2	5.00		8.35
TP3	5.00		6.04

Table 4.5 Properties of Ryukyu Limestone pillars tested at University of the Ryukyus

Test No	A_t / A_p	*E (GPa)*	σ_{max} *(MPa)*
RP1	1.00	28.2	41.1
RP2	4.00	12.4	22.4
RP3	9.00	9.12	6.12
RP4	16.0	28.0	25.0
RP5	36.0		

where C is the photo-elasticity coefficient, λ is the vacuum length, and σ_1 are σ_3 the principal stress components. A fringe pattern appears due to the optical interference of waves (e.g. Frocht, 1965). The number of fringe orders N is related to the retardation factor as

$$N = \frac{\Delta}{2\pi} \tag{4.2}$$

One can determine the state of stress at various points in the material from fringe patterns. The actual photo-elasticity devices are quite large and expensive, requiring a light source with well-known characteristics. However, it is quite difficult to find commercially available light sources in Japan these days, whereas they can be easily obtained from other countries such as India, China, and the USA.

Originally, polariscopes, which are based on the principle of photo-elasticity, were developed to display stress distributions and concentrations for educational purposes. It is also possible to produce hand-made polariscopes utilizing LED lights and polarizer films, while digital cameras with circular polarizers can be used as analyzers to capture static and dynamic images (Figure 4.7). The author developed such a system for photo-elasticity experiments, and the outcomes of these studies are presented in the following sections. Some compression experiments on photo-elasticity materials were conducted to determine their elastic modulus and Poisson's ratio. Figure 4.8 shows the strain–stress response of several polyurethane samples under uniaxial compression conditions. Table 4.6 summarizes the material properties of several polyurethane samples. Polyurethane-20 was used in most of the photo-elasticity experiments.

The existence of discontinuities is also of great importance for the stability of underground structures. Figure 4.9 shows the stress distributions in rock mass models consisting of layers and blocks. The presence of discontinuities significantly alters the stress distributions, and some discontinuities in the stress field are observed at the joints and layer interfaces. Figure 4.10 illustrates a more complex rock mass situation in a room and pillar mine model subjected to a concentrated load at the ground surface, as well as gravity load, which may provide insight

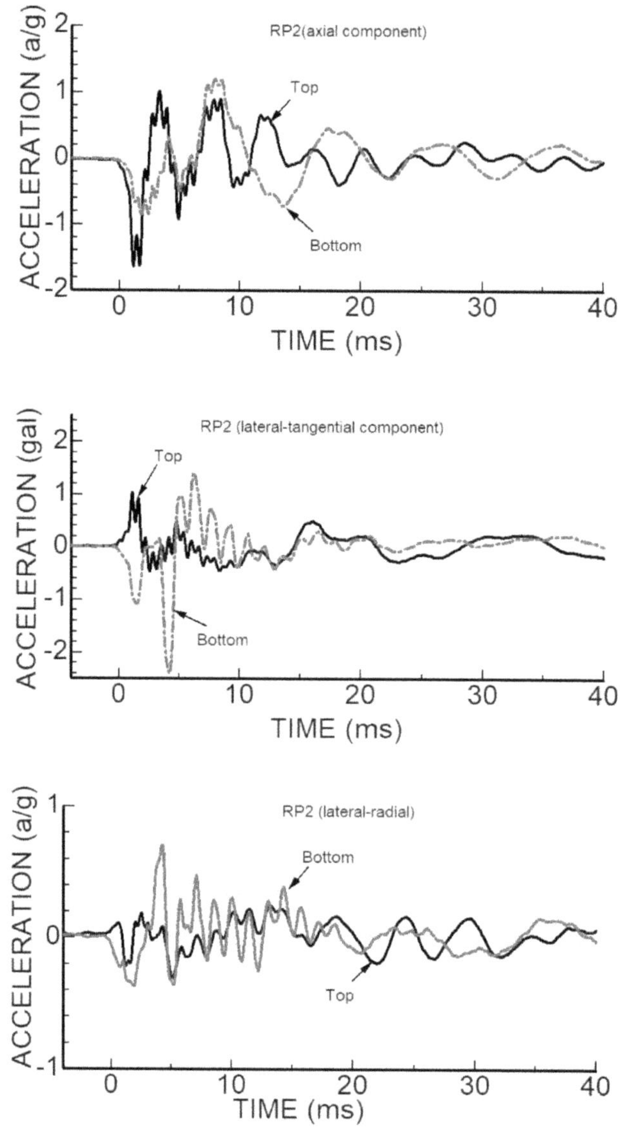

Figure 4.5 Acceleration responses of RP2 during failure. (From Tokashiki et al., 2004, 2005.)

into the stress distributions in abandoned mines beneath urbanized areas. Figure 4.11 shows the effect of backfilling on the stress distributions.

4.3 Shaking Table Tests

4.3.1 Model Tests Utilizing Non-breakable Model Materials

A series of experiments were conducted on model abandoned mines using non-breakable model materials (Aydan, 1994a; Aydan & Kawamoto, 1992; Aydan et al., 1994b, 2006; Geniş & Aydan, 2008). Figure 4.12a shows an illustration of the model abandoned mines. The main parameters

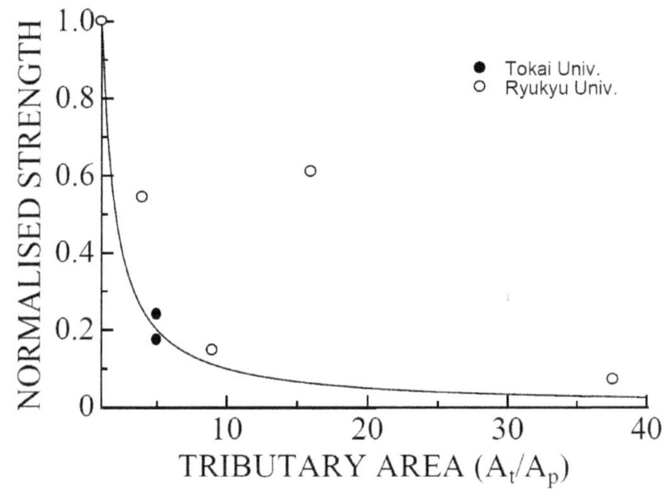

Figure 4.6 Comparison of pillar strength with the tributary area model. (From Tokashiki et al., 2004, 2005.)

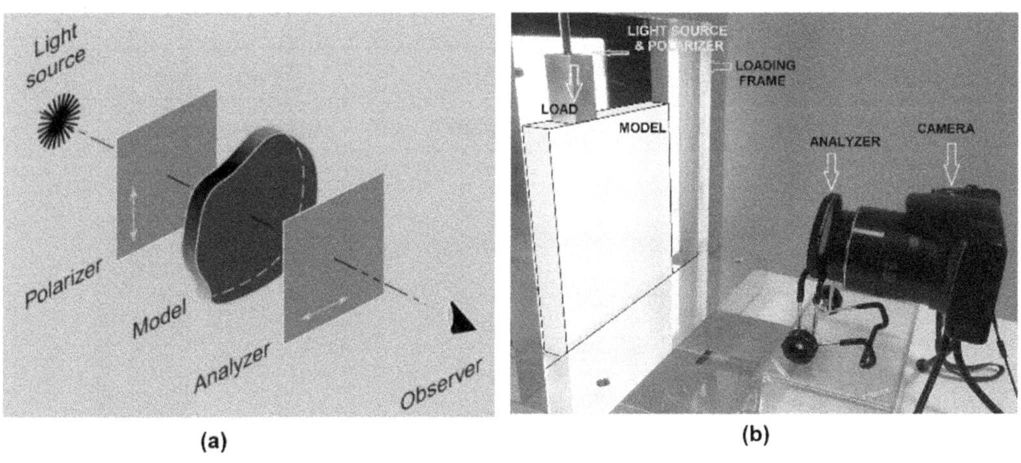

Figure 4.7 (a) Principle of photo-elasticity and (b) its practical implementation. (From Aydan, 2020; Aydan et al., 2021.)

of this series were to observe the effects of overburden ratio (H/h), excavation ratio (A_e/A_p), and measurement location on ground surface acceleration responses. Figure 4.13a shows the acceleration responses at the base, opening, and ground surface for an overburden ratio of 5 and excavation ratio of 3. As noted from the figure, the acceleration at the ground surface is at least twice that at the base of the model. Figure 4.13b shows the effect of the excavation ratio on the ground surface acceleration responses for an overburden ratio of 5. The acceleration increases as the excavation ratio increases. Figure 4.14a shows the acceleration responses at the ground surface above the opening and pillar for an overburden ratio of 5 and excavation ratio of 5 when the model was subjected to a sweeping test ranging from 3 to 100 Hz with an amplitude of 100 gals. As noted from the figure, the acceleration at the ground surface above

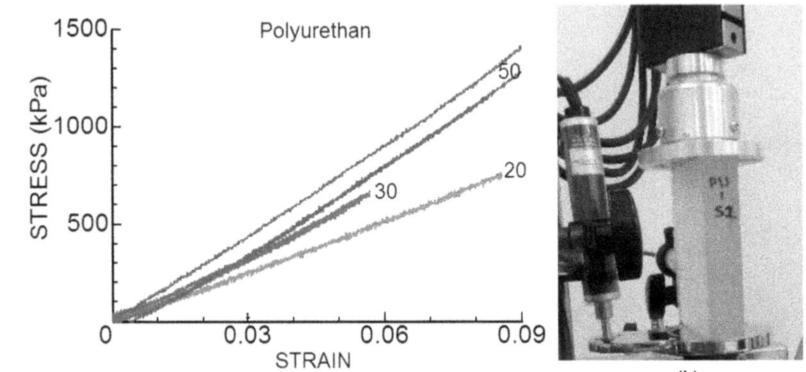

(a) (b)

Figure 4.8 (a) Strain–stress response of several polyurethane samples under uniaxial compression and (b) a view of an experiment. (From Aydan, 2020; Aydan et al., 2021.)

Table 4.6 Elastic modulus and Poisson's ratio of several polyurethane samples

Material	*Elastic Modulus (MPa)*	*Poisson's Ratio*	*Models*
Polyurethane-20	7.3	0.38	Faults, tunnels, layers, and blocks
Polyurethane-40	15.8	0.20	Frames

Figure 4.9 Stress distributions around underground openings in jointed complex rock mass. (From Aydan, 2020.)

Figure 4.10 Stress distributions around room and pillar mines with surface load in discontinuum. (From Aydan, 2020.)

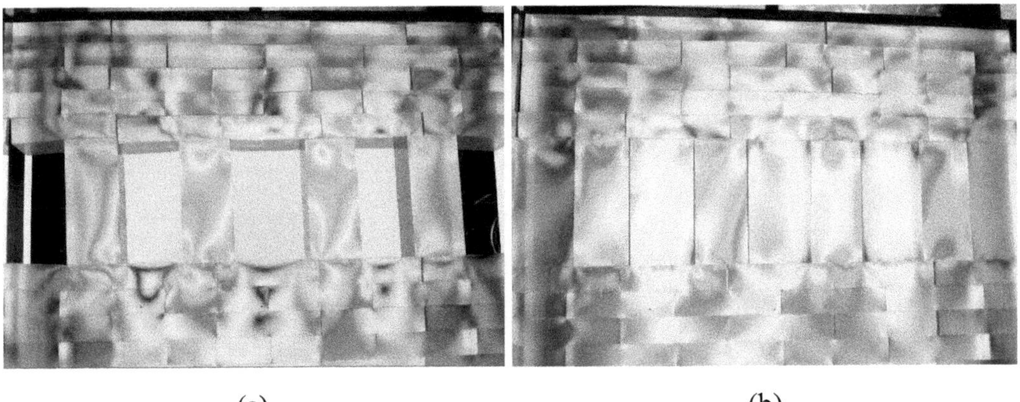

Figure 4.11 Stress distributions around room and pillar mines with/without backfilling. (a) Unfilled. (b) Backfilled.

the pillar is greater than that above the opening. The final example concerns the acceleration responses at the ground surface as a function of the overburden ratio for an excavation ratio of 5. These examples with non-breakable model materials clearly demonstrate what one should expect from acceleration responses at the ground surface for linear behavior of the ground during earthquakes.

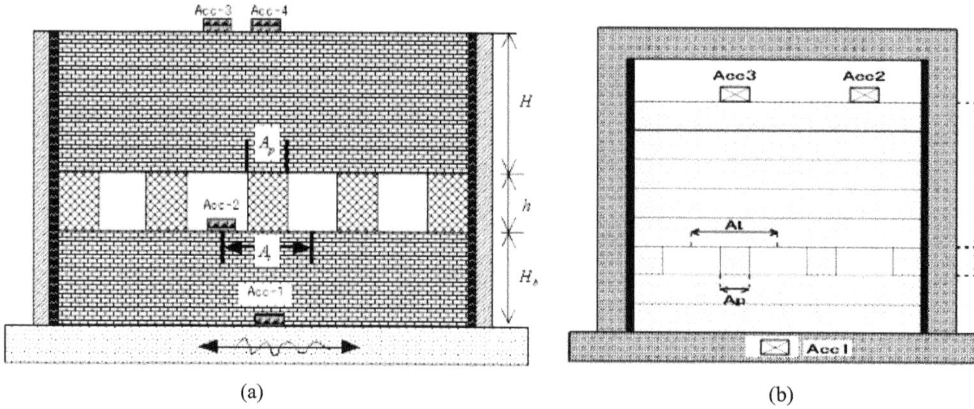

Figure 4.12 Illustration of the abandoned mines on shaking table. (a) Non-breakable model material. (b) Breakable model material. (From Aydan et al., 2013a.)

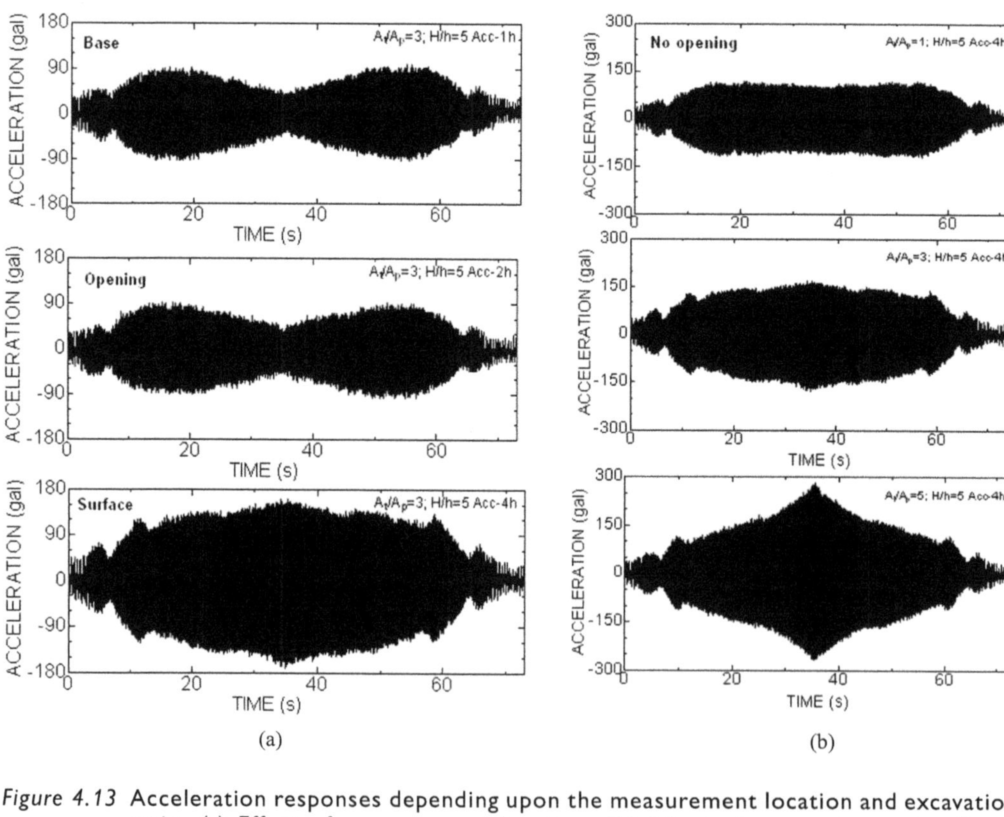

Figure 4.13 Acceleration responses depending upon the measurement location and excavation ratio. (a) Effect of measurement location. (b) Effect of excavation ratio. (From Aydan et al., 2013a.)

4.3.2 *Model Tests Utilizing Breakable Model Materials*

Some laboratory model tests on abandoned lignite mines were carried out. The material used in modeling rock layers consisted of $BaSO_4$, ZnO, and Vaseline oil, which were used in base friction model tests. The strength of this material depends upon the compaction pressure, allowing

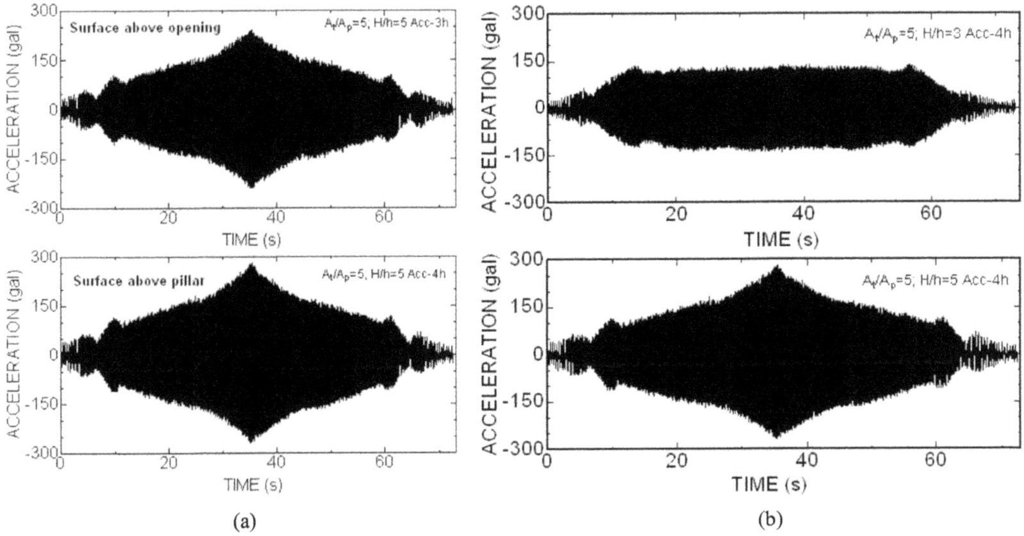

Figure 4.14 Acceleration responses as a function of measurement position and overburden ratio. (a) Effect of position at ground surface. (b) Effect of overburden ratio. (From Aydan et al., 2013a.)

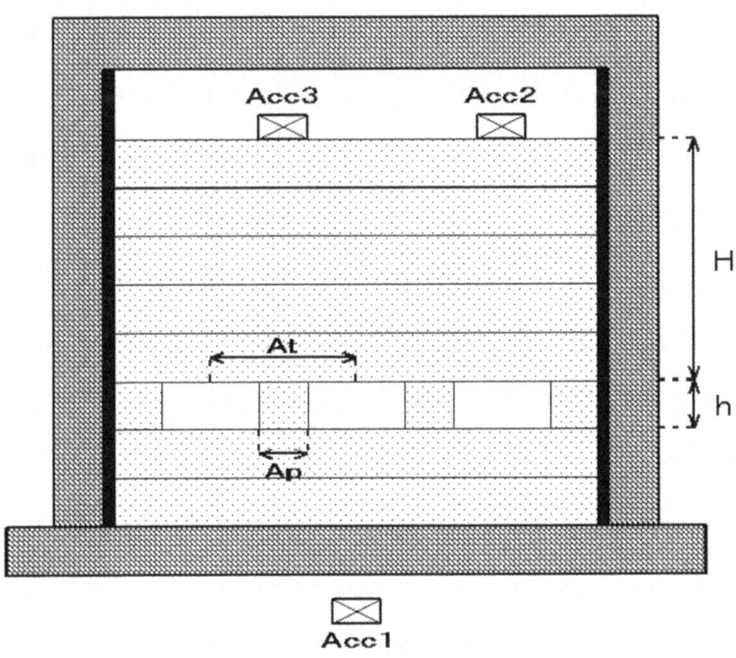

Figure 4.15 Model setup (Ohta et al., 2005).

for possibility of achieving failure in the models. Figure 4.15 shows the frame used in the model tests and the model setup. The model is fixed on a uniaxial shaking table apparatus and is shaken. Figure 4.16 shows the ground acceleration at failure versus the overburden ratio for

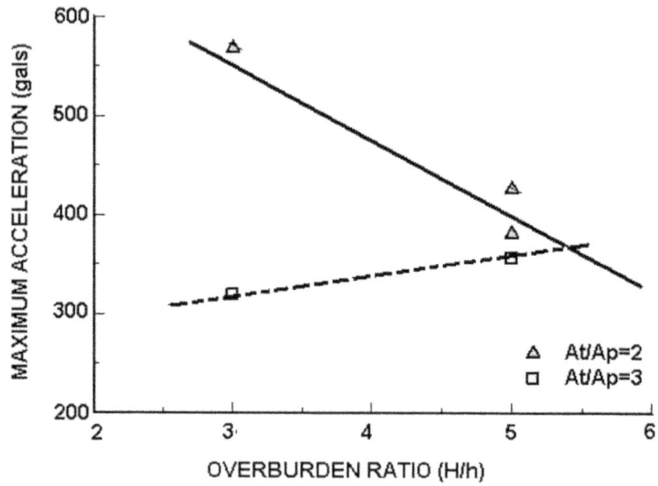

Figure 4.16 Maximum acceleration-to-overburden ratio (Ohta et al., 2005).

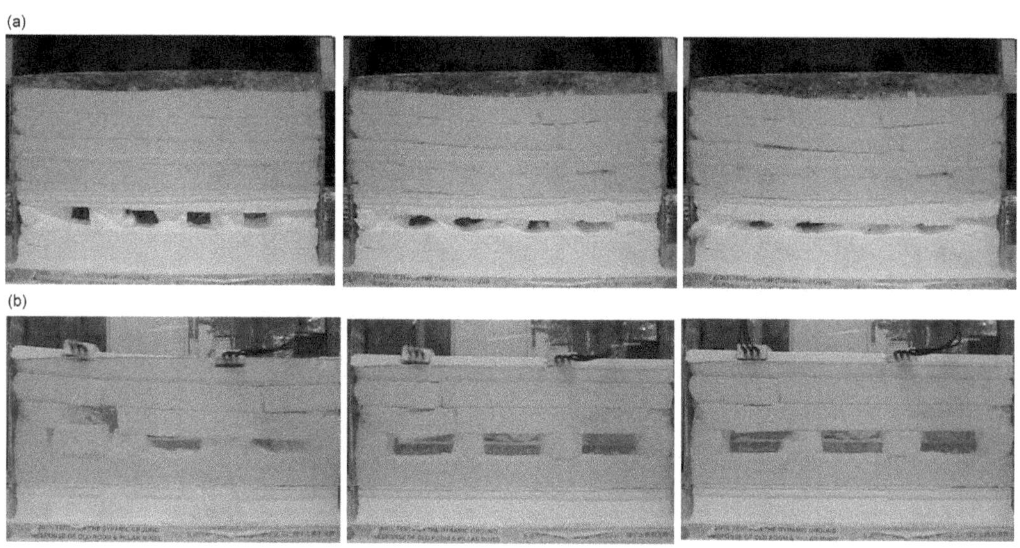

Figure 4.17 Some views of models of abandoned lignite mines during shaking table tests. (a) Compressive failure of pillars. (b) Failure of roof layers (Ohta et al., 2005; Aydan et al., 2010).

different excavation (tributary area) ratios of the models. In these particular model tests, failure occurred due to the compression failure of pillars for the excavation ratio (A_t/A_p) of 2, while failure was attributed to the roof layers for the excavation ratio (A_t/A_p) of 3, as shown in Figure 4.17.

Aydan et al. (2006) reported that shallow room and pillar abandoned mines fail due to the collapse of the roof layer, resulting in a sinkhole, while deeper room and pillar abandoned mines

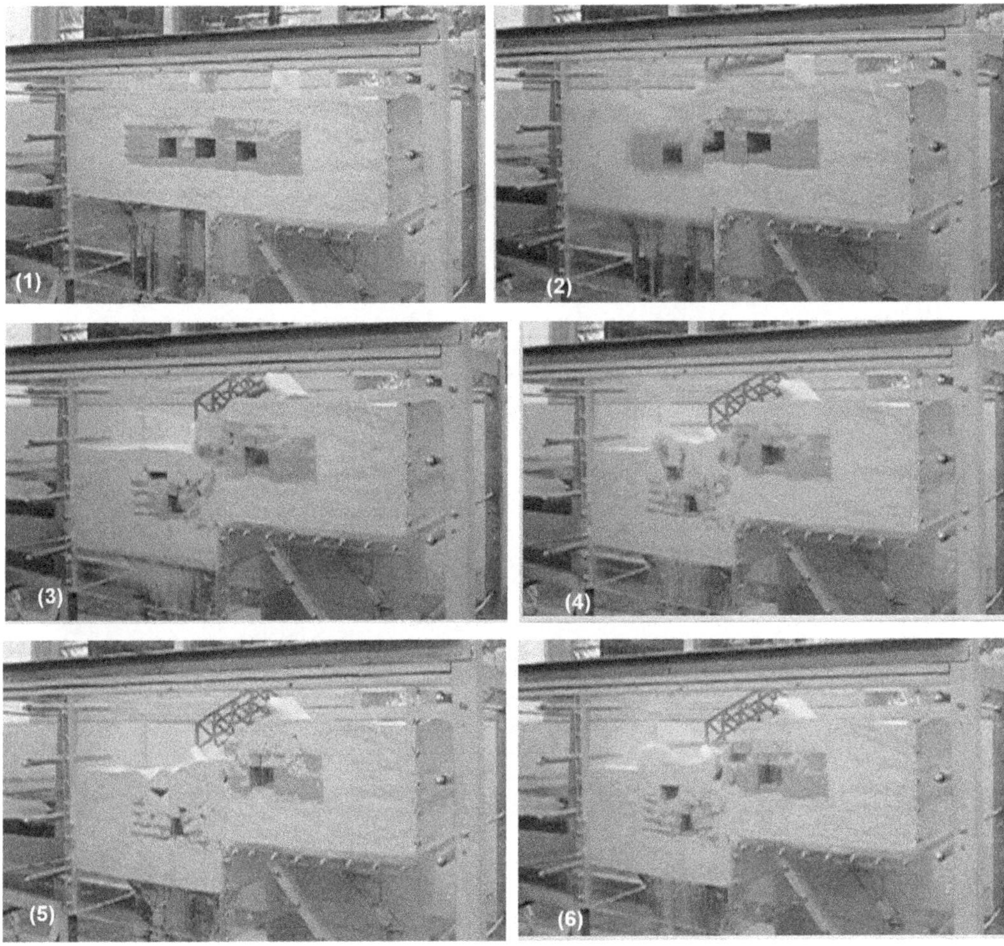

Figure 4.18 Views of the behavior of underground openings during faulting.

fail due to pillar failure, resulting in ground subsidence, as seen in Figure 4.17. A possible explanation for this observation, along with the reported cases (i.e., Nishida et al., 1984; Walton and Cobb, 1984), is that the stress acting on pillars increases as depth increases, while bending stresses in roof layers are higher at shallow depths.

4.4 Model Tests on the Effect of Faulting on Room and Pillar Mines

Shallow and large intra-plate earthquakes involve surface rupturing resulting from earthquake faults. Although there are currently no reports on the effect of earthquake faulting on abandoned mines worldwide, except for shaking-induced damage (Aydan, 2015), faulting may cause damage in underground abandoned mines. Figure 4.18 shows some model test results addressing this issue.

Figure 4.19 A view of the model setup mounted on a shaking table. (From Aydan and Geniş, 2015.)

4.5 Effect of Tunneling beneath Abandoned Room and Pillar Mines

A box with inner dimensions of 500 mm wide and high, and 50 mm thick, is used; this model device can be mounted on a shaking table, as shown in Figure 4.19. A circular cavity with a diameter of 50 mm can be created after preparing the model, which may collapse if the surrounding material yields under the induced stress state. The model can be subjected to base shaking after the cavity is created, as desired. The device is equipped with a laser profiler, a laser displacement sensor, and accelerometers. Displacement responses can be measured before, during, and after the experiments.

Two sets of model experiments were carried out to investigate the effect of backfilling on room and pillar-type abandoned lignite mines above a circular cavity, which may collapse after the cavity's creation or during shaking. The model ground consists of sandstone and shale fragments.

4.5.1 Unfilled Room and Pillar Mine Model

A room and pillar mine with an exploitation ratio of 50% was created. The model cavities were 50 mm wide and 20 mm high. A circular cylinder simulating a tunnel boring machine (TBM)

Figure 4.20 Views of the model before and after excavation. (a) Before excavation and (b) After excavation and subsequent collapse. (From Aydan and Geniş, 2015.)

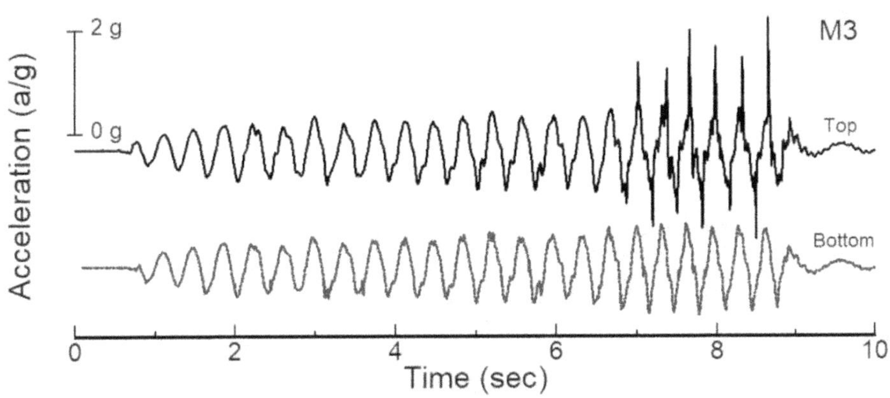

Figure 4.21 Applied waveform subjected to model base. (From Aydan and Geniş, 2015.)

excavation was subsequently withdrawn to simulate the excavation process. Depending on the ground characteristics, the surrounding ground may collapse. In this study, the surrounding ground was chosen such that it could yield if the cavity is not supported. Figure 4.20 shows the model before and after the excavation process. As noted from Figure 4.20, the room and pillar type abandoned mine is highly deformed and has moved towards the collapsed cavity. Subsidence occurred at the ground surface in relation to the volume of the collapsed cavity, and the rooms of the abandoned mine moved towards the collapsed cavity.

Then, the base of the model was subjected to horizontal shaking with a waveform shown in Figure 4.21 in order to see the effect of shaking with a base acceleration up to 0.9*g* on the ground subsidence profile as well as on the highly deformed mine cavities. As seen in Figure 4.22, the subsidence increased and the influenced region became larger.

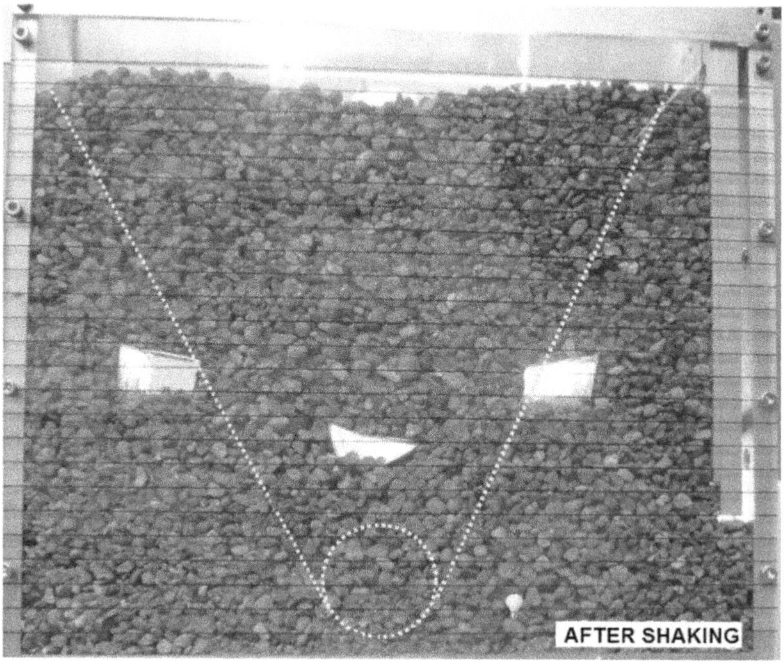

Figure 4.22 Applied waveform and a view of the model after shaking. (From Aydan and Geniş, 2015.)

4.5.2 Back-Filled Room and Pillar Mine Model

A backfilled room and pillar mine model was prepared to investigate the effect of backfilling on the collapse of an underground cavity beneath the abandoned mine and/or the subsequent base shaking. The excavation and shaking procedures were applied in the same way as in the unfilled model test. Figure 4.23 shows views of the backfilled room and pillar mine before and after the collapse of the newly created cavity. After excavation and the collapse of the cavity beneath the backfilled room and pillar mine model, the model was subjected to base shaking up to 0.9*g*. As noted from the figure, a cavity appears just below the backfilled cavity above the collapsed cavity. Although some small movements occurred in the created cavity, the overall deformation was quite negligible. The surface profiles of the model were measured, and it was noted that there was almost no difference in the measured ground surface profiles, implying that the backfilling was quite effective despite the collapse of the newly excavated circular cavity.

4.6 Model Tests on Shafts of Abandoned Room and Pillar Mines

Aydan (2004) performed shaking table experiments on the response of non-cohesive backfill material in vertical shafts with a square shape, following observations of the subsidence of backfill material in an abandoned shaft in the Komatsu lignite field of Yamoto Town during the 2003 Miyagi Hokubu Earthquake and one of its large aftershocks on September 4, 2003. Figure 4.24 shows pictures of shaking table model experiments on the subsidence of backfilled shafts with non-cohesive material to explain the mechanism of subsidence during ground shaking.

(a) (b)

Figure 4.23 Views of the model before and after excavation. (a) Before excavation. (b) After excavation and subsequent collapse. (From Aydan and Geniş, 2015.)

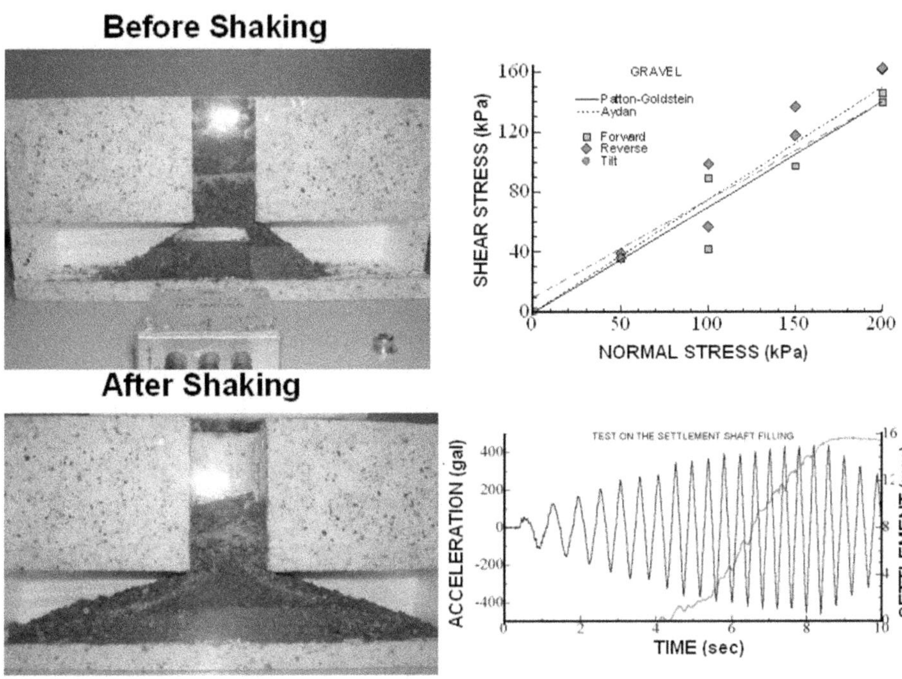

Figure 4.24 Views of models before and after shaking of backfilled shafts and their subsidence response during shaking, together with yield function and measured responses. (Arranged from Aydan, 2004.)

If the similitude law is based on the yielding strength of granular material and interface friction with shaft walls, the observed phenomenon is expected to correspond to the same phenomenon observed in actual situations, except for the geometrical differences between the model and actual mines. When shaking is strong enough to cause the yielding of the backfill material, it

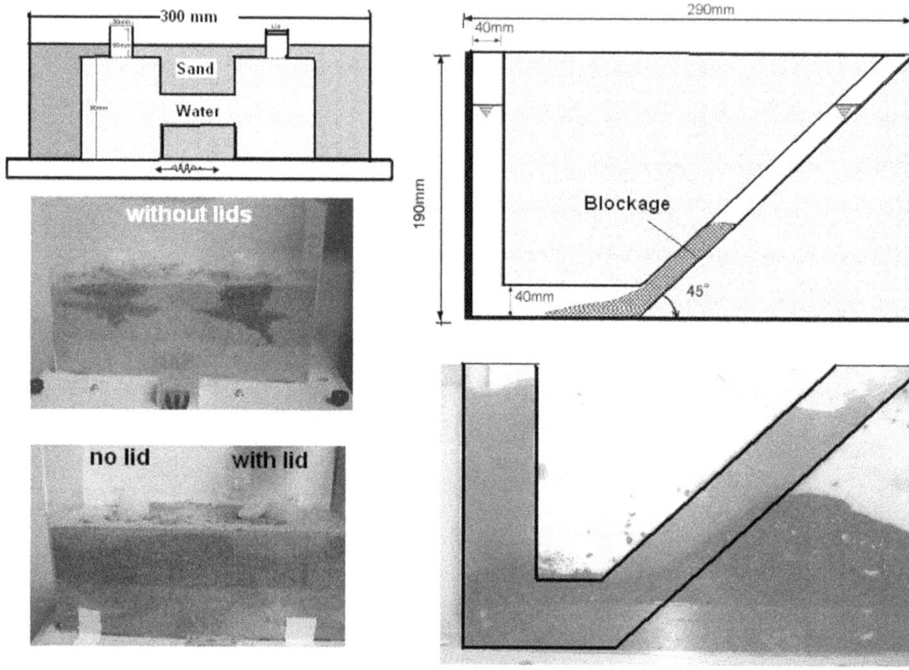

Figure 4.25 Motion of groundwater and its effects on abandoned mine models during shaking. (Arranged from Aydan, 2004.)

would flow to adjacent openings until the slope of the granular backfill material becomes equal to its dynamic repose angle for the applied maximum ground acceleration.

Aydan (2004) also performed shaking table experiments on model abandoned mines immersed in underground water, following observations of damage to abandoned mines in the Komatsu lignite field of Yamoto Town caused by the 2003 Miyagi Hokubu Earthquake. In these experiments, the movement of groundwater in vertical and inclined shafts, with or without blockage due to fallen material from the roof, and the effect of the impact force of groundwater on shaft plugs resulting from shaking were investigated, as seen in Figure 4.25. The shaking table experiments clearly showed that the motion of groundwater within a mine is fundamentally similar to the sloshing phenomenon observed in tanks. When ground motions become stronger, the groundwater may induce impact forces on shaft plugs, which may even result in their displacement. In addition, such effects are also expected on roof layers above the collapsed chimneys within the mines. The anticipated effects would include the collapse of roof layers near the ground surface, leading to their failure and the spouting of groundwater from the resulting sinkhole.

Chapter 5

Tests on Backfilling of Abandoned Room and Pillar Mines

Backfilling of abandoned room and pillar mines is primarily a countermeasure against possible collapses, subsidence, or sinkholes. This chapter describes some experimental studies on backfilling using different materials and discusses their effectiveness (Aydan et al., 2013c, 2018; Aydan & Tokashiki, 2013).

5.1 Characteristics of Backfilling Materials

5.1.1 Granular Materials

A steel cylindrical container with an internal diameter of 60 mm and a height of 60 mm was used in experiments for investigating the compression behavior of granular and cohesive backfilling materials. Granular backfilling material was No. 4 quartzite sand. The direct shear test and compaction response of this granular backfilling material are shown in Figure 5.1. As noted from the figure, granular backfill material becomes denser after each loading and unloading cycle, and residual (plastic) strains occur. Furthermore, the bearing capacity of the backfill material increases, and no softening occurs due to the confinement of the container. This implies that the granular backfilling material can be effective provided that it is confined laterally. (Aydan et al., 2013c, 2018; Aydan and Tokashiki 2013).

5.1.2 Cohesive Materials

Cohesive backfilling material was chosen to be plaster. The physical and mechanical properties of plaster are given in Table 5.1. Figure 5.2 shows the strain–stress response of the plaster at the age of 7 days under uniaxial compression.

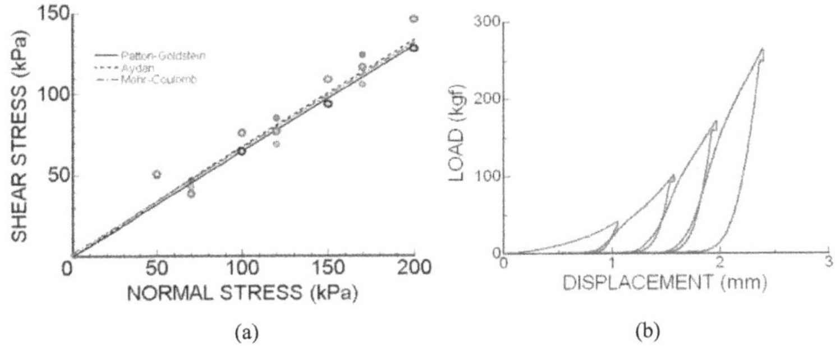

Figure 5.1 Failure criteria and compression response for granular sand No.4. (a) Failure criteria. (b) Compression response. (From Aydan et al., 2018.)

DOI: 10.1201/9781003517214-5

Table 5.1 Physical and mechanical properties of cohesive backfilling material

Sample No.	*Age (days)*	*Unit Weight (kN/m^3)*	*V_p (km/s)*	*Elastic Modulus (MPa)*	*UCS (kPa)*	*BUT (kPa)*
Plaster FM1	7	13.50	1.209	69.9	524	
Plaster FM2	7	13.66	1.273	41.32	616	
Plaster FM3	7	13.30	1.336			
Brazilian FM1	7	13.68	1.390			30.16

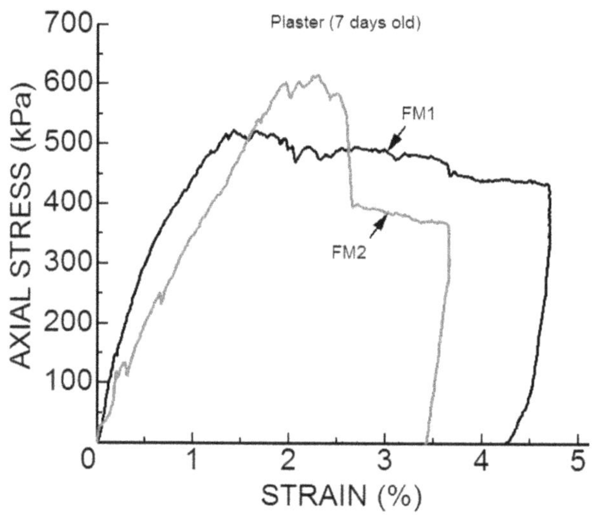

Figure 5.2 Strain–stress relation for plaster at the age of 7 days. (From Aydan et al., 2018.)

5.2 Experiments on the Effect of Backfilling Technique

A steel cylindrical container with an internal diameter of 60 mm and a height of 60 mm was used in preliminary experiments for investigating the supporting effect of granular backfilling on lignite, mudstone, and sandstone samples from abandoned lignite mines of Mitake Town. Load, displacement, and acoustic emission (AE) were monitored continuously using a setup as shown in Figure 5.3. The same steel container was used to measure the stress–strain response of the backfilling material in compression experiments.

Pillars were in a prismatic shape, and the excavation ratio ranged between 70% and 83%, which is quite similar to actual excavation ratios in abandoned lignite mines in Mitake Town.

A 10 mm thick cylindrical acrylic container with an internal diameter of 100 mm and a height of 250 mm was used in large-scale experiments to investigate the supporting effect of granular and cohesive backfilling on pillar samples of lignite from abandoned lignite mines in Nagakute Town and tuff from Oya Town (Tochigi Prefecture). The size of the pillars was 46–50 mm in diameter and 96–100 mm in height. The pillars were sandwiched between two 100 mm high cylindrical rock blocks with a diameter of 98 mm. The top and bottom blocks were made of Oya tuff for Oya tuff samples and Ryukyu limestone for experiments on Nagakute lignite. In some

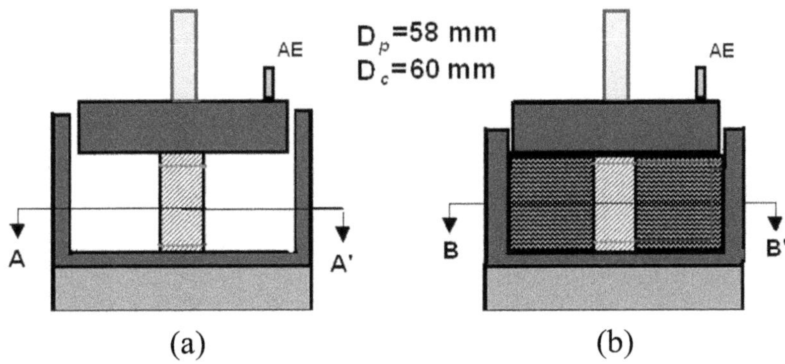

Figure 5.3 Experimental setup for short-term small-scale experiments for (a) unfilled and (b) backfilled states. (From Aydan et al., 2013c.)

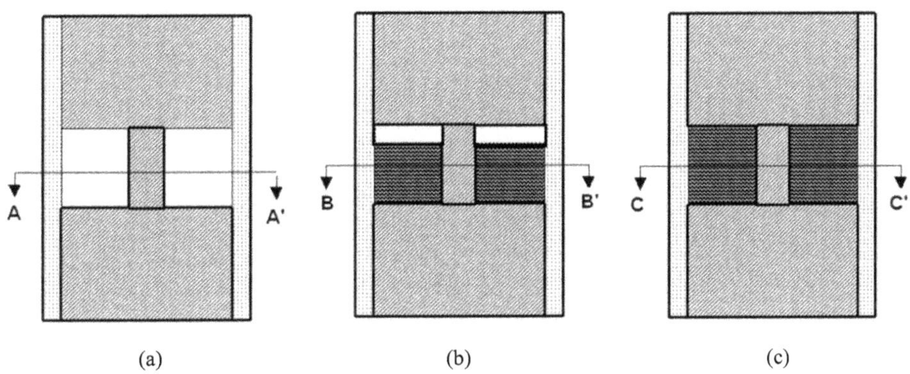

Figure 5.4 Backfilling states in experiments. (a) Non-backfilled. (b) Partially-backfilled. (c) Fully backfilled. (From Aydan et al., 2018.)

experiments on rectangular prism pillars of Oya tuff, Ryukyu limestone blocks were used as the top and bottom blocks. Load, displacement, AE, and accelerations were monitored continuously. Besides the type of backfilling material, the degree of backfilling was investigated in some of the experiments, namely, non-backfilled, partially backfilled, and fully backfilled (Figure 5.4).

5.2.1 *Tests on Backfilling Mitake Lignite Pillars*

Figure 5.5 shows the stress, strain, and AE responses of granular backfill material during cyclic compression experiments. As noted from the figure, the granular backfill material becomes denser after each loading and unloading cycle, and residual (plastic) strains occur. It is also interesting to notice that the cumulative AE response is associated with straining, and a somewhat Kaiser effect is noted from the measured responses even in this granular backfill material. Furthermore, the bearing capacity of the backfill material increases, and no softening occurs due to the confinement of the container.

Next, Figures 5.6 and 5.7 show the stress, strain, and AE responses of an unfilled lignite sample and a filled lignite sample from Mitake Town with granular backfill material during cyclic compression experiments, respectively. As noted from the figures, the unfilled lignite sample

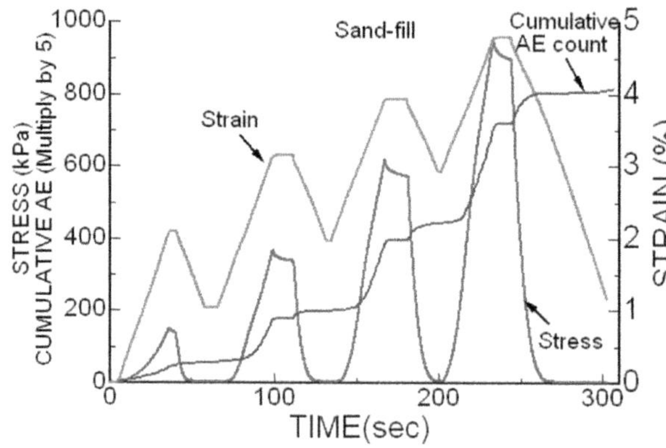

Figure 5.5 Stress, strain, and cumulative AE count responses of granular backfill material during cyclic compression. (From Aydan et al., 2013c.)

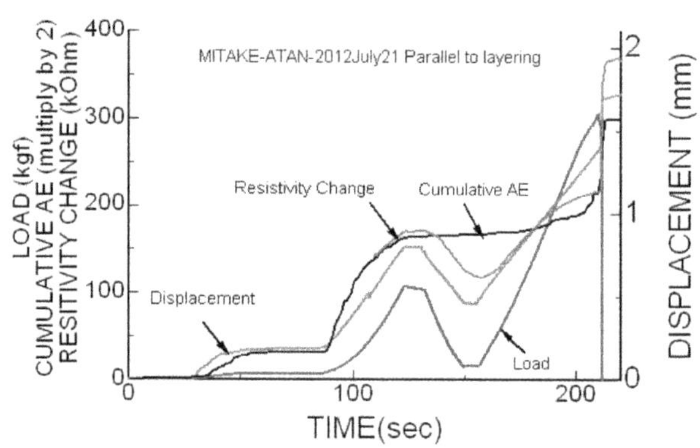

Figure 5.6 Stress, strain, and cumulative AE count responses of backfilled samples during cyclic compression (Aydan et al., 2013c).

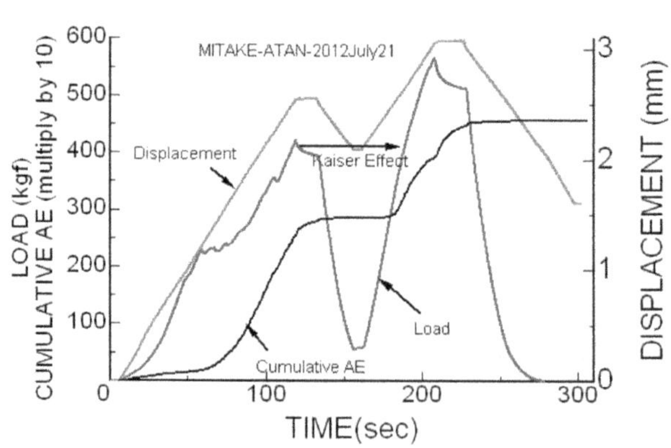

Figure 5.7 Stress, strain, and cumulative AE count responses of backfilled sample during cyclic compression. (From Aydan et al., 2013c.)

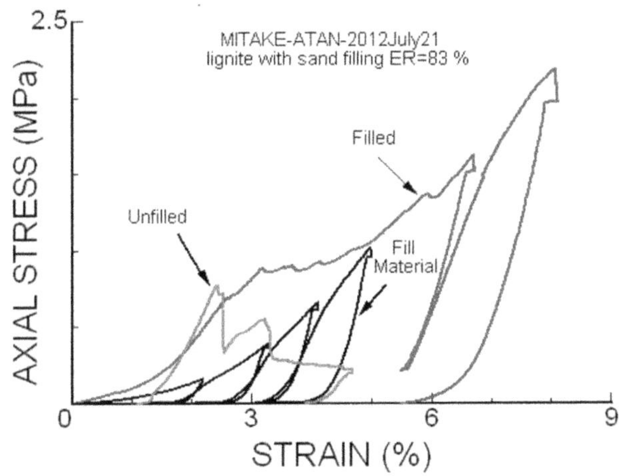

Figure 5.8 Comparison of strain–stress responses of unfilled samples, backfilled samples, and backfill material during cyclic compression. (From Aydan et al., 2013c.)

fails in a brittle manner, while the backfilled sample continues to sustain higher loads. Furthermore, permanent straining is noted after each loading–unloading cycle, and the cumulative AE response is closely associated with straining, with the Kaiser effect noted from the measured AE responses.

The stress–strain responses of the unfilled sample, the backfilled sample, and the backfill material itself are compared in Figure 5.8. It is interesting to note that the bearing capacity is increased by about 1.2 times compared with that of the unfilled sample at the same strain level, and the behavior of the sample backfilled with granular material is elasto-plastic of the strain hardening type. After a certain amount of straining, the overall strain–stress behavior depends upon that of the backfill material. Despite the backfill material being granular non-cohesive frictional material, the bearing capacity of the backfilled material was of great significance. Three experiments were carried out for each case, and the results were basically similar to each other. Sharp decreases in strain–stress responses are due to the relaxation behavior of the tested samples.

Similar experiments were carried out on pillar samples of mudstone and sandstone from the abandoned lignite mines in Mitake Town. Figures 5.9 and 5.10 show the average strain–stress relations for unfilled and backfilled pillars and backfill material. It is interesting to note that the bearing capacity increases by about 1.2–1.3 times compared with that of the unfilled sample at the same strain level. The behavior of the backfilled sample is elasto-plastic with slight softening just after the failure of the pillar and exhibits strain hardening after a certain amount of straining. Nevertheless, the residual strength is almost the same as that of the unfilled pillar or greater. Furthermore, the bearing capacity of the backfilled material is of great significance, despite the backfill material being a granular non-cohesive frictional material.

5.2.2 *Experiments on Nagakute Lignite Pillars*

Figures 5.11–5.13 show the stress, strain, and AE responses of the unfilled lignite pillar and lignite pillars backfilled with granular and cohesive (NSK: Nendo-Suna-Kira) filling materials during cyclic compression, respectively. As noted from the figures, the unfilled lignite sample

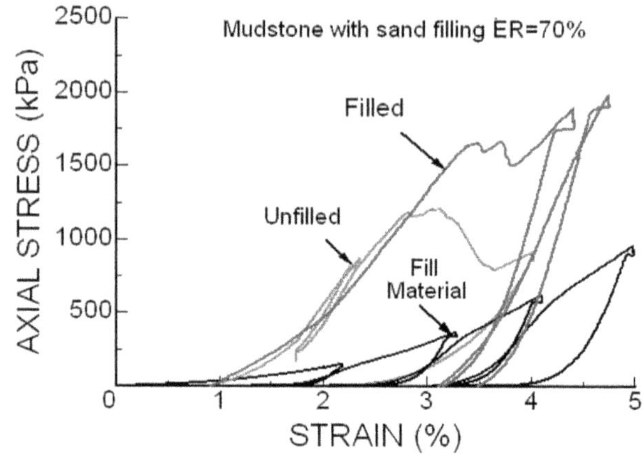

Figure 5.9 Comparison of strain–stress responses of unfilled samples, backfilled samples, and backfill material during cyclic compression. (From Aydan et al., 2013c.)

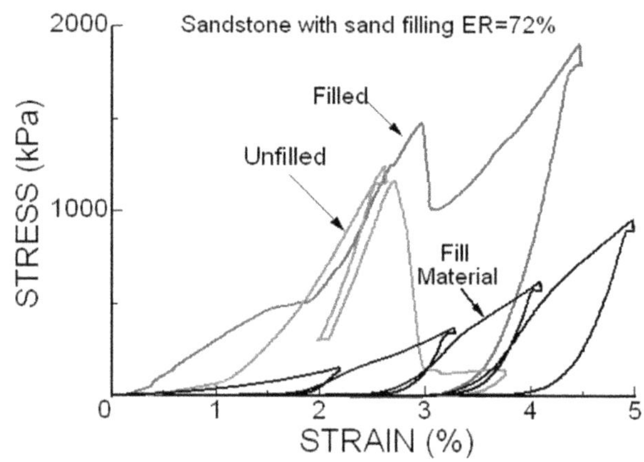

Figure 5.10 Comparison of strain–stress responses of unfilled samples, backfilled samples, and backfill material during cyclic compression. (From Aydan et al., 2013c.)

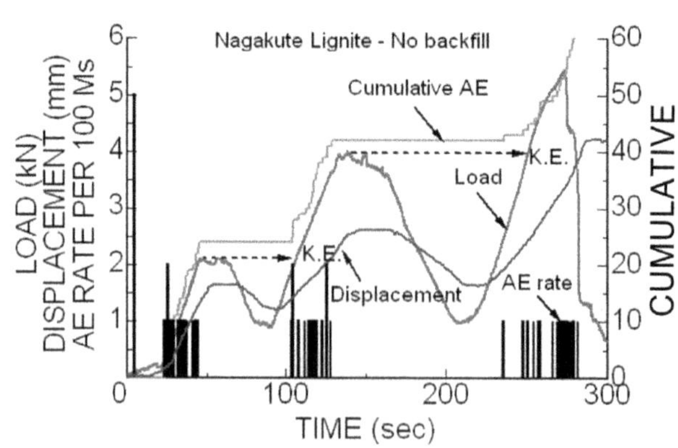

Figure 5.11 Stress, strain, and cumulative AE count responses of an unfilled lignite pillar during cyclic compression. (From Aydan et al., 2013c.)

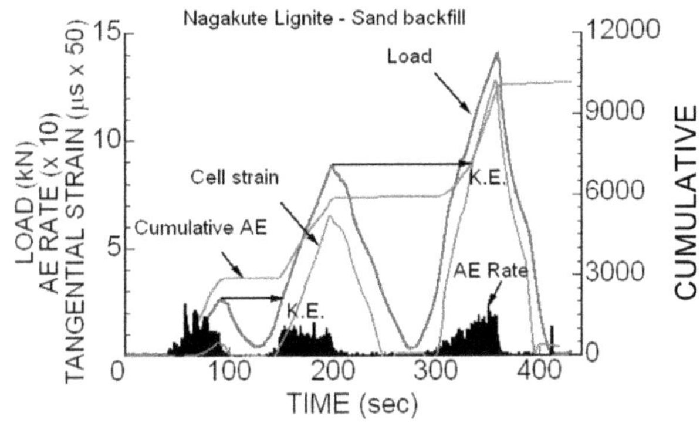

Figure 5.12 Stress, strain, and cumulative AE count responses of a Nagakute lignite sample backfilled with sand during cyclic compression. (From Aydan et al., 2013c.)

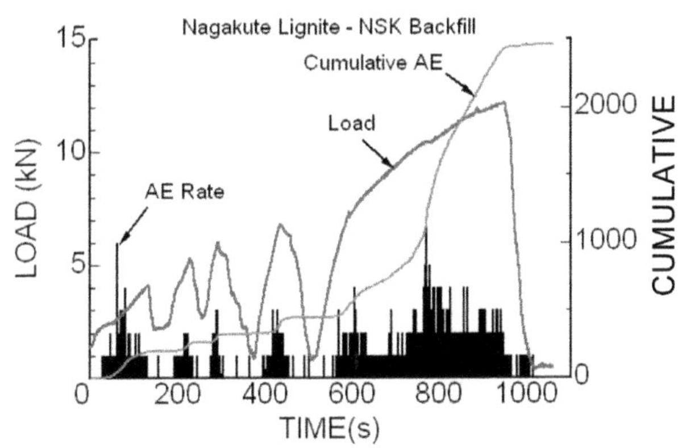

Figure 5.13 Stress, strain, and cumulative AE count responses of a Nagakute lignite pillar backfilled with cohesive NSK backfill material during cyclic compression. (From Aydan et al., 2013c.)

fails in a brittle manner, while the backfilled sample continues to sustain higher loads. Furthermore, permanent straining is observed after each loading–unloading cycle, and the cumulative AE response is closely associated with straining. Somewhat of a Kaiser effect is noted from the measured responses, as seen in the small-scale short-term experiments. The tangential strain of the acrylic container is also shown in Figure 5.12. The tangential strain begins after a certain strain level, which is closely associated with the deformation and initiation of yielding of the pillars.

Figure 5.14 compares the average strain–stress relations for unfilled pillars and those backfilled with granular and cohesive materials. It is noteworthy that the bearing capacity of backfilled pillars increases by about 1.3–1.5 times compared with that of the unfilled sample at the same strain level. Furthermore, the behavior of backfilled pillars is elasto-plastic without any softening, even after the failure of the pillars. The bearing capacity of the pillar backfilled with NSK backfill material is greater than that of the pillar backfilled with granular material.

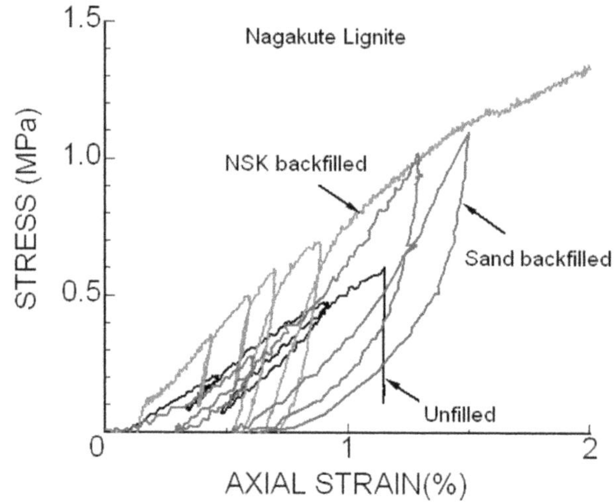

Figure 5.14 Comparison of strain–stress responses of unfilled pillars and backfilled pillars with granular and cohesive backfill materials during cyclic compression. (From Aydan et al., 2013c.)

5.2.3 *Experiments on Oya Tuff Pillars*

There are many abandoned and active underground quarries in Oya Town, Tochigi Prefecture. Quarrying has been active in this area for more than 100 years. As briefly mentioned in the introduction, significant sinkhole formations have occurred in Oya Town from time to time. There is growing attention regarding whether to fill the abandoned underground quarries.

Figures 5.15–5.17 show the stress, strain, and AE responses of the unfilled tuff pillar and tuff pillars backfilled with granular and cohesive (NSK) filling materials during cyclic compression, respectively. As noted from the figures, the unfilled lignite sample fails in a brittle manner, while the backfilled sample continues to sustain higher loads. Furthermore, permanent straining is observed after each loading–unloading cycle, and the cumulative AE response is closely associated with straining; a Kaiser effect is noted from the measured responses. Experiments on Oya tuff are conducted under both dry and saturated conditions. As the experiments on pillars with NSK backfilling material were carried out at 28 days of curing in water, it is expected that the strength of Oya tuff would be lower than that under dry conditions. The experiment showed that the uniaxial compressive strength of Oya tuff is reduced to 30%–50% of the uniaxial compressive strength under dry conditions. The responses are shown in Figure 5.15 for a tuff sample under saturated conditions.

In Figure 5.16, the tangential strain of the acrylic cell is also shown. As noted clearly from this figure, the tangential strain of the cell drastically increased as soon as the pillar started to fail and exhibited dilatant behavior. Nevertheless, the confinement effect of the backfilling material, assisted by the constraint of the acrylic cell, increases the bearing capacity of the pillar despite slight softening behavior following the yielding of the pillar.

Figure 5.18 compares the average strain–stress relations for unfilled pillars and pillars backfilled with granular and cohesive backfill materials. It is interesting to note that the bearing capacity of backfilled pillars is increased by about 1.3–1.5 times compared with that of the unfilled sample at the same strain level. Furthermore, the behavior of backfilled pillars is elastoplastic, with slight softening following the yielding of the pillars and strain hardening thereafter.

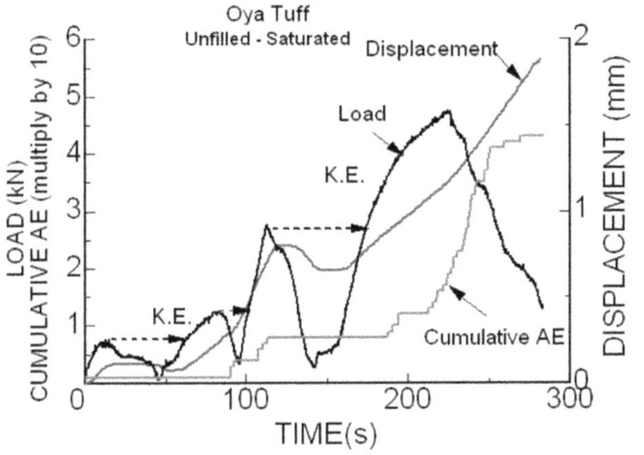

Figure 5.15 Stress, strain, and cumulative AE count responses of an unfilled saturated Oya tuff pillar during cyclic compression. (From Aydan et al., 2013c.)

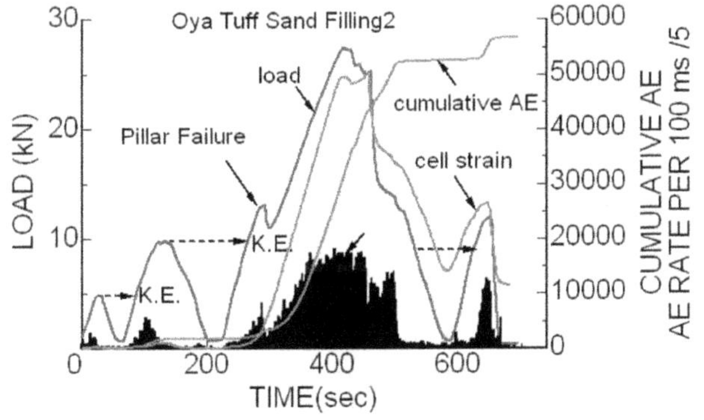

Figure 5.16 Stress, strain, and cumulative AE count responses of an Oya tuff pillar backfilled with granular backfill material during cyclic compression. (From Aydan et al., 2013c.)

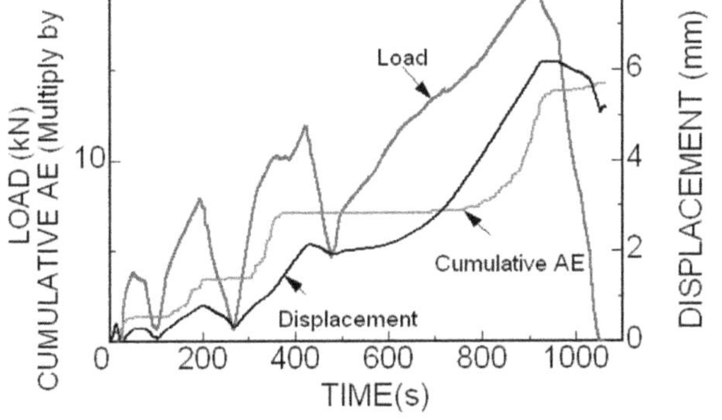

Figure 5.17 Stress, strain, and cumulative AE count responses of an Oya tuff pillar backfilled with cohesive NSK backfill material during cyclic compression. (From Aydan et al., 2013c).

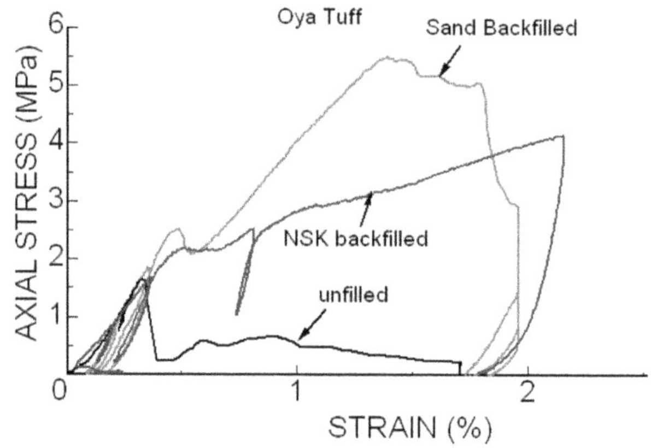

Figure 5.18 Comparison of strain–stress responses of unfilled pillars and backfilled pillars with granular and cohesive backfill materials during cyclic compression. (From Aydan et al., 2013c.)

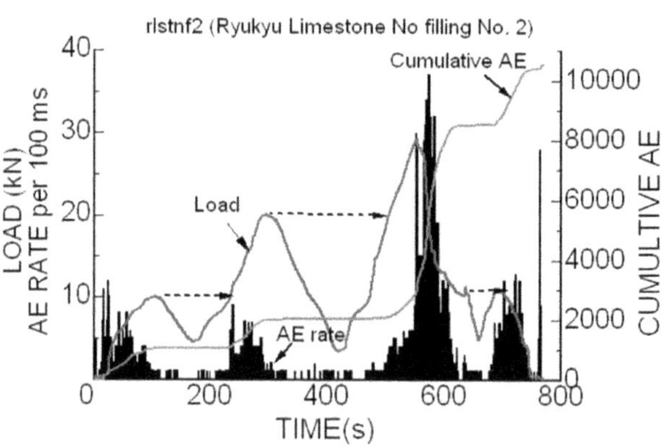

Figure 5.19 Stress, strain, and cumulative AE count responses of an unfilled Ryukyu limestone pillar during cyclic compression. (From Aydan et al., 2013c.)

Although the cohesive backfill material provides better resistance for pillars, the difference between the average strain–stress response of the pillar (saturated) backfilled with NSK backfilling material and the pillar (dry) backfilled with granular sandy material is due to the strength difference of the tuff pillars. The tuff pillar is dry in the case of the sand-backfilled case, while it is saturated in the case of the NSK backfilled case.

5.2.4 *Experiments on Ryukyu Limestone Pillars*

Ryukyu limestone is widely distributed in the Ryukyu Islands, and this formation contains various sizes of solution cavities. These solution (karstic) cavities cause occasional sinkhole problems in Ryukyu limestone regions. Therefore, some experiments were carried out to investigate the effect of backfilling on the bearing capacity of the Ryukyu limestone formation.

Figures 5.19–5.21 show the stress, strain, and AE responses of unfilled limestone pillars and limestone pillars backfilled with granular and cohesive (NSK) filling materials during cyclic

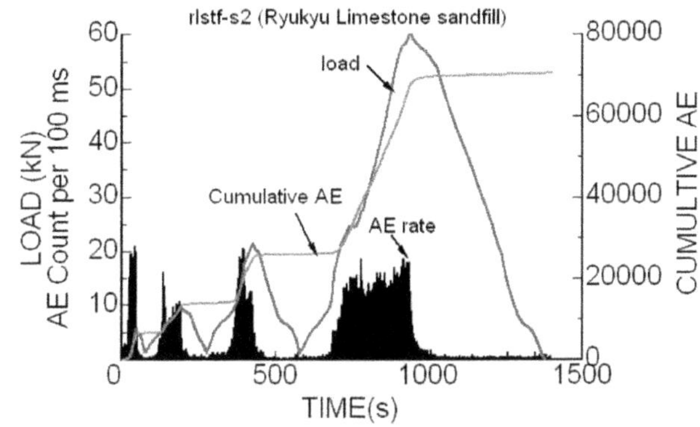

Figure 5.20 Stress, strain, and cumulative AE count responses of Ryukyu limestone backfilled with granular backfill material during cyclic compression. (From Aydan et al., 2013c.)

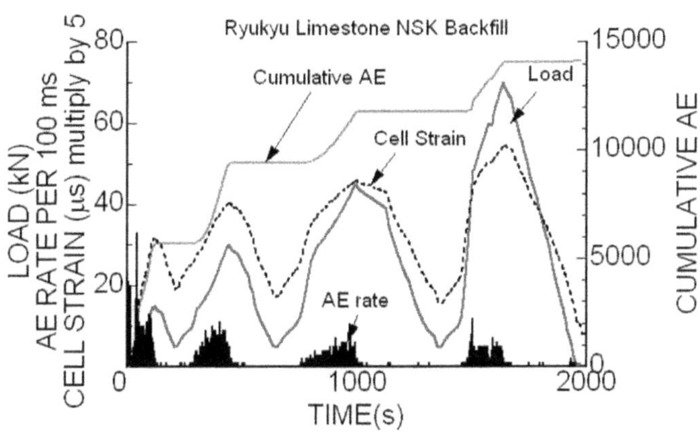

Figure 5.21 Stress, strain, and cumulative AE count responses of a Ryukyu limestone pillar backfilled with cohesive NSK backfill material during cyclic compression. (From Aydan et al., 2013c.)

compression, respectively. Although Ryukyu limestone is a much stronger rock compared with lignite and tuff, the overall responses and observations are quite similar to those of experiments on lignite and tuff samples. As noted from the figures, the unfilled limestone sample fails in a brittle manner, while the backfilled samples continue to take up higher loads. Furthermore, permanent straining is noted after each loading–unloading cycle, and cumulative AE response is closely associated with straining. Kaiser effects are clearly observed in both unfilled and backfilled samples.

Figure 5.22 compares the average strain–stress relations for unfilled pillars and pillars backfilled with granular and cohesive backfill materials. It is interesting to note that the bearing capacity of backfilled pillars is increased by about 1.2–1.3 times compared with that of the unfilled sample at the same strain level. Furthermore, the behavior of backfilled pillars is elasto-plastic without any softening after the yielding of the pillar. When the limestone pillar is backfilled with NSK backfilling material, the overall strength of the pillars greatly increases, and the

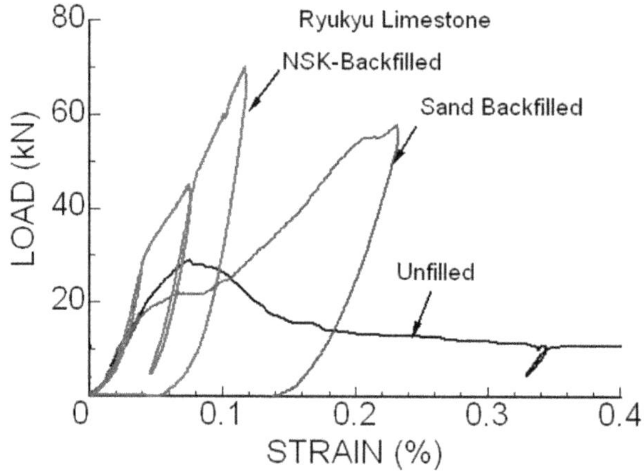

Figure 5.22 Comparison of strain–stress responses of unfilled pillars and backfilled pillars with granular and cohesive backfill materials during cyclic compression. (From Aydan et al., 2013c.)

experiments were terminated to prevent the bursting of the acrylic cell, which could lead to undesirable accidents.

5.3 Creep Tests

Several experiments on lignite samples from Mitake Town regarding long-term behavior subjected to atmospheric degradation and saturation conditions were carried out. It is well known that lignite is vulnerable to variations in water content, starting to crack due to shrinkage, and its strength decreases upon saturation. While one sample (ATC3) was directly exposed to the environment, the other sample (ATC1) was backfilled after 36 hours of exposure to atmospheric degradation. Both unfilled and backfilled samples were fully saturated at 1,360 hours (about 2 months) after the commencement of the experiments. Figure 5.23 shows the views of the unfilled and backfilled samples. During the long-term experiments, axial displacement, AEs, and environmental conditions such as temperature, humidity, and air pressure were monitored continuously. The monitoring devices were entirely battery-operated, and the creep devices were of the cantilever type, so they were not susceptible to power failures.

Figures 5.24 and 5.25 show the measured displacement and AE responses over time. Both experiments were terminated at the same time when the ATC3 sample failed. It is noteworthy that the displacement and cumulative AE count of the sample exposed to the atmosphere increased with elapsed time. Following the saturation of the sample at 1,360 hours from the commencement of the experiment, the sample failed almost 2 days after saturation. On the other hand, the rate of AE occurrences in the ATC1 sample was almost non-existent, except during backfilling, loading, and saturation procedures. The AE occurrences were drastically reduced following the backfilling of the sample. There were almost no AE occurrences after saturation, while the unfilled sample failed after saturation. The main mechanisms for the reduction of AE occurrences in the backfilled sample can be attributed to the prevention of exposure to variations in environmental parameters such as temperature and humidity, which govern water content variation and subsequent degradation of lignite, and the reduction of deviatoric stress, which may be related to the well-known Kaiser effect. Therefore, these two mechanisms may be utilized to verify the

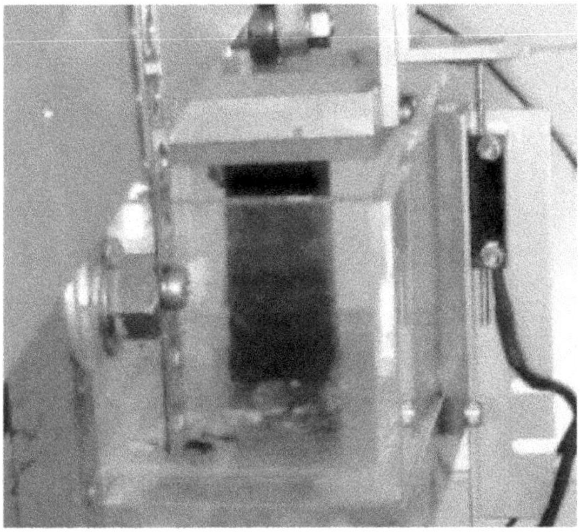

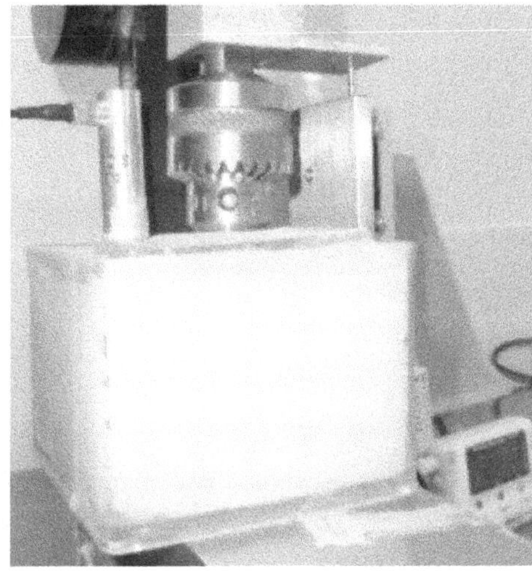

Figure 5.23 Views of unfilled and backfilled lignite samples. (a) Unfilled (ATC3). (b) Backfilled (ATC1). (From Aydan et al., 2013c.)

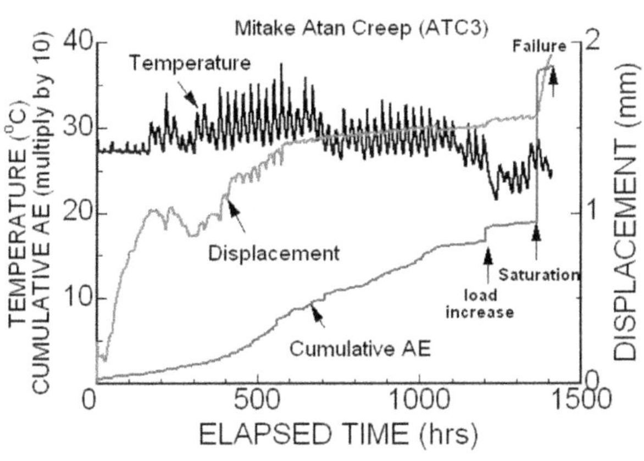

Figure 5.24 Displacement and cumulative AE count responses of the ATC3 sample. (From Aydan et al., 2013c.)

effect of backfilling in abandoned lignite mines, as well as other types of backfilled cavities, through in-situ monitoring of AE responses.

5.4 Tests and Techniques for Evaluation of Mechanical Effect of Backfilling

The in-situ evaluation of the effect of backfilling on the characteristics of the surrounding ground is also an important aspect. Several techniques are possible. One of them is the utilization of the velocity of p-waves and s-waves of the ground before and after backfilling. Table 5.2 provides the improvement of p-wave velocity of the ground before and after backfilling. There is no doubt that backfilling increases the overall p-wave velocity of the ground.

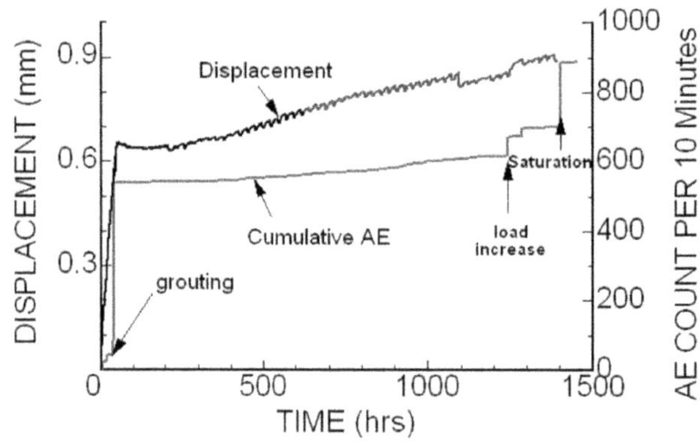

Figure 5.25 Displacement and cumulative AE count responses of the ATC3 sample. (From Aydan et al., 2013c.)

Table 5.2 Variation of p-wave velocity of pillar models before and after backfilling

Rock	*Pillar Ratio*	*Pillar-Dry (km/s)*	*Pillar-Dry Container*	*Pillar-Filled*	
				Sand	*Plaster*
Shimajiri sandstone	0.373	1.610	1.452	1.452	1.74
Shimajiri tuff	0.421	1.308	1.290	1.279	1.745
Shimajiri siltstone	0.367	1.775	1.763	1.811	1.774
Oya tuff	0.454	2.347	2.011	2.025	2.127

5.4.1 *First Series: Non-Backfilled and Fully Backfilled with Plaster*

A number of tests on rock pillars made of Mitake Siltstone, Oya Tuff, Shimajiri Mudstone, and Shimajiri Tuff were carried out. In these experiments, the main purpose was to investigate the effect of backfilling. The samples were either unfilled or fully backfilled with plaster. Experimental results are shown in Figure 5.26. The main observations from the experimental results shown in Figure 5.26 are as follows:

- i The overall stiffness of backfilled samples is higher than that of unfilled samples
- ii The peak strength of backfilled samples is higher than that of the unfilled samples
- iii The residual strength of backfilled samples is much higher than that of the unfilled samples. In some cases, the yielding strength shows strain hardening behavior, while the unfilled samples show strain softening behavior.

5.4.2 *Second Series: Degree of Backfilling with Quartzite Sand (No.4)*

The second series of experiments was carried out to investigate the degree of backfilling. Three different situations were considered:

- i Non-backfilled,
- ii Partially backfilled up to 70% of the pillar height, and
- iii Fully backfilled.

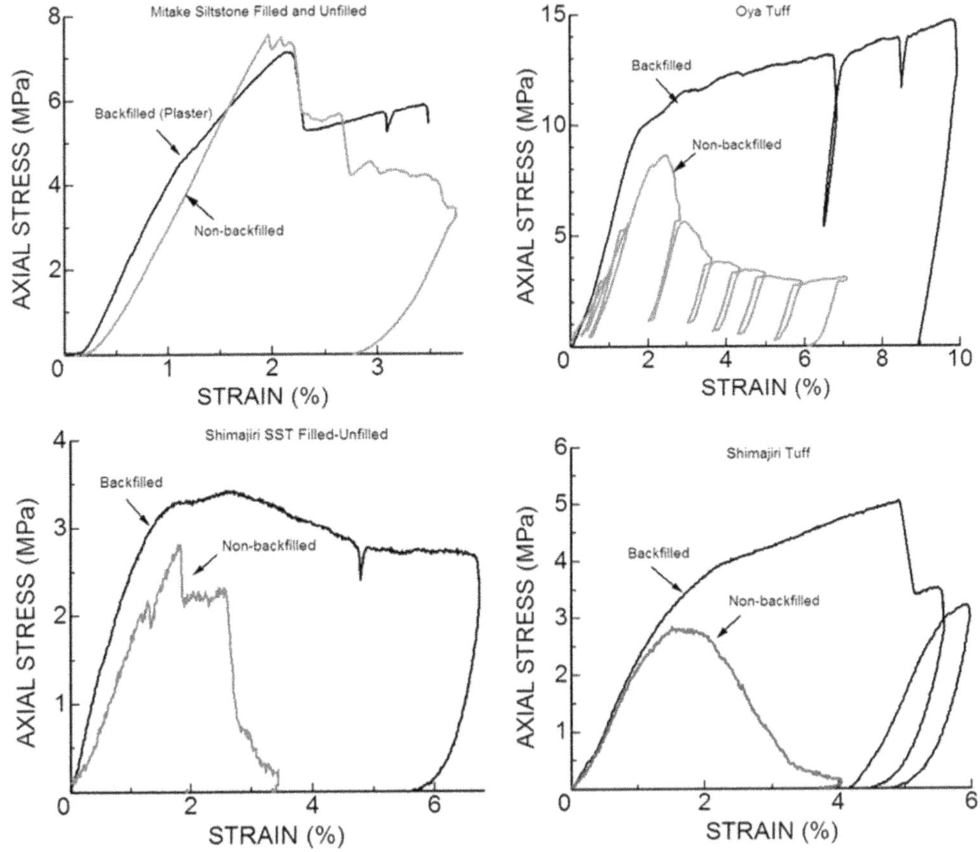

Figure 5.26 Strain–stress responses of unfilled and backfilled samples. (From Aydan et al., 2018.)

The backfill material in the second series was quartzite sand, whose properties are shown in Figures 5.1 and 5.2.

The experimental results are shown in Figure 5.27. As noted from the experimental results, the overall stiffness of the fully backfilled sample is higher than that of the other samples. Although the peak strength and residual strength of partially backfilled samples are higher than those of non-backfilled samples, the residual strength tends to decrease as the strain becomes larger. On the other hand, the residual strength of the fully backfilled sample increases with strain following the initial softening stage. This implies that partial backfilling is not effective in preventing the settlement of the ground surface.

5.5 Tests on Backfilling under Dynamic Conditions

The third series of experiments was carried out to investigate the response of pillars under three different backfilling states during shock tests. The backfilling situations were the same as those of the second series of experiments, which were as follows:

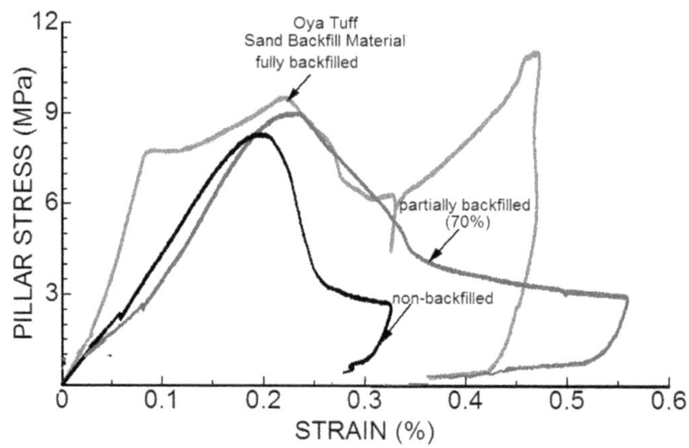

Figure 5.27 Effect of backfilling states on strain–stress responses of the pillars. (From Aydan et al., 2018.)

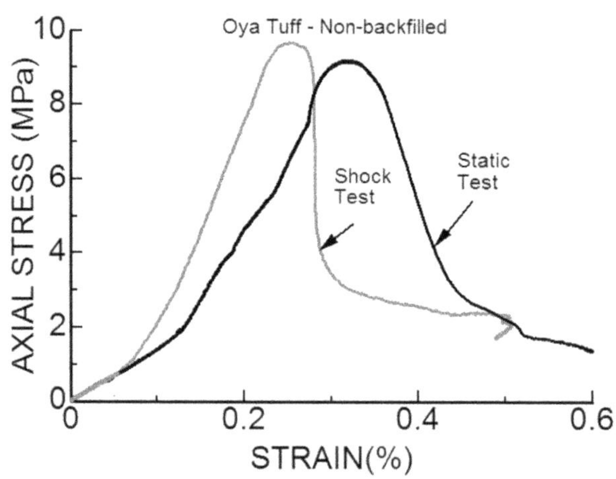

Figure 5.28 Effect of loading condition on strain–stress responses of the pillars in a non-backfilled situation. (From Aydan et al., 2018.)

i Non-backfilled,
ii Partially backfilled up to 70% of the pillar height and
iii Fully backfilled.

The load was imposed as shocks on the samples. Figure 5.28 compares the response of Oya Tuff without backfilling under static and shock loads. As noted from the figure, the overall stiffness and strength of the pillar sample subjected to shock loading are slightly higher than those of the sample tested under static loading. Furthermore, the residual strength of the sample subjected to shock loading is also slightly higher than that of the sample tested under static conditions.

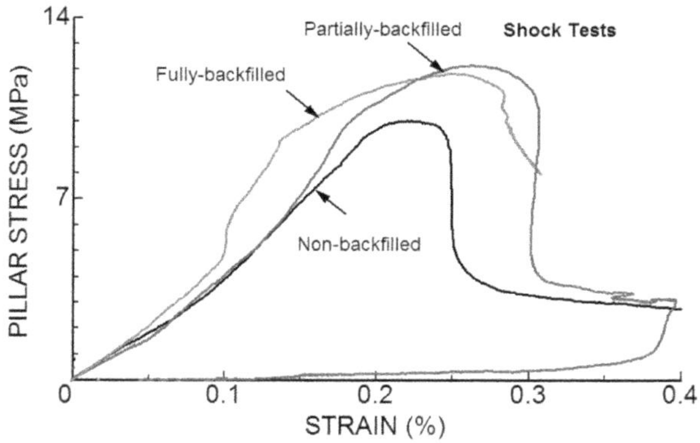

Figure 5.29 Strain–stress responses of the pillars under different backfilling conditions subjected to shock loads. (From Aydan et al., 2018.)

Figure 5.29 shows the strain–stress responses of the Oya Tuff pillar samples with different degrees of backfilling subjected to shock loads. It is also noted that the overall stiffness increases as a function of the degree of backfilling, similar to the samples tested under static conditions.

As the samples were equipped with accelerometers, the acceleration response of the samples at the time of failure for non-backfilled and fully backfilled situations is shown in Figure 5.30. As noted from the figures, the accelerations are higher for the non-backfilled sample compared with the fully backfilled sample. The maximum amplitude of the acceleration is almost four times that of the fully backfilled sample. Furthermore, the accelerations are not symmetric with respect to the time axis, as noted previously by Aydan et al. (2011).

5.6 Conclusions

In this chapter, the outcomes of an experimental program on the supporting effect of backfilling on the response and stability of abandoned mines, quarries, and karstic caves, as well as its verification in actual backfilling applications in Mitake Town (Mitake lignite), Takenoyama (Nagakute lignite), and Gushikawa Castle remains (Ryukyu limestone), are presented. The backfilling materials are granular and cohesive, and the rock samples include Mitake lignite, mudstone, sandstone, Nagakute lignite, Oya tuff, and Ryukyu limestone. Short- and long-term laboratory experiments were carried out, and some of the major conclusions are as follows:

- The bearing capacity of backfilled pillars increases by about 1.2–1.5 times compared with that of the unfilled sample at the same strain level.
- Unfilled samples fail in a brittle manner, while backfilled samples continue to sustain higher loads. Furthermore, the behavior of backfilled pillars is elasto-plastic, with none or slight softening following the yielding of the pillars.
- Permanent straining is noted after each loading–unloading cycle, and cumulative AE response is closely associated with straining. The Kaiser effect is clearly observed in both unfilled and backfilled samples.
- Long-term creep tests indicated that AE occurrence is almost non-existent for backfilled samples following saturation, while unfilled samples fail after saturation.

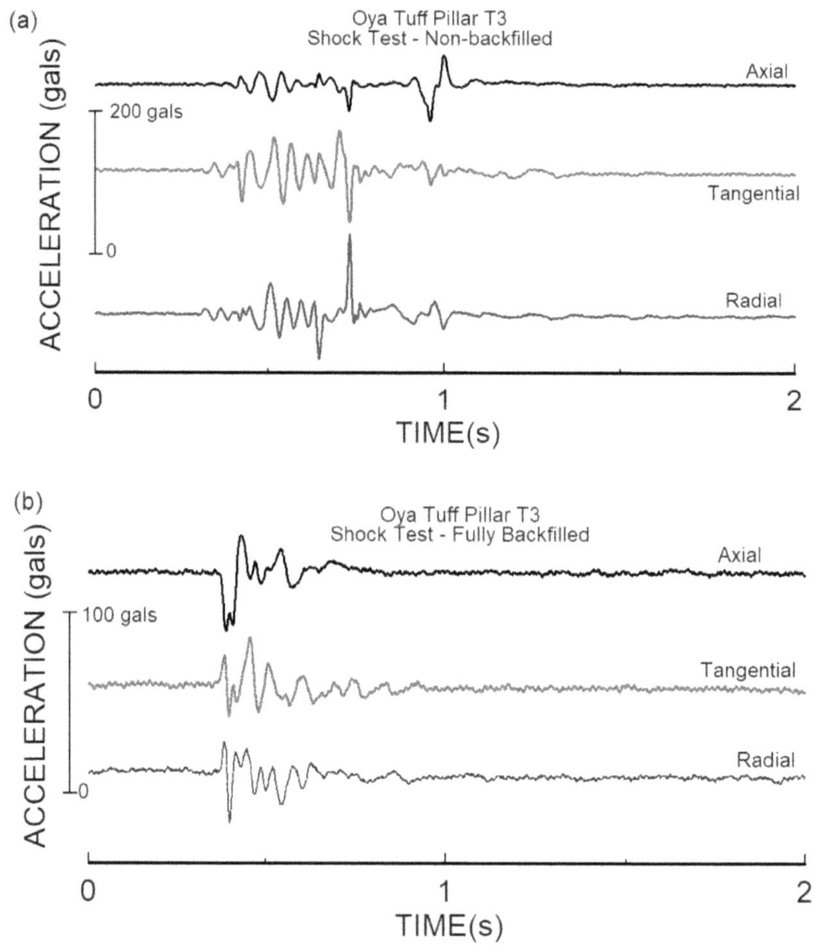

Figure 5.30 Acceleration responses of the (a) non-backfilled and (b) fully backfilled samples. (From Aydan et al., 2018.)

- The main mechanisms for the reduction of AE occurrence in the backfilled sample can be attributed to the prevention of exposure to variations in environmental conditions such as temperature and humidity, which govern the water content variation and subsequent degradation of lignite, as well as the reduction of deviatoric stress, which is related to the well-known Kaiser effect.
- These two mechanisms were utilized in the verification of the effect of backfilling in abandoned lignite mines, as well as other types of backfilled cavities. Long-term monitoring of AE responses before and after the backfilling operation clearly confirmed that the conceptual model could be used to verify the effect of backfilling.
- The reduction of groundwater velocity due to backfilling should also be considered an additional reason for the reduction of AE occurrences.

Chapter 6

Static and Dynamics Stability Assessment of Abandoned Room and Pillar Mines

6.1 Empirical Methods based on Rock Mass Classifications

Although there are many rock classifications, RMR and the Q-System were chosen for the assessment of the stability of the abandoned mine. The ratings of the rock mass of the abandoned mine have already been presented in Chapter 3 and are also given in Table 3.7. The span of the opening ranges between 2 and 6 m. For the rates given in Table 3.7 and the range of opening span, the results are plotted in the stability charts proposed by Bieniawski (1984) and Barton (1995), as shown in Figure 6.1. As noted from the figure, the RMR system predicts that the openings should have already collapsed totally many decades ago. Although roof collapses were observed within the abandoned mine, the abandoned mine is still stable in broad terms. It seems that the RMR system is somewhat over-conservative regarding the stand-up time of the underground cavities. On the other hand, the estimations by the Q-System seem to be more reasonable, as the support requirement would be necessary. However, the required support system would be light.

6.2 Simple Analytical Methods

The stability assessment methods for the roof of shallow underground openings are generally based on the bending theory of beams or arching theory used in structural mechanics (Figure 6.2). Particularly, the arching theory has been popular in mining engineering, and many formulations have been developed with consideration of various modes of failure (i.e., Aydan, 1989; Kawamoto et al., 1991).

6.2.1 *Beam Theory*

The limit of roof span (L) normalized by roof rock layer thickness (h_r) under its dead weight and topsoil with a thickness of (h_s) can be obtained in the following form:

$$\frac{L}{h_r} = \sqrt{\beta \frac{\sigma_t}{(\gamma_r h_r + \gamma_s h_s)}} \tag{6.1}$$

where σ_t is the tensile strength of the roof layer. The value of β would take the following values: 1/3, 2/3, and 2 for cantilever, simple, and built-in beams. Cantilever beam situations may arise when the roof of an opening is crossed by a fault, which is a common occurrence in abandoned room and pillar mines.

DOI: 10.1201/9781003517214-6

(a)

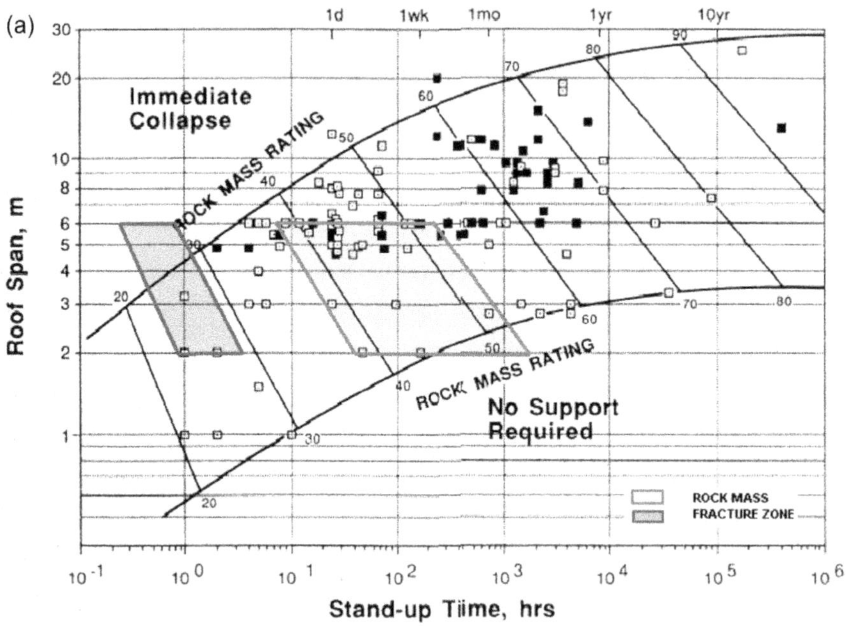

(b)

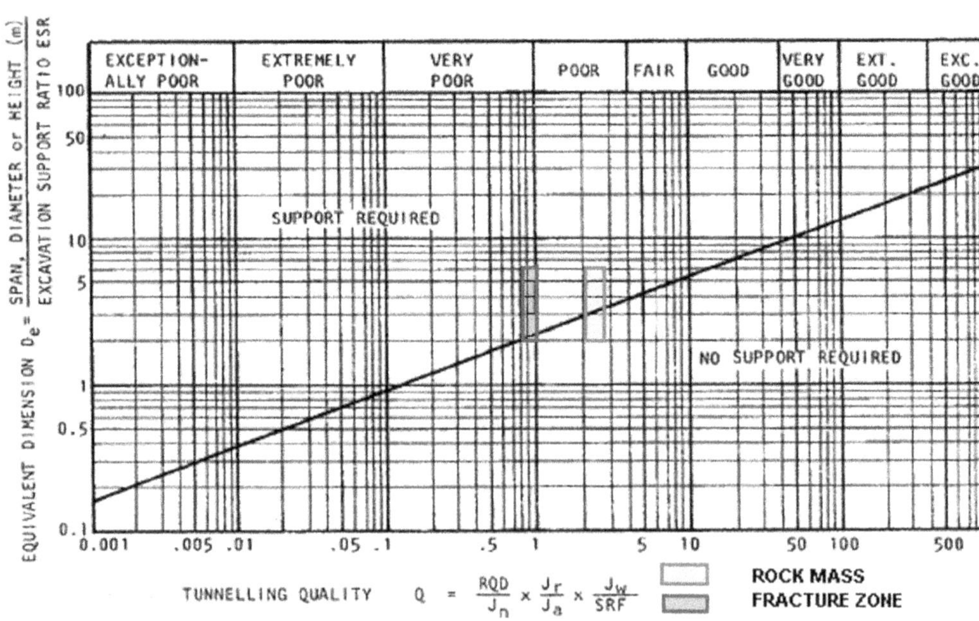

Figure 6.1 Comparison of observed and suggested stand-up time or stable opening span. (a) RMR. (b) Q-system. (From Aydan et al., 2006.)

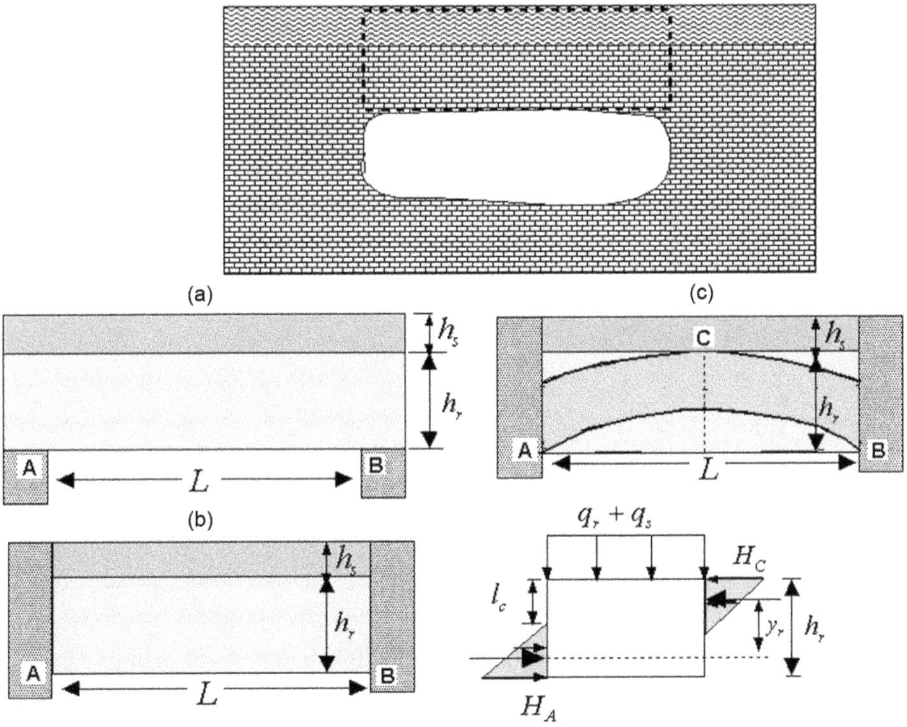

Figure 6.2 Illustrations of mechanical models for theories of bending and arching of beams. (a) Simply supported beam, (b) built-in beam, and (c) arching model. (From Aydan, 1989.)

6.2.2 *Arching Theory*

The arching theory for assessing the stability of roofs of shallow underground openings is generally based on three-hinged beams. As discussed by Aydan (1989), the rotation of the rock layer in which the arch is formed should be initiated for arch action to take place. The failure modes of an arch are varied and include crushing at the crown and/or abutments, vertical or horizontal sliding at abutments, and sliding along an existing discontinuity within the arch. Detailed derivations for each failure mode are presented in Aydan's (1989) doctoral thesis and in the article by Kawamoto et al. (1991). The classical arching theory assumes that the failure mode of the arch occurs through crushing at the crown or abutments, and the final form for the limit of roof span (L) normalized by rock roof layer thickness (h_r) under its own weight and topsoil with a thickness of (h_s) is as follows:

$$\frac{L}{h_r} = \sqrt{\beta \frac{\sigma_c}{(\gamma_r h_r + \gamma_s h_s)}} \tag{6.2}$$

where σ_c is the compressive strength of the roof layer. The value of β would be 4/3 when the crack length (l_c) is zero. However, the maximum resistance of the arch would be attained when the value of β is 3/2 after some manipulations concerning the crack length in the arch. The effect of concentrated loads can be taken into account if necessary.

6.2.3 *Effect of Long-Term Creep and Degradation on Roof Stability*

The stability of roof layers can be checked against various forms of failure. Although roof layers may crack due to bending stresses, their ultimate failure would be governed by the arching phenomenon (although shear failure may be conceivable, it is not generally likely to occur) within the roof layers. If the fiber stress exceeds the compressive strength at the crest of the arch within the roof layer, the roof layer will collapse. Under such conditions, the following relation can be derived (i.e., Aydan, 1989):

$$\sigma_c = \frac{3}{8}\rho g L\left(\frac{L}{t}\right) \tag{6.3}$$

where L and t are the span and thickness of the roof layer, respectively. The effect of degradation can be imposed on the thickness of the roof layer, while the effect of long-term strength can be imposed on the compressive strength of the roof layers. Figure 6.3 shows some computational results for the effects of creep loading and degradation on mudstone and sandstone roofs.

6.2.4 *Tributary Area Method for Pillar Stability*

The tributary area is a simple yet effective method for assessing the overall stability of pillars. Let us assume that the geometry of a representative pillar and its overburden load can be modeled as shown in Figure 6.4. From the assumed geometry, it is easy to show that the average pillar stress is

$$\sigma_p = \rho g H \frac{A_t}{A_p} \text{ or } \sigma_p = \sigma_V \frac{A_t}{A_p} \tag{6.4}$$

where ρ is the unit weight of rock, g is the gravitational acceleration, H is the overburden, A_t is area supported by the pillar, A_p is the area of the pillar, and $\sigma_V = \rho g H$

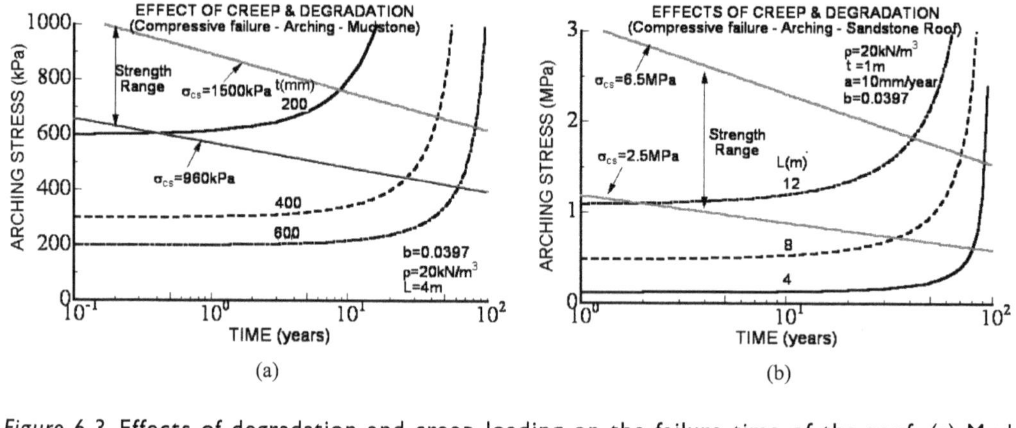

Figure 6.3 Effects of degradation and creep loading on the failure time of the roof. (a) Mudstone roof. (b) Sandstone roof. (Aydan et al., 2006).

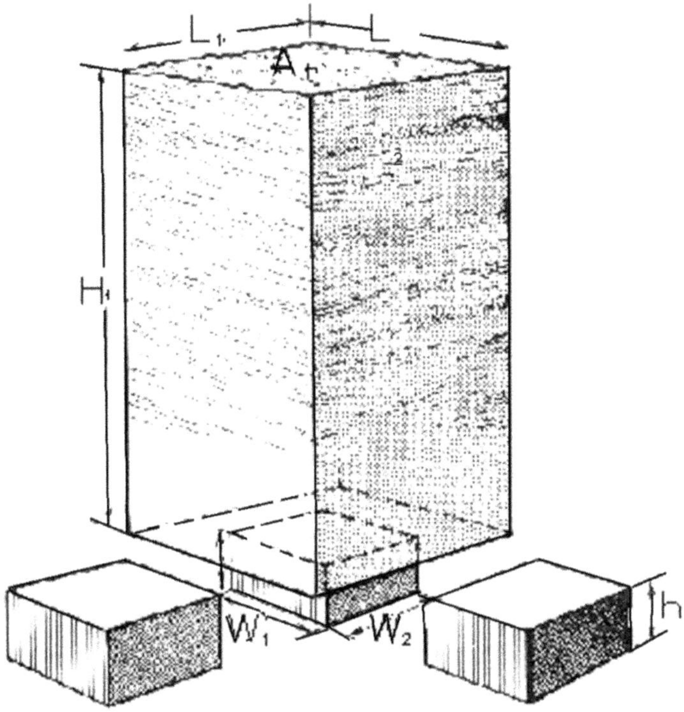

Figure 6.4 Model for pillar stability analyses.

6.2.4.1 *Effect of degradation*

The effect of degradation of rock due to cycles of wetting and drying causes flaking of rock near the surface, which then falls under the effect of gravity. This process repeats itself endlessly, subsequently causing a reduction in the area of the pillar. As a result, the normalized stress on the pillar can be expressed as

$$\frac{\sigma_p}{\sigma_{p0}} = \frac{1}{1 - \frac{\Delta L \times (L_{10} + L_{20})}{A_{p0}} + \frac{\Delta L^2}{A_{p0}}} \tag{6.5}$$

The degradation process may be approximated using some functional forms. For the simplicity, we assume that degradation is a linear function of time and is given in the following form:

$$\Delta L = a \cdot t \tag{6.6}$$

where constant a is called the degradation rate. Since the uniaxial strength of the lignite seam ranges between 4 and 6 MPa, the computed results for the long-term stability of the abandoned mine, varying the degradation rate and tributary area ratio, are shown in Figure 6.5. As expected, the computed results imply that if the degradation rate is higher or the tributary area ratio is larger, the failure time of the pillar becomes shorter.

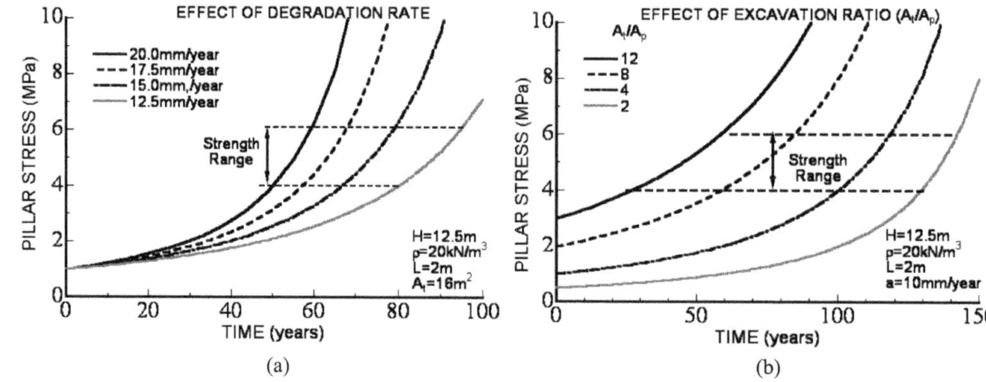

Figure 6.5 Effects of degradation on tributary area and the failure time of pillars (a) Effect of degradation rate. (b) Effect of excavation ratio (tributary area). (Aydan et al., 2006).

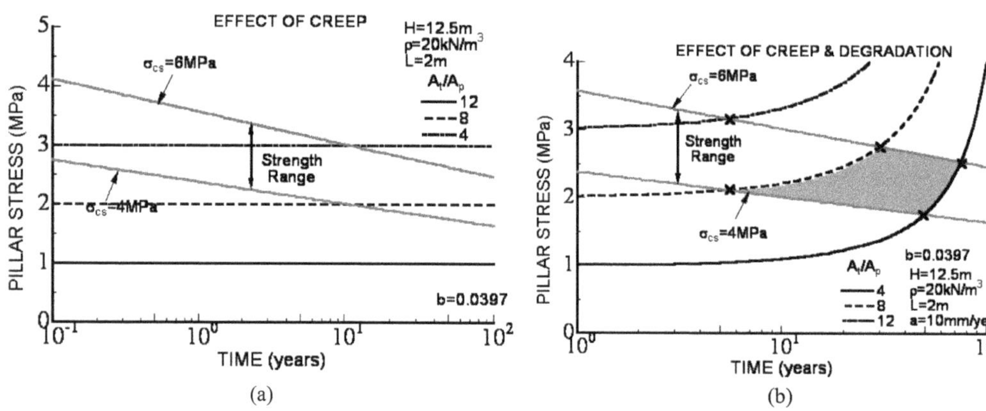

Figure 6.6 Effects of degradation and creep loading on the failure time of pillars (a) Effect of creep loading only. (b) Effect of creep loading and degradation. (Aydan et al., 2006).

6.2.5 Effect of Long-term Creep and Degradation

As explained in Section 3, the long-term strength of rock under creep loading would be mobilized. The long-term strength of rock depends on the level of the sustained loading with respect to its strength in short-term tests. Figure 6.6a shows the computed results for the failure time of pillars considering their long-term strength. From the results, it can be inferred that the failure time would be shorter if the tributary area (excavation ratio) increases. This situation may be observed when the abandoned mines are fully submerged. However, it should be noted that the long-term strength properties for saturated conditions should be used in such cases. Figure 6.6b shows the combined effects of long-term strength and degradation rate. Compared to the failure time of the pillar under creep loading only, the failure time of the pillars becomes shorter. The shaded range likely corresponds to the actual situation of the abandoned mine that the authors have been investigating (Aydan et al., 2005a,b,c). For the lower bound strength range, stability problems of the pillars would occur after 5.6–50 years while it would be after 30–75 years for the upper bound strength range.

6.2.5.1 *Effect of earthquakes*

However, there is currently no available approach to consider the effect of earthquakes on the stability of abandoned room and pillar mines. Although appropriate consideration would require solving the equation of motion, Aydan et al. (2005a, 2006, 2010) and Ohta et al. (2005) proposed a simple approach based on the seismic coefficient method used in the seismic design of structures in earthquake engineering. Figure 6.7a illustrates the main concept of this approach. The earthquake force acting on a typical pillar is determined through the horizontal seismic coefficient, in conjunction with the tributary area concept. When the pillar is in full contact with the overburden layers, the stress on a pillar may be expressed as follows (Aydan et al., 2005a, 2006; Ohta et al., 2005).

$$\sigma_p = \rho g H \frac{A_t}{A_p}\left[1 + 6\alpha_H \frac{H}{w}\cdot\frac{y}{w}\right] \tag{6.7}$$

where $\rho, g, H, A_t, A_p, w, \alpha_H$, and y are density, gravity, overburden height, area supported by a pillar, pillar area, pillar width (e.g., w_1 or w_2), horizontal seismic coefficient, and distance

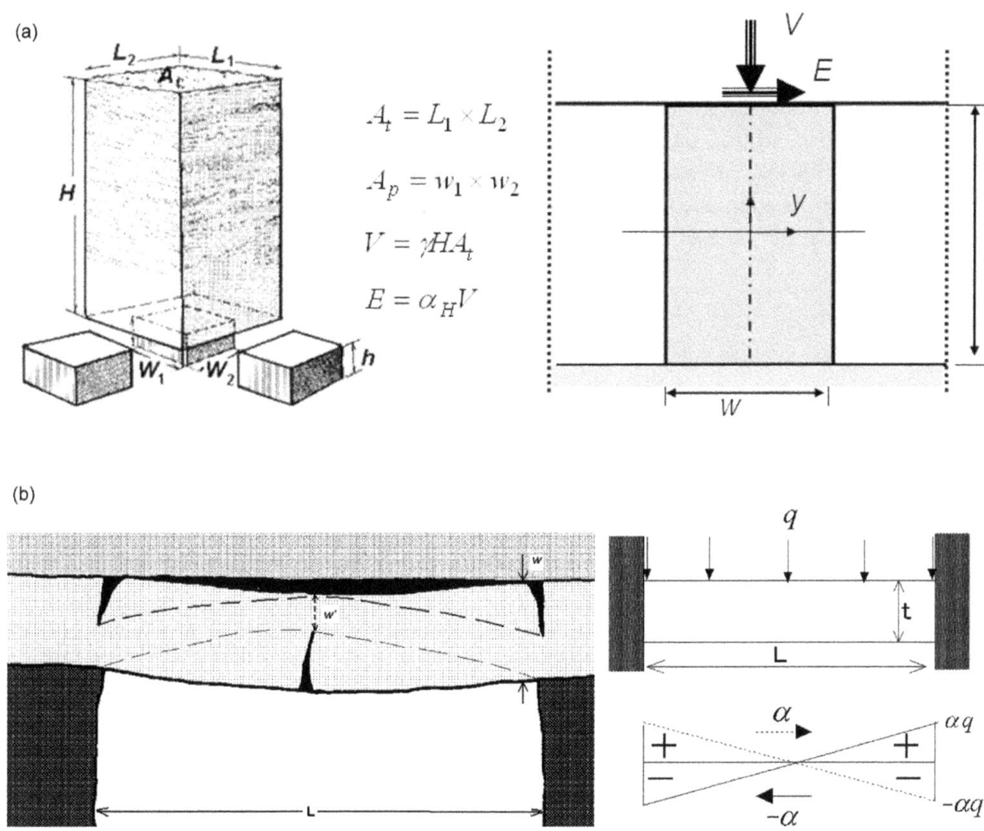

Figure 6.7 Mechanical model for roof stability and modeling of earthquake load for pillars (a) and roofs (b). (From Aydan et al., 2006; Ohta et al., 2005.)

from the center of the pillar, respectively. The maximum compressive and tensile stress occurs alternately at the sides of the pillar, depending on the direction of shaking. If the tensile strength between the roof layer and the pillar is assumed to be nil, and yielding occurs to a certain depth from the side of the pillar, the maximum compressive stress on the pillar side can be obtained as follows:

$$\sigma_p = \sigma_{po}\frac{1}{(1-\xi)^2}\left[1+\frac{6}{(1-\xi)}\left(\alpha_H\frac{H}{2w}-\xi\right)\right] \tag{6.8}$$

where $\xi = \frac{e}{w}$, $e = w - w^*$, $A_p^* = w^* \cdot w^*$, $\sigma_{po} = \rho g H \frac{A_t}{A_p}$, and w^* is the effective pillar width. The relation between the normalized yield distance and seismic coefficient α_H for no-tension conditions between the roof layer and pillar is obtained from Eq. (6.8) after some manipulations as follows:

$$\alpha_H = \frac{w}{3H}[5\xi + 1] \tag{6.9}$$

The horizontal seismic coefficient at the collapse of the pillar may be determined under the condition that the maximum compressive stresses from Eqs. (6.8) and (6.9) are the same. In Figure 6.8a, H and h represent the overburden height and pillar height, respectively. Model tests on room and pillar mines on a shaking table indicated that the roof layers also fail by bending. Therefore, the load condition under horizontal shaking is assumed to consist of gravitational load inducing bending stresses and linearly varying axial stress along the roof axis, transitioning from tension to compression due to horizontal shaking. Based on this assumption, the horizontal seismic coefficient at the time of roof layer failure by bending is obtained as follows:

$$\alpha_H = \frac{\sigma_t}{\rho g t(A_t/A_p - 1)} - \frac{t}{2H}(A_t/A_p - 1) \tag{6.10}$$

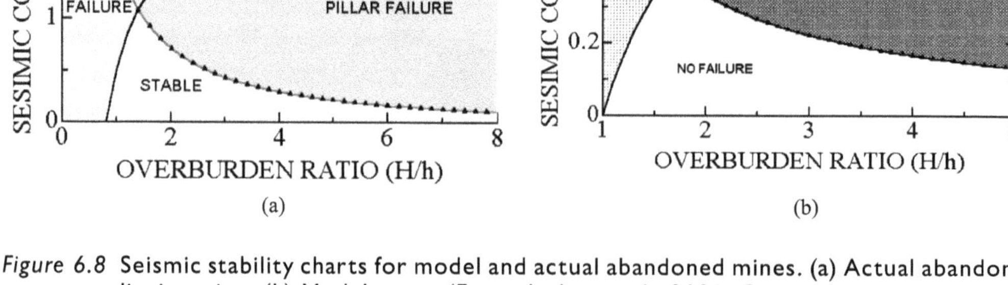

Figure 6.8 Seismic stability charts for model and actual abandoned mines. (a) Actual abandoned lignite mine. (b) Model tests. (From Aydan et al., 2006; Ohta et al., 2005.)

where t and σ_t are the thickness and tensile strength of the roof layer, respectively. Figure 6.8 shows a computed diagram of the relations between the overburden ratio and seismic coefficient for various failure modes with the chosen parameters depicted in the same figure. As noted from the figure, there are four different regions in the diagram: Region 1: roof layer and pillar are stable; Region 2: roof layer is stable, pillar is unstable; Region 3: roof layer is unstable, pillar is stable; Region 4: roof layer and pillar are both unstable. These results indicate that shallow mines are prone to roof failure, while deeper mines are more susceptible to pillar failure during actual strong ground motions in earthquakes.

6.2.6 Effect of Earthquakes on Roof

The stability of the roof of shallow room and pillar mines subjected to earthquake loads can also be analyzed using a similar mechanical model presented above. The seismic stability of shallow sinkholes is evaluated as a built-in beam subjected to horizontal and vertical seismic loads (e.g., Aydan et al., 2010; Aydan & Tokashiki, 2011). Based on experimental observations, the horizontal seismic load will cause tension on one side while resulting in compression on the other side (Figure 6.7b):

$$\alpha_H = \frac{\dfrac{\sigma_t}{\gamma t} - \left(1 + \dfrac{t^*}{t}\right)\dfrac{1}{2}\left(\dfrac{L}{t}\right)^2}{\left(\dfrac{L}{t}\right)\left(1 + \dfrac{t^*}{t}\right)\left[1 + \dfrac{\beta}{2}\left(\dfrac{L}{t}\right)\right]} \tag{6.11}$$

where

γ: unit weight of roof material
t: roof thickness
t^*: surcharge load height
L: opening width
β: ratio of vertical seismic coefficient (α_V) to horizontal seismic coefficient (α_H).

6.2.7 Effect of Groundwater

Groundwater in abandoned mines can be considered a factor in preventing the degradation of pillars and roof layers. Furthermore, the effective vertical stress acting on the pillar would decrease due to the fluid pressure support provided by groundwater, and pillar stress would be expressed as follows (Figure 6.9):

$$\sigma'_p = \frac{1}{1-e}(\gamma_r H - \gamma_w H_w) \tag{6.12}$$

where $\gamma_r, \gamma_w, H, H_w$, and e are unit weight of rock, water, overburden, water depth, and excavation ratio, respectively (Aydan & Ito, 2015). The excavation ratio is defined as the area of excavation (A_e) over the typical total area (A_t) (Aydan et al., 2005a,b, 2006; 2011):

$$e = \frac{A_e}{A_t} \tag{6.13}$$

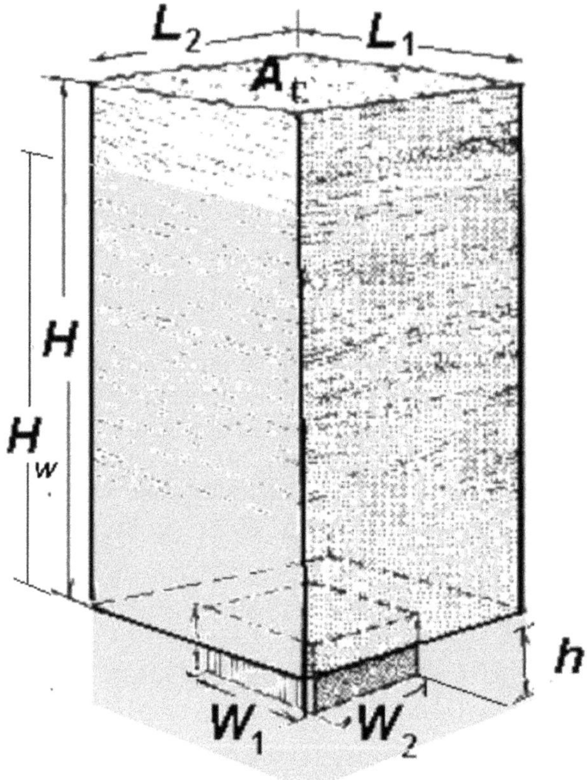

Figure 6.9 Modeling the effect of inundation on abandoned room and pillar mines.

If an abandoned mine is fully submerged and the water height is equivalent to the overburden height, the effective stress acting on the pillar would be decreased by about 43% (Aydan & Ito, 2015). However, as noted from Figure 3.20, the strength of the surrounding lignite layer decreases by about 60%–75% under fully saturated conditions (Aydan, 2016). In other words, groundwater would further decrease the overall stability of abandoned lignite mines.

6.3 Numerical Methods

6.3.1 Fundamentals and Formulation

For any structure or continuum body subjected to motion, the momentum conservation law must hold. Momentum is defined in integral form as given below (e.g., Aydan, 2017, 2021; Eringen, 1980; Maze, 1970).

$$\mathbf{p} = \int_{\Omega} \rho \mathbf{v} d\Omega \tag{6.14}$$

Preliminary relations

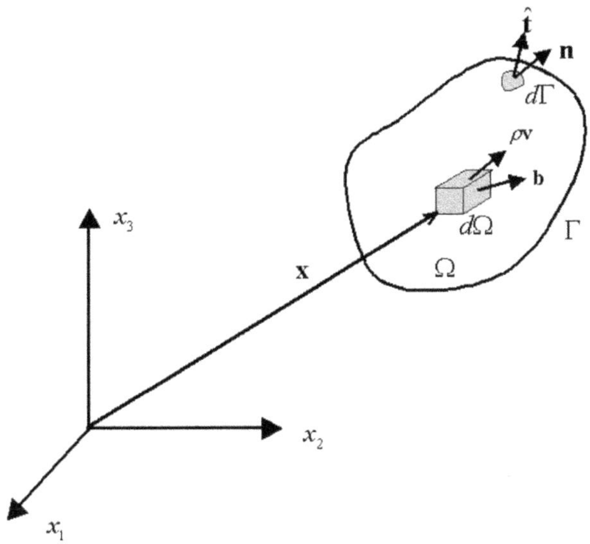

Figure 6.10 Illustration of momentum conservation law. (From Aydan, 2017.)

$$\int_{\Omega} \nabla \cdot \boldsymbol{\sigma} d\Omega = \int_{\Gamma} \boldsymbol{\sigma} \cdot \mathbf{n}\, d\Gamma; \frac{d(d\Omega)}{dt} = (\nabla \cdot \mathbf{x}) d\Omega; \mathbf{t} = \boldsymbol{\sigma} \cdot \mathbf{n} \tag{6.15}$$

Conservation of momentum is expressed in the following form, in view of Eq. (6.15), which is also known as the Reynolds transport theorem (Figure 6.10)

$$\frac{d}{dt} \int_{\Omega} \rho \mathbf{v} d\Omega = \int_{\Gamma} \mathbf{t} d\Gamma + \int_{\Omega} \mathbf{b} d\Omega \tag{6.16}$$

Equation (6.16) may be rewritten as

$$\int_{\Omega} \left(\frac{d(\rho \mathbf{v})}{dt} + (\rho \mathbf{v}) \nabla \cdot \mathbf{v} \right) d\Omega = \int_{\Omega} \nabla \cdot \boldsymbol{\sigma} d\Omega + + \int_{\Omega} \mathbf{b} d\Omega \tag{6.17}$$

Carrying out the derivation in Eq. (6.17), we have the following:

$$\int_{\Omega} \left(\left(\frac{d\rho}{dt} + \rho(\nabla \cdot \mathbf{v}) \right) \mathbf{v} \right) d\Omega + \int_{\Omega} \rho \frac{d\mathbf{v}}{dt} d\Omega = \int_{\Omega} (\nabla \cdot \boldsymbol{\sigma} + \mathbf{b}) d\Omega \tag{6.18}$$

The first term on the left-hand side disappears by virtue of the mass conservation law and takes the following form:

$$\int_{\Omega} \rho \frac{d\mathbf{v}}{dt} d\Omega = \int_{\Omega} (\nabla \cdot \boldsymbol{\sigma} + \mathbf{b}) d\Omega \tag{6.19}$$

Eq. (6.19) may be rewritten as

$$\int_{\Omega} \left[\rho \frac{d\mathbf{v}}{dt} - (\nabla \cdot \boldsymbol{\sigma} + \mathbf{b}) \right] d\Omega = 0 \tag{6.20}$$

To satisfy Eq. (6.20), the integrand must be zero, leading to the following relation:

$$\rho \frac{d\mathbf{v}}{dt} = \nabla \cdot \boldsymbol{\sigma} + \mathbf{b} \tag{6.21}$$

Furthermore, the derivation on the left-hand side may be related to acceleration or displacement vectors as follows:

$$\frac{d\mathbf{v}}{dt} = \mathbf{a} \text{ or } \frac{d\mathbf{v}}{dt} = \frac{d^2\mathbf{u}}{dt^2} \tag{6.22}$$

The seismic design of any structure and its foundations must be based on the solution of the equation of motion for a given dynamic loading condition, along with appropriate mechanical models of the materials constituting the structure and design specifications. The constitutive laws in the analyses should generally be elasto-visco-plastic, together with yield conditions appropriate for the materials involved (Figure 6.11). When structures are situated on or in rock mass, the characteristics of rock masses must be considered, which may require discontinuum-type modeling and consideration of geological discontinuities. While simple cases may be analyzed using analytical solutions, the use of numerical techniques such as finite difference, finite element, and/or boundary element methods is essential for general situations.

Since the loads acting on structures are transient, the utilization of hyperbolic-type solutions is necessary, taking into account inertia forces as well as damping characteristics. Nevertheless, the solution procedure may be significantly simplified, yielding a pseudo-static solution in which the transient load is applied like a static load in an incremental manner, and inertia forces are omitted. One such analysis is called push-over analysis, which accommodates the non-linear

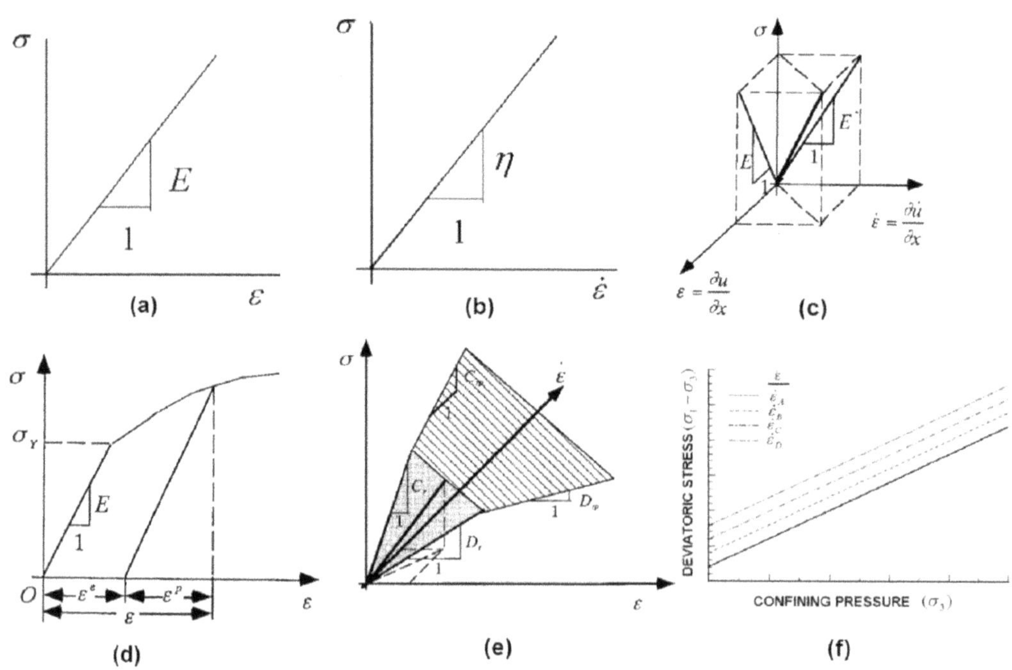

Figure 6.11 Illustrations of some constitutive models. (a) Elastic, (b) viscous, (c) visco-elastic, (d) elasto-plastic, (e) elasto-visco-plastic, (f) rate-dependent yield criterion. (From Aydan, 2021.)

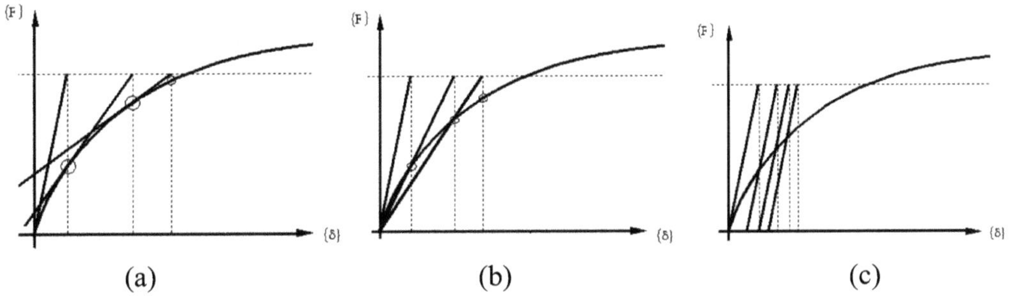

Figure 6.12 Methods for dealing non-linear behavior. (a) Tangential stiffness. (b) Secant stiffness. (c) Initial stiffness. (From Aydan, 2021.)

behavior and yielding of materials and/or components of the structure. In these analyses, various techniques are employed to deal with non-linear responses, including tangential stiffness, secant stiffness, and initial stiffness methods, as illustrated in Figure 6.12. If the overall response shows a hardening response, tangential stiffness may be used for quick solutions. However, if the system begins to exhibit softening behavior, the use of secant stiffness and initial stiffness methods becomes necessary. Particularly, the initial stiffness method yields absolute convergence.

The final form of the discretized equation of motion given in Eq. (6.21), irrespective of the method of solution (FDM, FEM, BEM), may be written in the following form:

$$\left[M\right]\left\{\ddot{\phi}\right\}+\left[C\right]\left\{\dot{\phi}\right\}+[K]\{\phi\}=\{F\} \tag{6.23}$$

The specific forms of matrices $[M]$, $[C]$, $[K]$, and vector $\{F\}$ in the equation above will only differ depending on the chosen method of solution and the dimensions of physical space. The viscosity matrix $[C]$ is associated with the rate dependency of geomaterials. However, in many dynamic solution schemes, the viscosity matrix $[C]$ is expressed in the following form using the Rayleigh damping approach as

$$\left[C\right]=\alpha[M]+\beta[K] \tag{6.24}$$

where α, β are called proportionality constants. This approach becomes very convenient in large-scale problems when using the central finite difference technique and mass lumping.

The natural frequencies of any structure or continuum body can be obtained from the following equation system by assuming that the damping matrix $[C]$ is nil and the external force $\{F\}$ is zero in Eq. (6.23).

$$\left[M\right]\left\{\ddot{U}\right\}+[K]\{U\}=\{0\} \tag{6.25}$$

Equation (6.25) is known as the equations system of free vibration.

The variable vector $\{U\}$ can be expressed as a harmonic function given as

$$\{U\}=\{\phi\}e^{iwt} \tag{6.26}$$

If Eq. (6.26) is inserted into Eq. (6.25), one can easily obtain the following equation system:

$$([K]-\omega^2[M])\{\phi\}=\{0\} \tag{6.27}$$

The solution of Eq.(6.27) is known as eigenvalue analysis, yielding ω^2 and $\{\phi\}$. ω^2 is known as eigenvalues and $\{\phi\}$ eigenvalue vector.

6.3.2 One dimensional (1D) Dynamic Multi-layer Shear Beam Approach

6.3.2.1 Fundamentals and theory of 1D multi-layer shearing approach

If the ground is modeled as shown in Figure 6.13, Eq. (6.21) takes the following form (Aydan, 1994, 2006, 2008; Aydan & Geniş, 2008; Aydan et al. 2008, 2012b):

$$\frac{\partial^2 u}{\partial t^2} = \frac{G}{\rho}\frac{\partial^2 u}{\partial x^2} + \frac{\eta}{\rho}\frac{\partial^2 \dot{u}}{\partial x^2} + \frac{b}{\rho} \tag{6.28}$$

where G, η, and b viscosity modulus, and body force, respectively. The FEM form of Eq. (6.28) can be obtained in analogy to Eq. (6.23) as

$$[M]\{\ddot{U}\}+[C]\{\dot{U}\}+[K]\{U\}=\{F\} \tag{6.29}$$

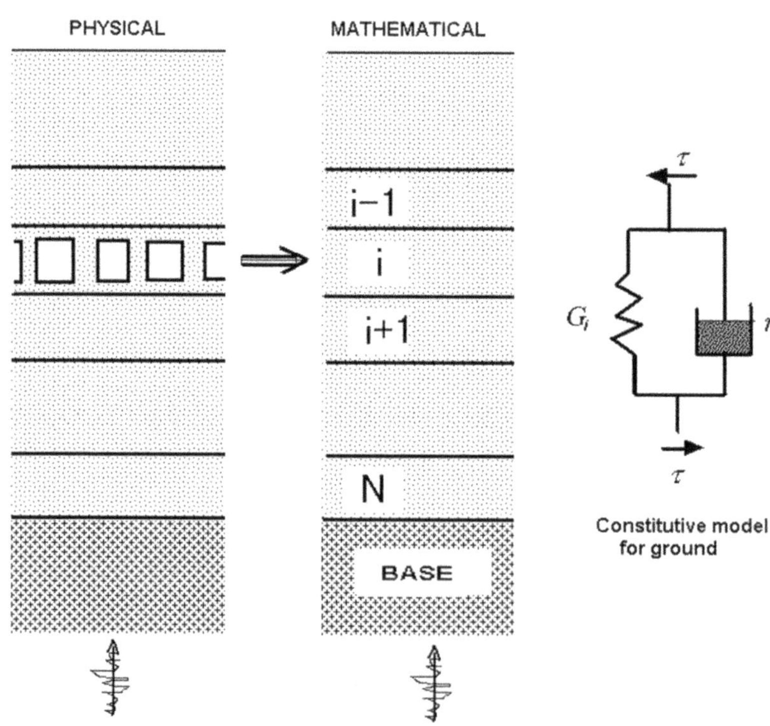

Figure 6.13 Layered half-space visco-elastic model. (From Aydan et al., 2013; Geniş & Aydan 2008.)

where [*M*] is the mass matrix, [*C*] is the viscosity matrix, and [*K*] is the stiffness, matrix while$\{U\}$ and $\{F\}$ are the nodal displacement and force vectors. Similarly, Rayleigh damping is assumed to take the following form:

$$[C]=\alpha[M]+\beta[K] \tag{6.30}$$

where α, β are damping coefficients.

In analyses, elasticity coefficients are replaced by p-wave or s-wave velocities. For example, the p-wave velocity is related to elasticity constants as follows:

$$V_p^2=\frac{E(1-\upsilon)}{\rho(1+\upsilon)(1-2\upsilon)} \text{ or } V_p^2=\frac{\lambda+2\mu}{\rho} \tag{6.31}$$

6.3.2.2 *Transfer matrix method*

The transfer matrix method is often employed to evaluate the natural frequencies of the ground. The method is illustrated in Figure 6.14, and the effect of damping is also considered.

6.3.2.3 *Analytical and numerical evaluation of vibration properties*

For a half-space with a height of H, one can show that the natural period of the ground can be obtained as given below (e.g., Aydan et al., 2004))

$$T_p=\frac{4H}{V_s}; V_s=\sqrt{\frac{G}{\rho}}. \tag{6.32}$$

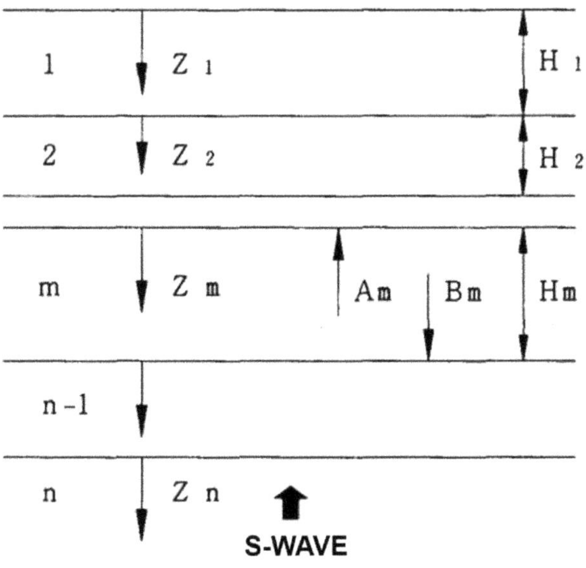

Figure 6.14 One-dimensional illustration of the multi-layer model in the transfer matrix method.

where G and ρ are the shear modulus and density of the layer. If the ground consists of n layers, the following relations hold between the equivalent vibration characteristics of the ground:

$$\bar{T}_p = \frac{4H}{\bar{V}_s}; \bar{V}_s = \sqrt{\frac{\bar{G}}{\bar{\rho}}}; \frac{1}{\bar{G}} = \sum_{i=1}^{N} \frac{\lambda_i}{G_i}; \bar{\rho} = \sum_{i=1}^{N} \lambda_i \rho_i; \lambda_i = \frac{h_i}{H}; \tag{6.33}$$

where h_i is the thickness of layer i. Furthermore, the equivalent shear modulus of the exploited layer (lignite, tuff, or limestone) can be given in the following form:

$$G_e = G_l \frac{A_p}{A_t} \tag{6.34}$$

where A_p and A_t are the pillar area and total area supported by a typical pillar. For example, if the modeled ground is assumed to consist of 3 layers, the equivalent shear velocity will be obtained as follows:

$$\bar{V}_s = \sqrt{\frac{\alpha G_1 G_2 G_3}{\alpha\lambda_1 G_2 G_3 + \lambda_2 G_1 G_3 + \alpha\lambda_3 G_1 G_2} \cdot \frac{1}{\lambda_1 \rho_1 + \alpha\lambda_2 \rho_1 + \lambda_3 \rho_3}} \tag{6.35}$$

where $\alpha = \frac{A_p}{A_t}$; $\lambda_1 = \frac{h_1}{H}$; $\lambda_2 = \frac{h_2}{H}$; $\lambda_3 = \frac{h_3}{H}$

If it is assumed such that $\lambda = \lambda_2$, $1 - \lambda = \lambda_1 + \lambda_3$, $G_1 = G_2 = G_3 = G$, $\rho_1 = \rho_2 = \rho_3 = \rho$, the equivalent shear wave velocity of the ground is obtained as follows:

$$\bar{V}_s = V_s \sqrt{\frac{\alpha}{\lambda + \alpha(1-\lambda)} \cdot \frac{1}{\alpha\lambda + (1-\lambda)}} \tag{6.36}$$

The normalized period by the equivalent period takes the following form:

$$\frac{\bar{T}}{T_p} = \sqrt{\frac{\lambda + \alpha(1-\lambda)}{\alpha} \cdot (\alpha\lambda + (1-\lambda))} \tag{6.37}$$

As noted from Eq. (6.37), if the excavation ratio (α) increases, the vibration period becomes longer, which implies that the abandoned mines may be affected by far-field earthquakes. Figures 6.15 and 6.16 illustrate the effect of tributary area and layer thickness on the equivalent natural frequencies of the ground.

Table 6.1 provides the properties of materials used in the finite element analyses conducted using the finite element code (1DFEM_BASEACC, Version 1.1) developed by Aydan et al. (2004a, b). The lignite mine was assumed to be 2 m thick with an excavation ratio of 50%. Figure 6.17a shows the acceleration response at the ground surface for sinusoidal shaking with various periods. The largest amplification occurs when the period of ground shaking coincides with the natural period of the domain, as expected. Figure 6.17b shows the maximum ground displacement response concerning the base with a wave period of 0.5 s. As expected, large relative displacement occurs at the level of the extracted lignite seam. The amplitude of the amplification entirely depends on the viscous properties of the layers and the frequency characteristics of the ground motions.

A series of natural frequencies of the ground on the abandoned lignite mines, with and without backfilling in Mitake Town, were computed using the matrix transfer method, as shown in Figures 6.18 and 6.19. As noted from the figures, the period of the ground becomes shorter as a result of the increase in the stiffness of the ground due to backfilling.

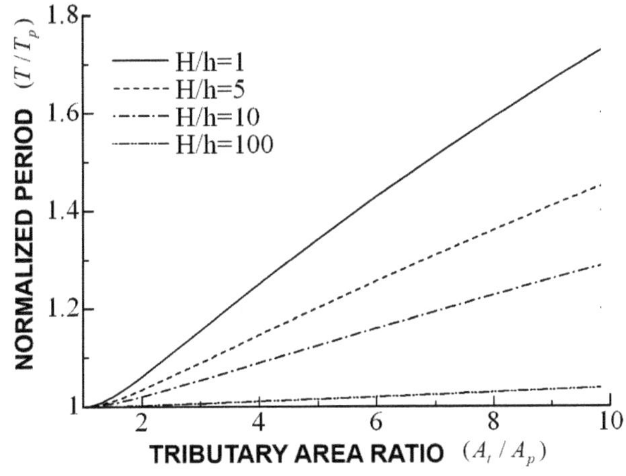

Figure 6.15 Effect of overburden on normalized natural period. (From Aydan et al., 2013a.)

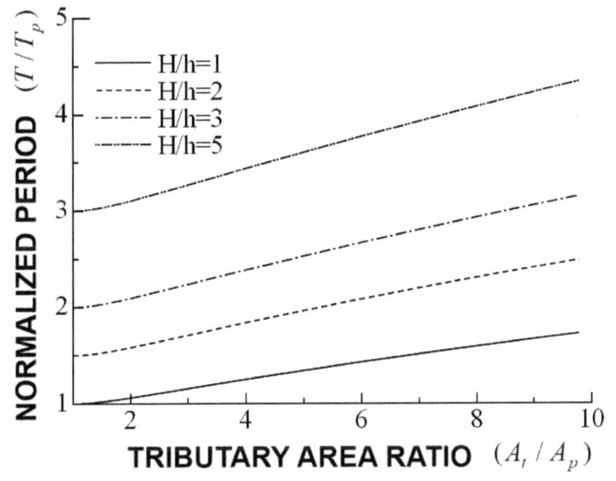

Figure 6.16 Variation of normalized natural period as a function of seam height. (From Aydan et al., 2013a.)

Table 6.1 Parameters used finite element analyses

Layers	*$(G/\rho)^{1/2}$ (m/s)*	*η/ρ (m^2/s)*	*(b/ρ) (m/s^2)*
Layer 1 (30–40 m)	320	32	10
Layer 2 (28–30 m)	160	16	10
Layer 3 (0–28)	320	32	10

The numerical model of Aydan was used to assess the shaking characteristics of the ground above the abandoned mine located at a depth of 16 m in Mitake Town for an exploitation ratio of 70%. Results are shown in Figure 6.20 for the abandoned lignite subjected to the base acceleration obtained according to the method proposed by Sugito et al. (2000) for the anticipated M9 Nankai–Tonankai–Tokai Earthquake. The material properties used in the analyses are given in Table 6.2.

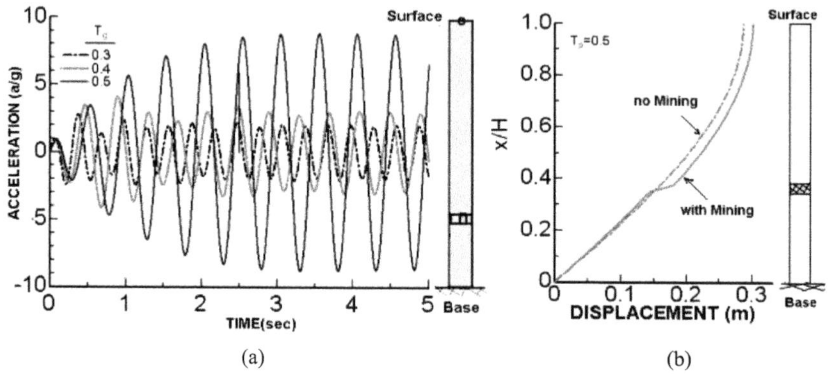

Figure 6.17 Computed responses for various wave forms. (a) Acceleration response. (b) Displacement response. (From Aydan et al., 2013a.)

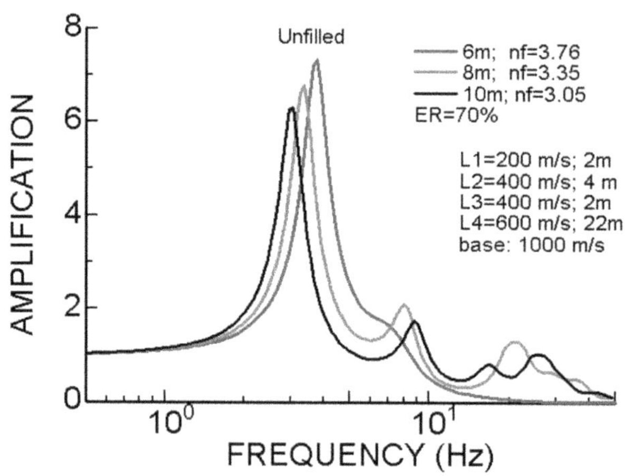

Figure 6.18 Computed natural frequency and amplification of an unfilled abandoned lignite mine. (From Aydan et al., 2013a.)

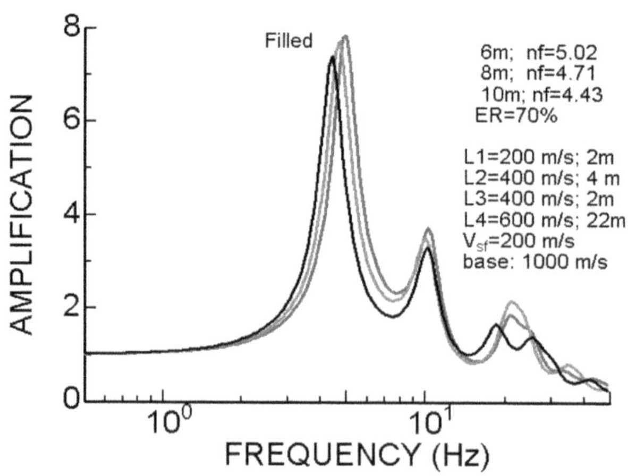

Figure 6.19 Computed natural frequencies of ground for backfilled lignite mine. (From Aydan et al., 2013a.)

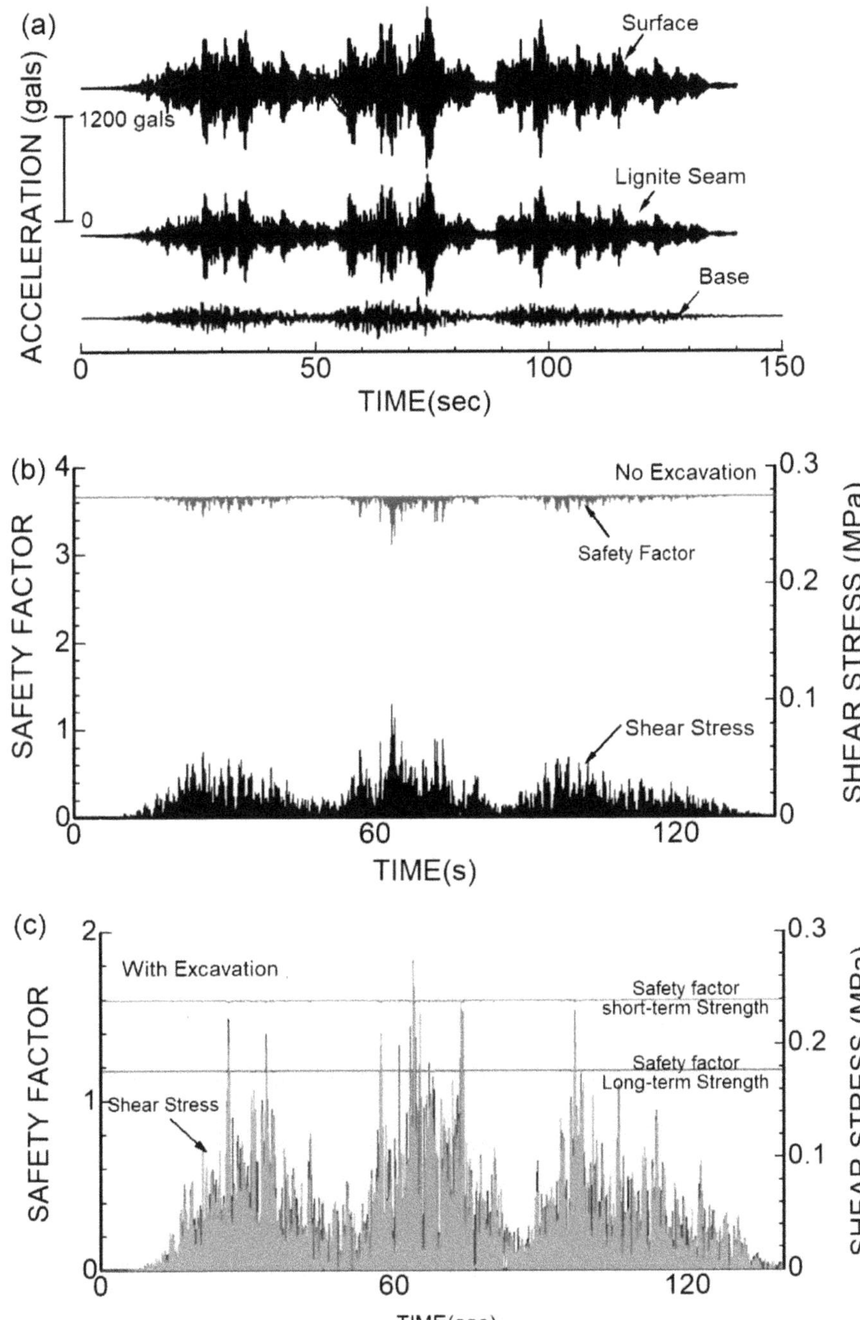

Figure 6.20 (a) The acceleration response at the ground surface and lignite seam, and (b,c) Extended shear layer model for estimating the safety of the mine during ground motions from the anticipated Nankai–Tonankai–Tokai triple earthquake. (From Aydan et al., 2013a.)

The maximum amplitude of the base acceleration is about 250 gals. The ground motions are amplified up to 3.7 times at the ground surface. This implies that the ground motions above the abandoned lignite mines can reach a very high level of shaking, which may be critical for surface structures.

6.3.2.4 *Effect of collapsed room and pillar mines on ground motions*

As mentioned in the introduction, long-wall mining is one of the commonly used extraction techniques for coal and lignite seams. The ground above the coal seam is forced to fail following the extraction of the coal. The broken rock mass above the mined areas is called goaf or gob, and its properties are greatly reduced compared to virgin ground. There are few reports on the in-situ investigations of the characteristics of the goaf, such as those by Fujii et al. (2011). The in-situ investigations showed that the reduction of wave velocity in goaf can be between 30% and 65% of that in virgin ground. As pointed out by Aydan et al. (2013a, c), there is no report on the possible ground amplifications above the goaf of old mine workings. The authors are currently performing three-dimensional ground response analyses for such grounds in addition to preliminary shear layering dynamic analyses. In this article, we report the results of analyses using the shear layering dynamic analysis method. Analyses were carried out for two ideal situations: virgin ground and ground with a goaf at a depth of 30 m, which is relatively shallow (Figure 6.21). Material properties are shown in Table 6.3, and the

Table 6.2 Parameters used in finite element analyses

Layers	$(G/\rho)^{1/2}$ (m/s)	η/ρ (m²/s)	(b/ρ) (m/s²)
Layer 1 (36–31 m)	900	45	19
Layer 2 (18–31 m)	600	30	19
Layer 3 (16–18)	400	20	14
Layer 4 (5–16 m)	500	25	19
Layer 5 (0–5)	300	15	19

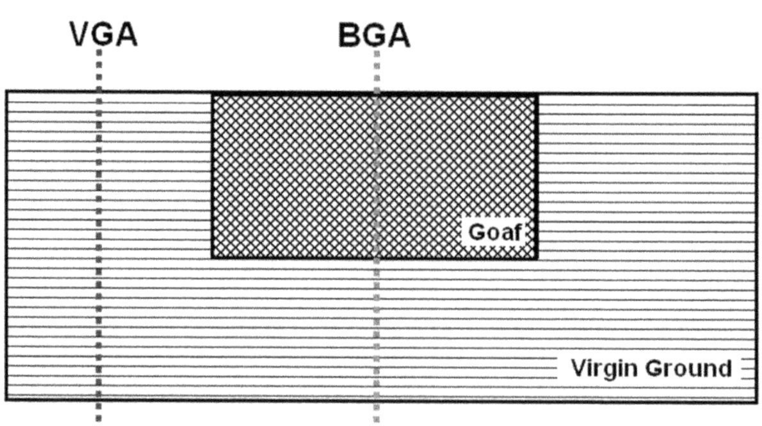

Figure 6.21 Illustration of ground conditions of old mine workings with longwall mining technique (Aydan et al., 2012a).

Table 6.3 Parameters used finite element analyses

Layers	$(G/\rho)^{1/2}$ (m/s)	η/ρ (m^2/s)	(b/ρ) (m/s^2)
Virgin ground	900	45	25
Goaf	300	15	18

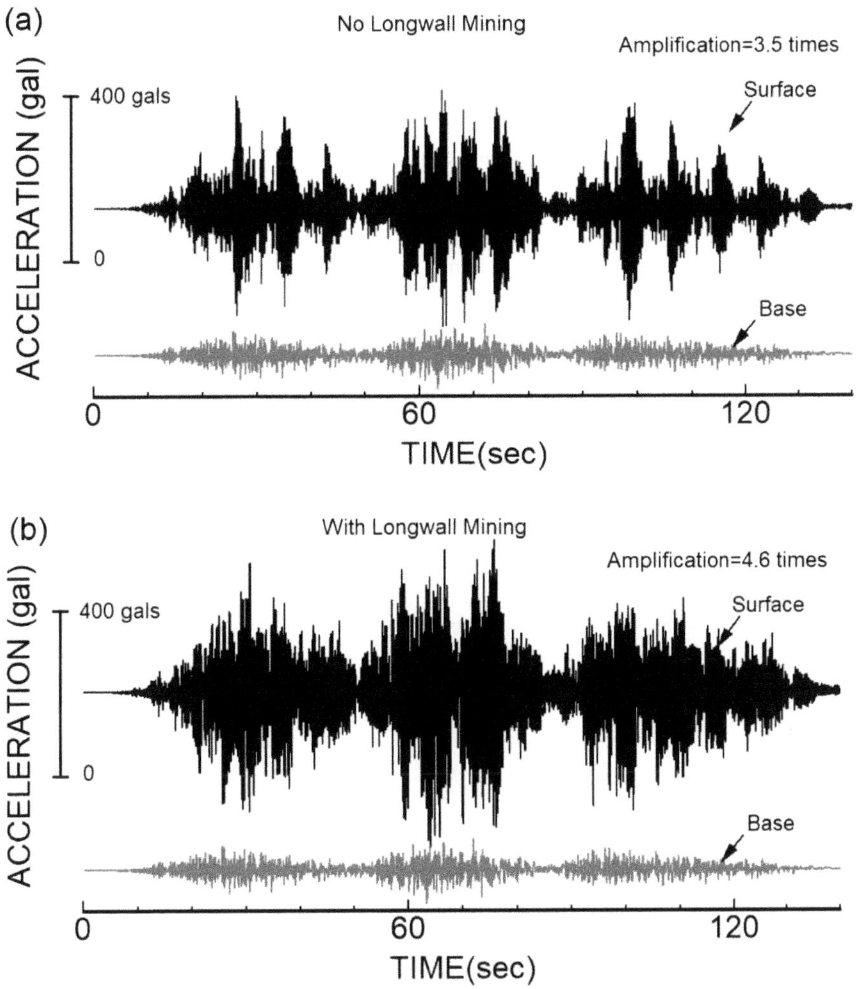

Figure 6.22 Acceleration response of ground surface for (a) virgin ground and (b) the mined ground (Aydan et al., 2012a).

ground motion was assumed to be due to the M9 anticipated Nankai–Tonankai–Tokai Earthquake, as used in the previous sections.

Figure 6.22 shows the computed acceleration responses for the ground surface of old mine workings without and with longwall mining. As expected, ground amplification occurs due to the free surface, even in the absence of mining. However, the ground amplification at the surface above the goaf is significantly larger compared to that of the virgin ground.

6.3.3 Two-dimensional Classic Elasto-Plastic Finite Element Method Analyses

6.3.3.1 Finite element method for static loading

Figure 6.23 shows the distribution of minimum principal stress (with tension as positive) for an abandoned room and pillar mine beneath Kyowa Secondary School in Mitake Town, Gifu Prefecture, Japan, under static loading condition for material properties given in Table 6.4. Although the maximum pillar stresses are slightly higher than those computed using the tributary area method, the quick stability assessment using the tributary area method should be quite acceptable.

6.3.3.2 2D Tunneling beneath abandoned mines

Depending on the deposition of lignite and coal seams, there may be single or multiple seams. The thickness and quality of the seams of coal or lignite are also governing factors for their exploitation. Furthermore, the overall mechanical properties of the seams and surrounding rock may decrease due to long-term creep-type loading as well as physical degradation processes. Therefore, these conditions are considered in detailed two-dimensional elasto-plastic numerical analyses (Rocscience, 1998) presented in this section (from Aydan & Geniş, 2015). The specific geological conditions are similar to those in the Tokai Region of Japan. In a typical numerical

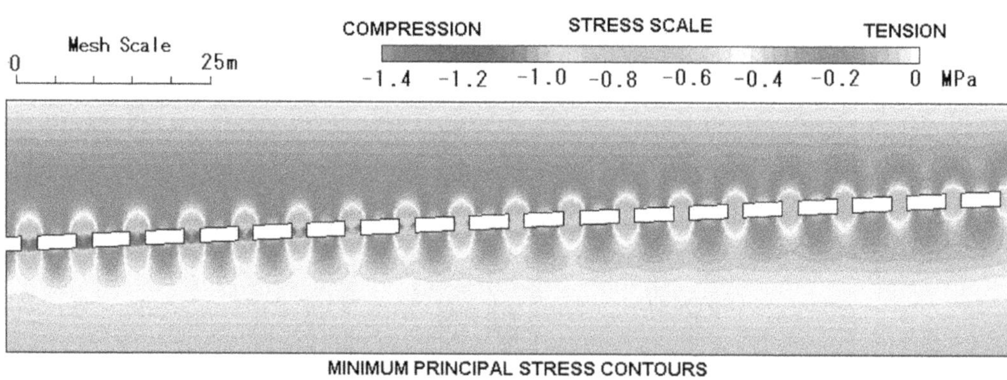

Figure 6.23 Contours of minimum principal stress beneath the Kyowa Secondary School. (From Aydan et al., 2013d.)

Table 6.4 Material properties of layers

Layer	*γ kN/m³*	*E MPa*	*ν*	*c MPa*	*φ (°)*
Topsoil	19	270	0.35	0.0	38
Upper Mst-Sst	19	750	0.3	0.7	25
Lignite	14	400	0.3	0.66	45
Lower Mst-Sst	19	1,073	0.3	1.00	45
Chert	19	3,647	0.3	3.00	45

analysis, gravity load is imposed on the whole domain, then abandoned mines are excavated, and finally, a new tunnel beneath the abandoned mine is excavated. An elastic-perfectly plastic constitutive model, along with the Mohr–Coulomb type yield criterion, is adopted in the numerical analyses.

6.3.3.2.1 SINGLE EXPLOITED SEAM

Firstly, the effect of the distance between the exploited seam and the crown of a circular cavity with a diameter of 14 m and 44 m below the ground surface upon its excavation is investigated. Figure 6.24 shows the mesh used in the numerical analyses. Displacement and safety factor contours for elastic response are shown in Figure 6.25. As noted from the figure, the displacement contours differ above and below the abandoned lignite mine. While the displacement contours above the abandoned lignite mine are almost vertical, the displacement contours between the excavated circular cavity and the abandoned lignite mine incline at an angle of approximately 60° to the horizontal. Figure 6.26 shows a close-up view of the displacement contours together with the plastic zone formation around the circular cavity.

If the strength properties of Pc1, lignite seam, and Pc2 layers are reduced as shown in Table 6.5, the yielding zone around the circular cavity, as well as above and below the lignite seam, increases in size, as shown in Figure 6.27. The percentage of yielding at the nodes shown in Figure 6.27 and other figures hereafter represents the ratio of the number of yielded elements to the total number of elements connected to the node used in contouring.

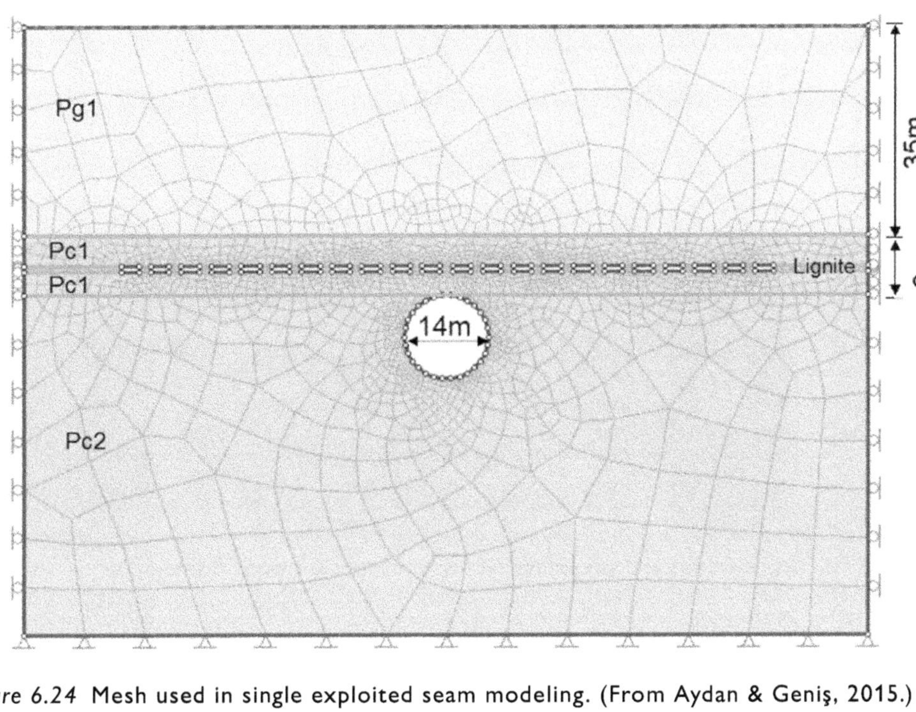

Figure 6.24 Mesh used in single exploited seam modeling. (From Aydan & Geniş, 2015.)

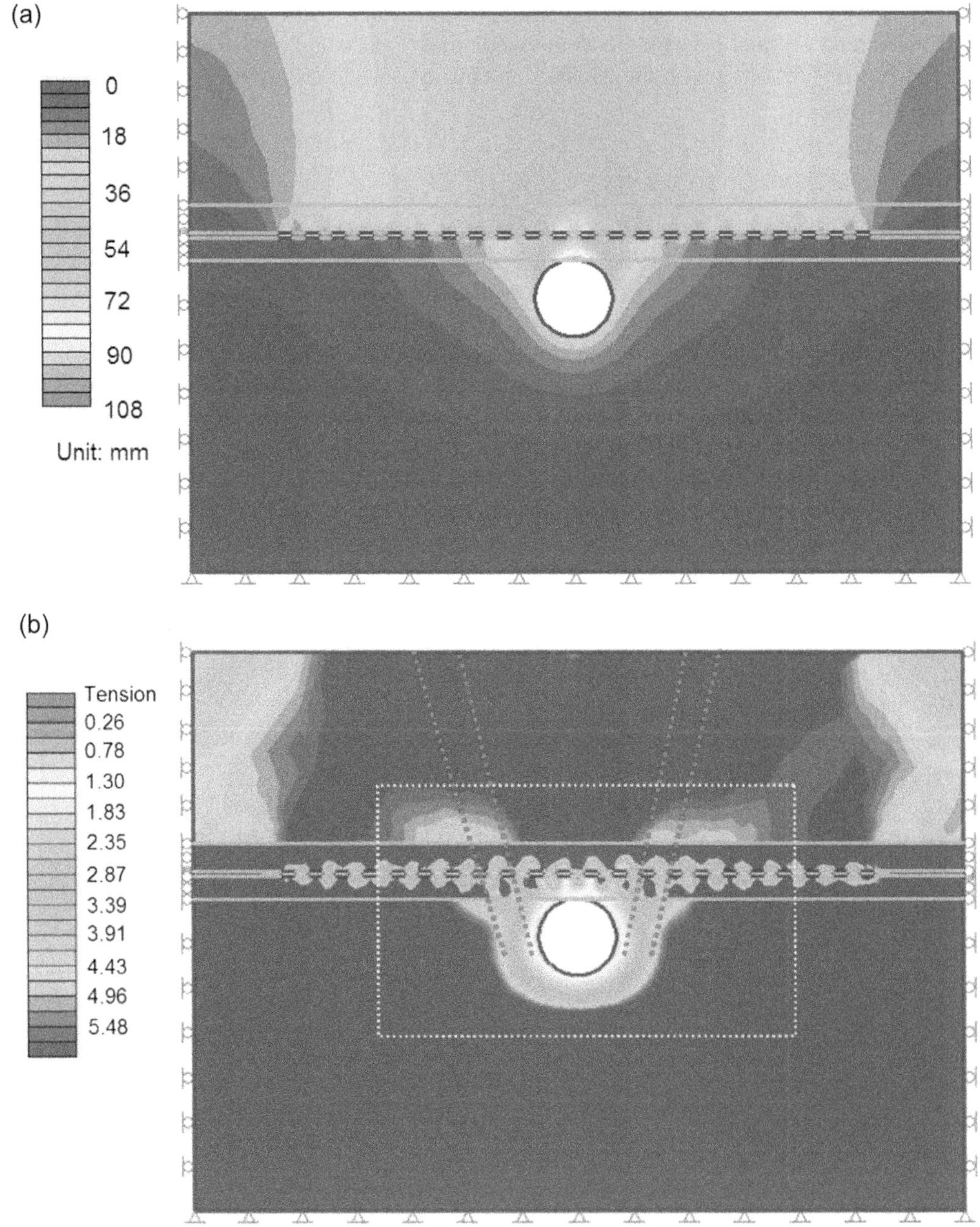

Figure 6.25 (a) Displacement and (b) safety factor contours for elastic analysis.

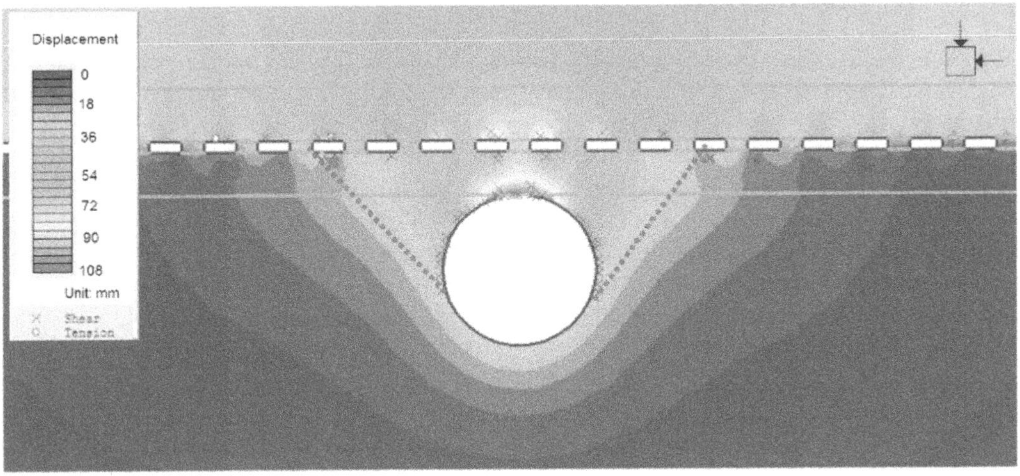

Figure 6.26 Displacement contours and yield zone. (From Aydan & Geniş, 2015.)

Table 6.5 Material properties used numerical analyses

Layer	γ *kN/m³*	*c kPa*	ϕ *(°)*	σ_t *kPa*	*E MPa*	ν
Pg1	16	50	35	50	80	0.3
Pc1	18	240	25	240	110	0.4
Lignite	14	160	30	160	180	0.26
Pc2	18	300	25	300	400	0.4

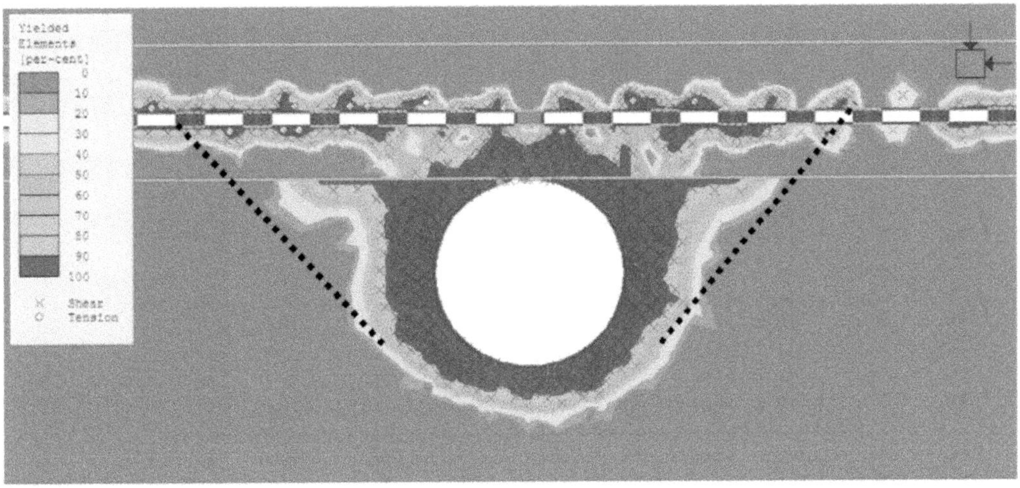

Figure 6.27 Yield zone for reduced properties. (From Aydan & Geniş, 2015.)

6.3.3.2.2 MULTIPLE SEAMS WITH A SINGLE SEAM EXPLOITED

The next computational examples involve two lignite seams, focusing on the exploitation of a single seam and both seams. The distance between the top lignite seam and the crown of the circular cavity is 56 m, while the distance between the lower lignite seam and the crown of the circular cavity is 26 m. It should be noted that the overburden height of the circular cavity is larger than in the previous case. Figure 6.28 shows the mesh used in the numerical analyses, along with the material properties outlined in Table 6.6. First, the top lignite seam is excavated, followed by the excavation of the circular cavity. Figures 6.29 and 6.30 illustrate the displacement contours and plastic zone formation before and after the excavation of the circular cavity. The displacement contours for the abandoned mine excavation are quite similar to those in the

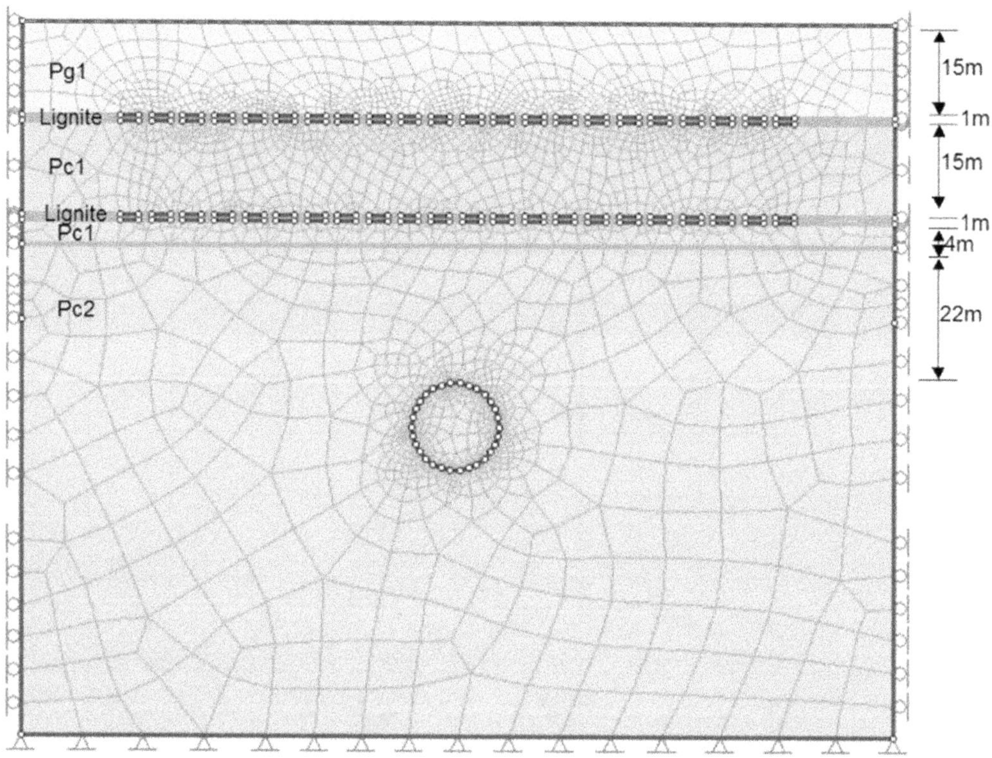

Figure 6.28 Mesh used in numerical analyses of the model with two lignite seams. (From Aydan & Geniş, 2015.)

Table 6.6 Material properties used numerical analyses

Layer	γ *kN/m*^{3}	*c kPa*	ϕ (°)	σ_t *kPa*	*E MPa*	ν
Pg1	16	50	35	50	80	0.3
Pc1	18	560	25	560	110	0.4
Lignite	14	660	30	660	180	0.26
Pc2	18	1,000	25	1,000	400	0.4

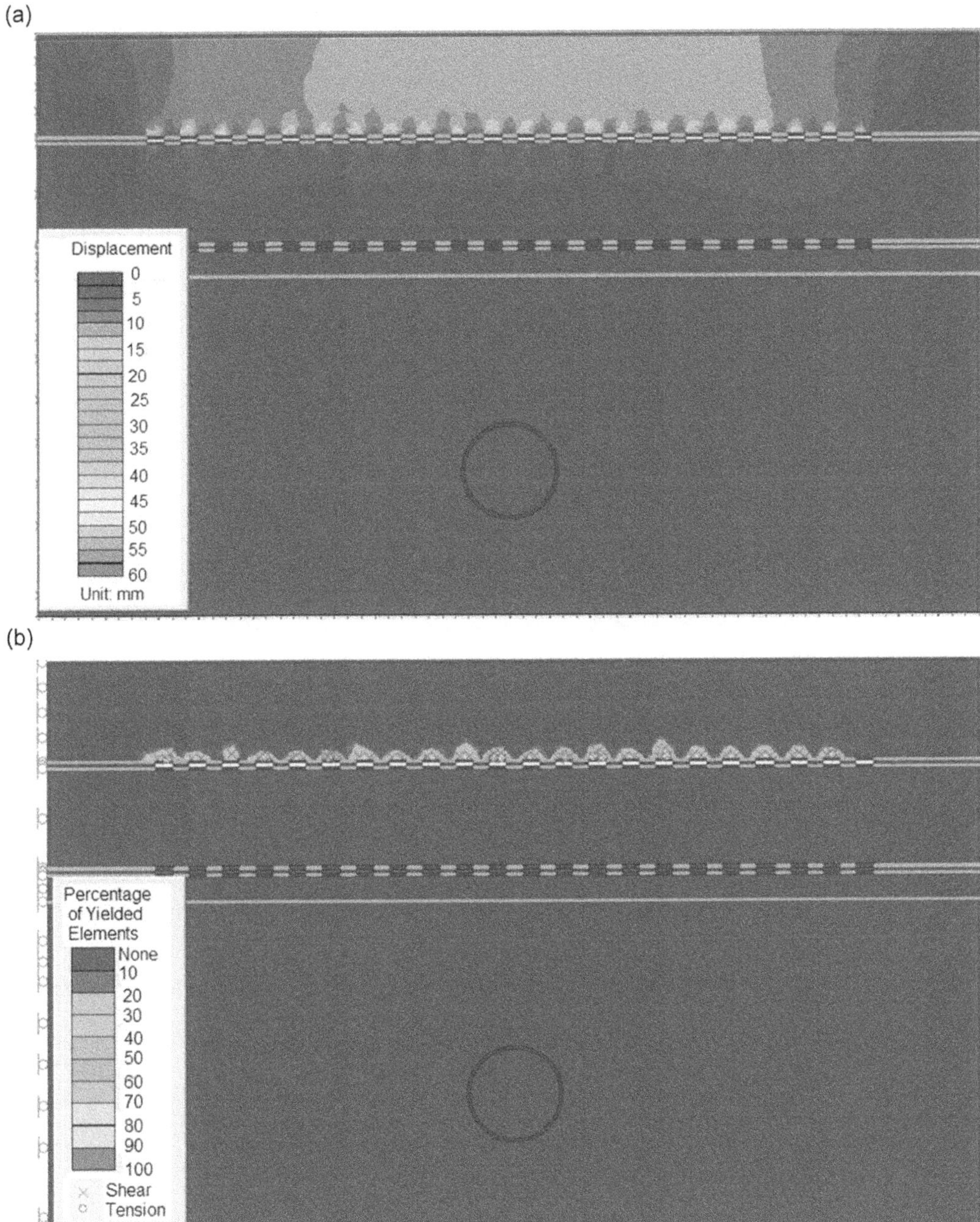

Figure 6.29 (a) Displacement contours and (b) yield zones due to single seam extraction. (From Aydan & Geniş, 2015.)

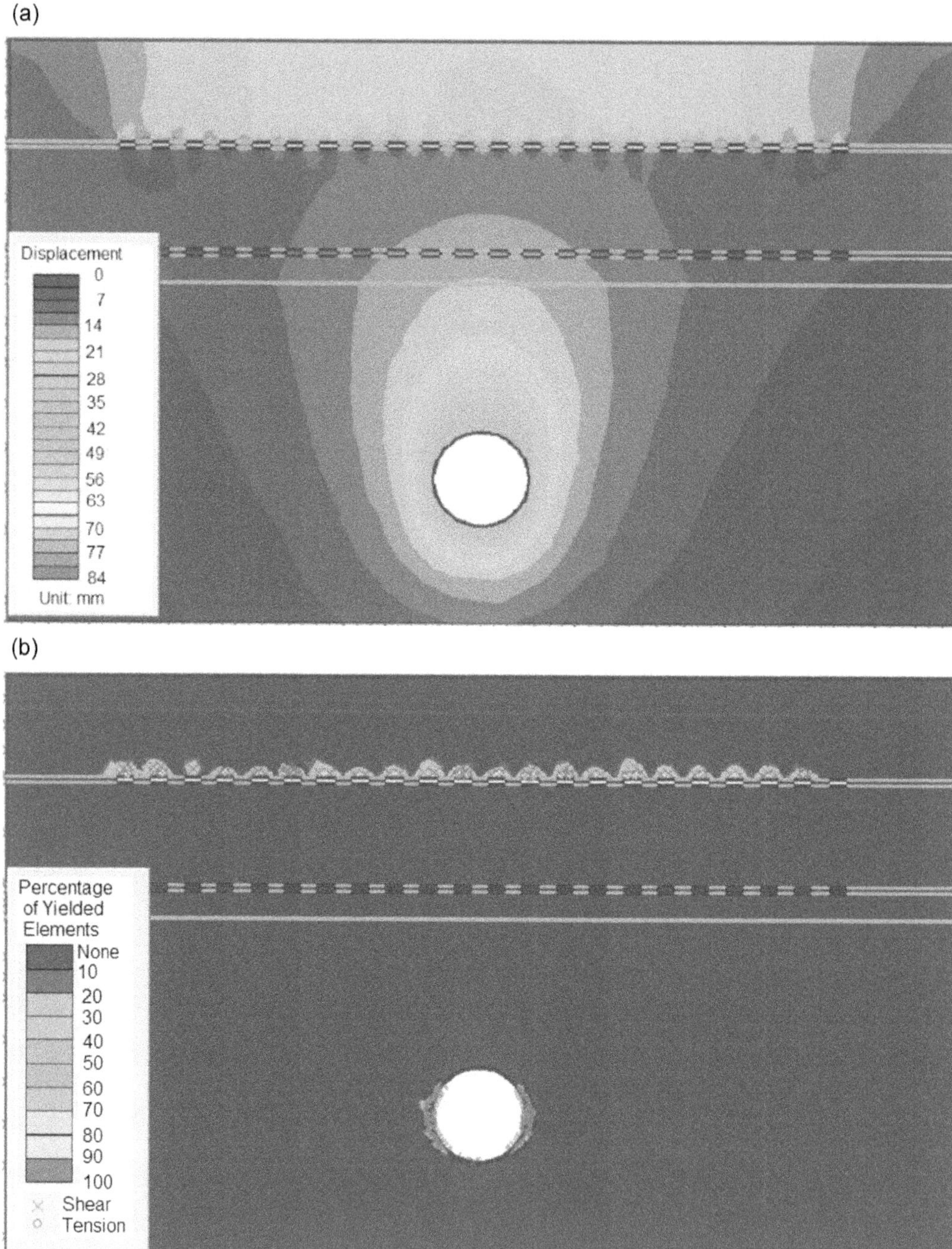

Figure 6.30 (a) Displacement contours and (b) yield zones due to tunnel excavation. (From Aydan & Geniş, 2015.)

previous case. However, yielding occurs in the top layer above the lignite seam due to tensile stresses. As noted from Figure 6.30a, the effect of excavating the circular cavity has very little influence on the displacement above the abandoned mine. It is also interesting to observe that the displacement contours have an ovaloid shape.

Yielding occurred in the layer just above the top lignite seam, as seen in Figure 6.30b. This yielding is attributed to tensile stresses. Following the excavation of the circular cavity, some yielding occurred near the perimeter of the cavity, while there was almost no further propagation of the yield zone in the vicinity of the lignite seam (Figure 6.30b).

6.3.3.2.3 MULTIPLE SEAMS WITH TWO SEAMS EXPLOITED

In addition to the exploitation of the top lignite seam, the bottom lignite seam was also exploited. Figure 6.31 shows the displacement contours and yielding in the analyzed example. The displacement field exhibits a similar response to that observed after the exploitation of the top lignite seam. The exploitation of the bottom seam did not induce any yielding in the pillars and layers in the vicinity of the lower lignite seam, likely due to the higher strength of the layers above and below the lower lignite seam.

Figure 6.32 illustrates the displacement contours and yield zone formation following the excavation of the circular cavity after the exploitation of both the top and lower lignite seams. The results are essentially similar to those from the previous case involving the exploitation of the top lignite seam.

6.3.4 Two-dimensional Discrete Finite Element Analysis for Collapsing Mines

6.3.4.1 Pillar failure

A series of analyses using Discrete Finite Element Method (DFEM) was conducted to evaluate the supporting effect of backfilling on a fractured pillar of an abandoned lignite mine (Aydan et al., 2016b; Aydan, 2018). The overburden was 6 m, the pillar height was 4 m, and the width was 2 m. Table 6.7 provides the material properties used in the analyses. Figure 6.33 shows the finite element meshes used in the analyses. The pseudo-dynamic version of DFEM analyses was employed, and computations were carried out up to 12 steps.

Figure 6.34 illustrates the deformed configuration of non-backfilled and backfilled pillars at computation step 12. Figure 6.35 displays the settlement of the pillar. As noted from the figures, when the pillar is not supported by backfilling, sliding occurs along the fracture plane, and a vertical crack appears in the middle of the pillar. However, if the pillar is supported by backfilling, the sliding and separation of fracture planes are restrained, stabilizing the pillar with backfilling material. Although the backfilling material is relatively soft, its supporting effect is remarkable.

6.3.4.2 Abandoned room and pillar analyses adjacent to Cliffs

An abandoned lignite mine with an overburden of 4 m next to a steep cliff is considered, as shown in Figure 6.36. The pillar height was assumed to be 2 m with a width of 1 m. The computations were conducted up to 12 steps, and the deformed configuration of the abandoned mine for two cases is shown in Figure 6.37. The material properties used were the same as

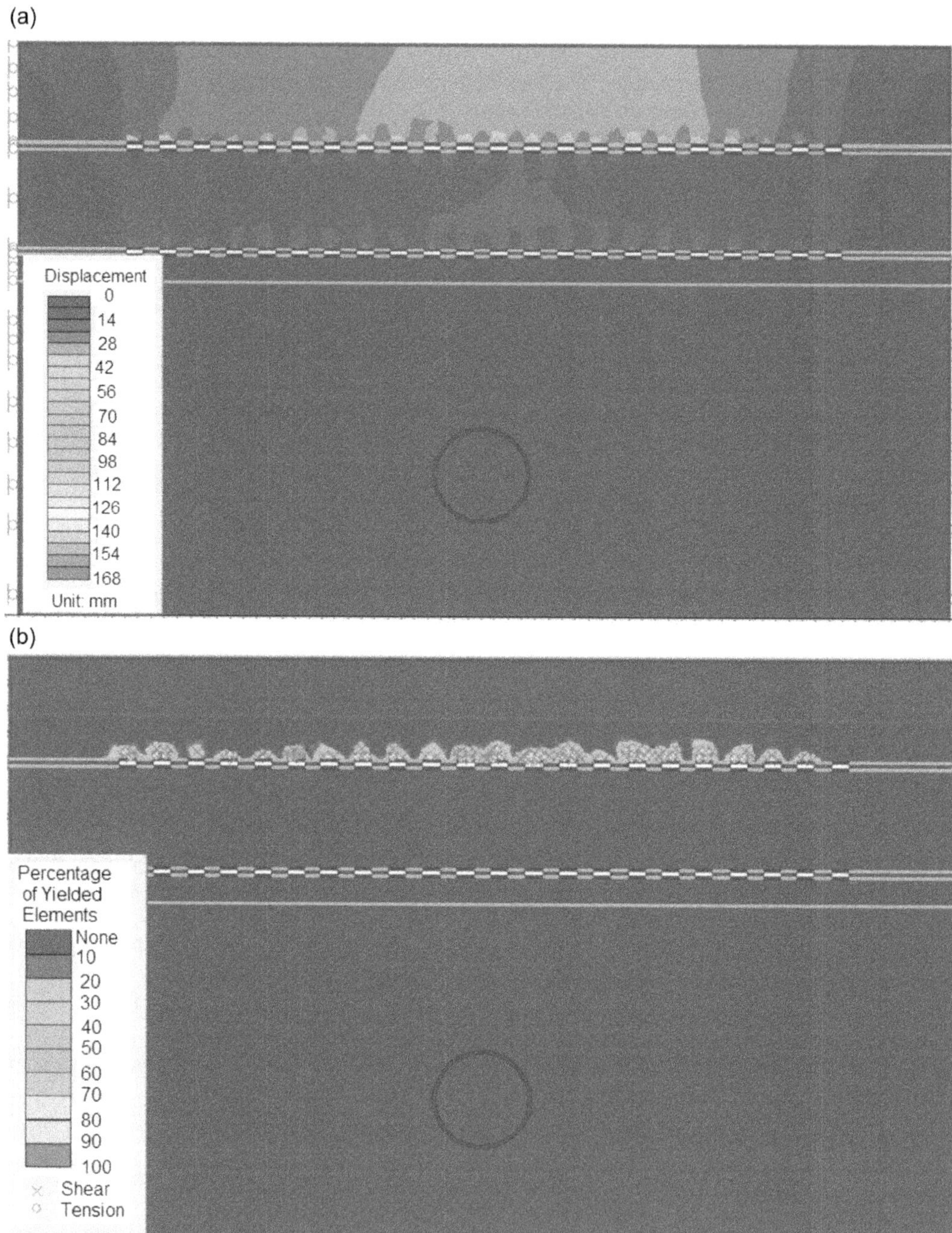

Figure 6.31 (a) Displacement contours and (b) yield zones due to multi-seam extraction. (From Aydan & Geniş, 2015.)

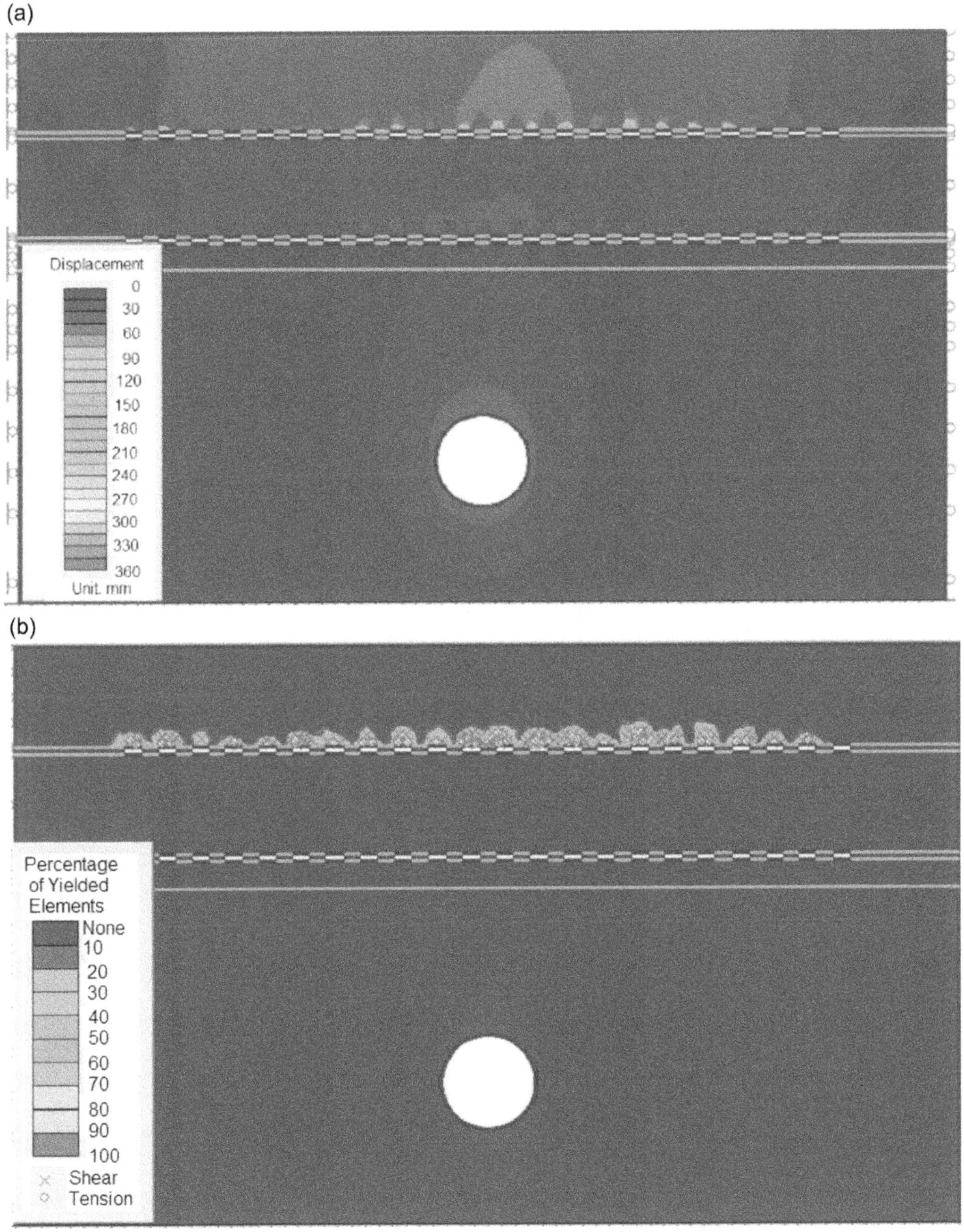

Figure 6.32 (a) Displacement contours and (b) yield zones due to tunnel excavation. (From Aydan & Geniş, 2015.)

Table 6.7 Material properties used in DFEM analyses

Material	*$\lambda(E_n)$ (MPa)*	*$\mu(E_s)$ (MPa)*	*γ (kN/m³)*	*c (kPa)*	*ϕ (°)*	*σ_t (kPa)*
Top layer	200	200	22			
Lignite	200	200	14			
Lower layer	300	300	22			
Backfilling	100	100	14			
Fracture plane	200	10	-	10	10	5

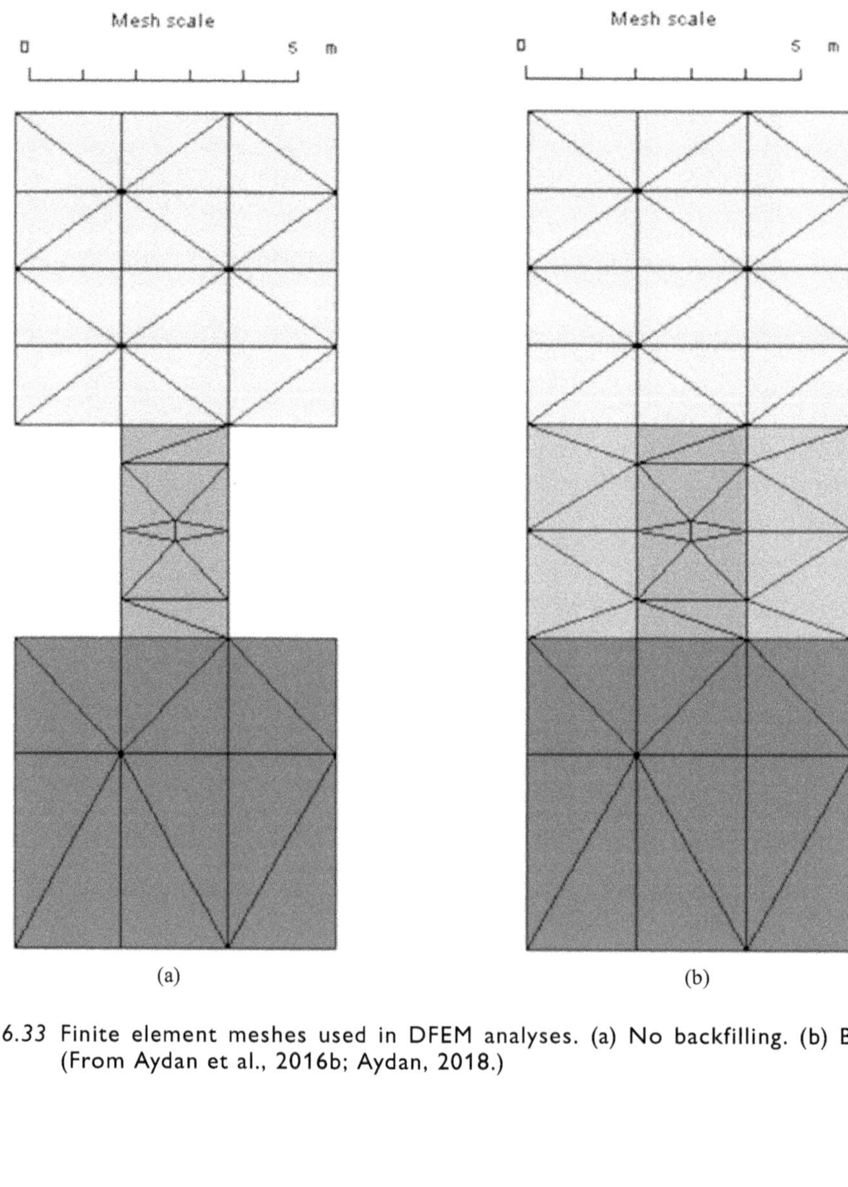

Figure 6.33 Finite element meshes used in DFEM analyses. (a) No backfilling. (b) Backfilled. (From Aydan et al., 2016b; Aydan, 2018.)

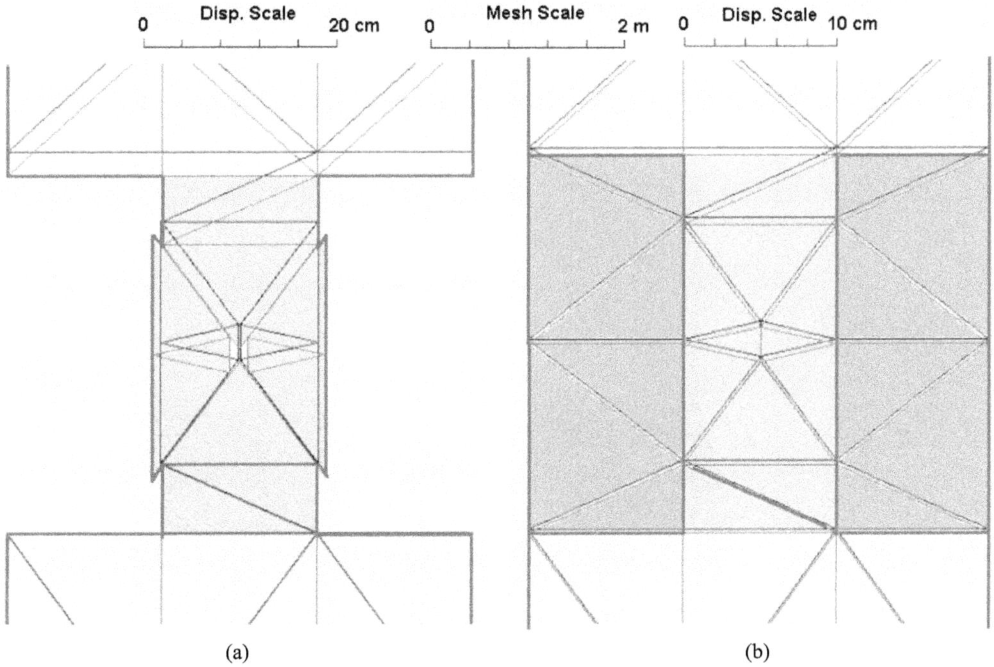

Figure 6.34 Deformed configurations of the pillar for (a) non-backfilling and (b) backfilling cases. (From Aydan, 2018.)

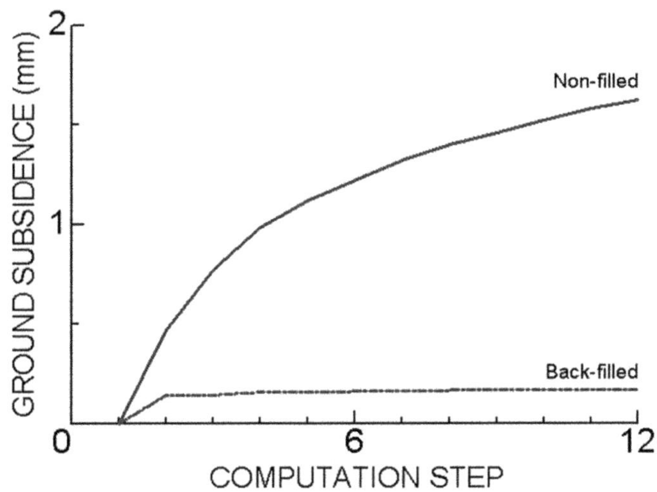

Figure 6.35 Settlement of the pillar for non-backfilling and backfilling cases. (From Aydan et al., 2016b; Aydan, 2018.)

(a)

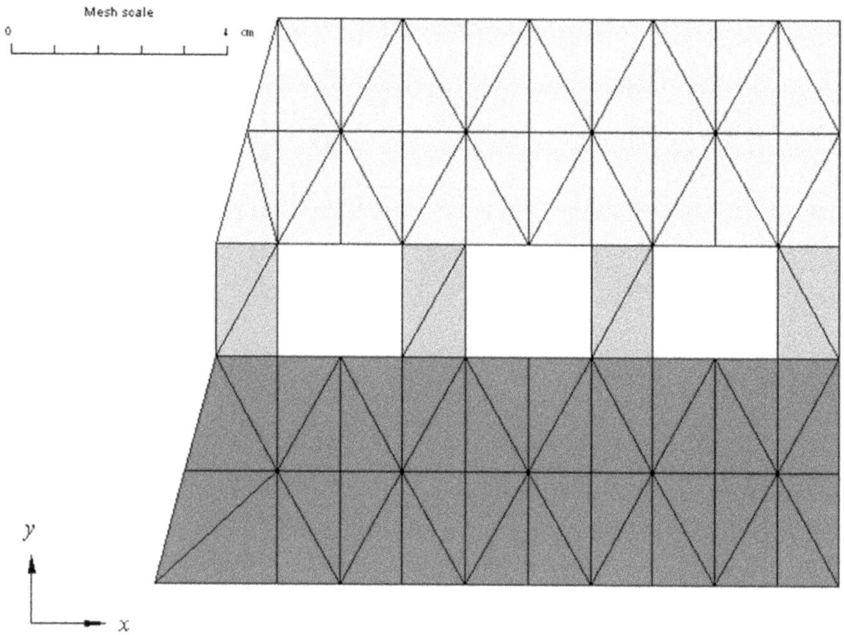

(b)

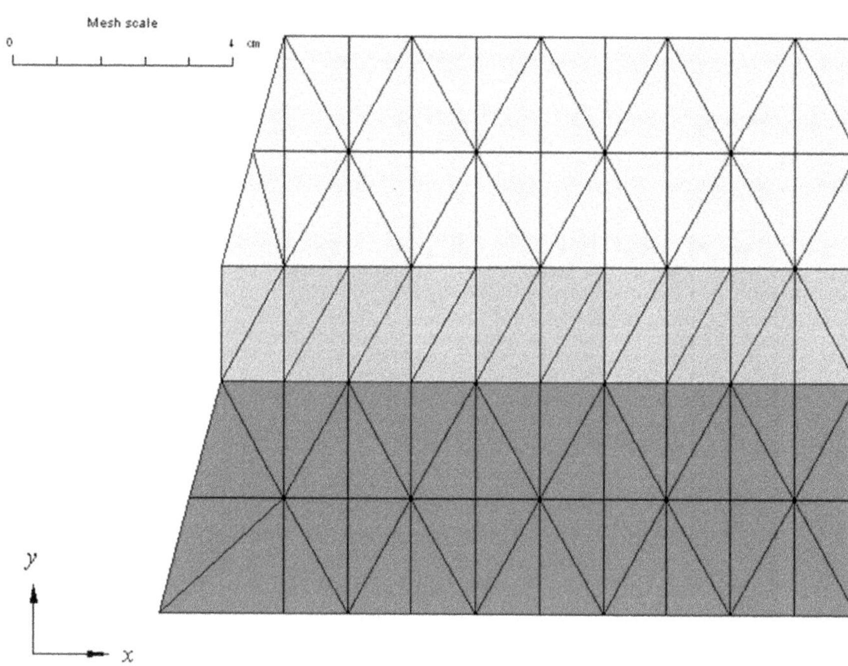

Figure 6.36 Finite element meshes used in DFEM analyses. (a) Non-backfilling. (b) Backfilled. (From Aydan et al., 2016b; Aydan, 2018.)

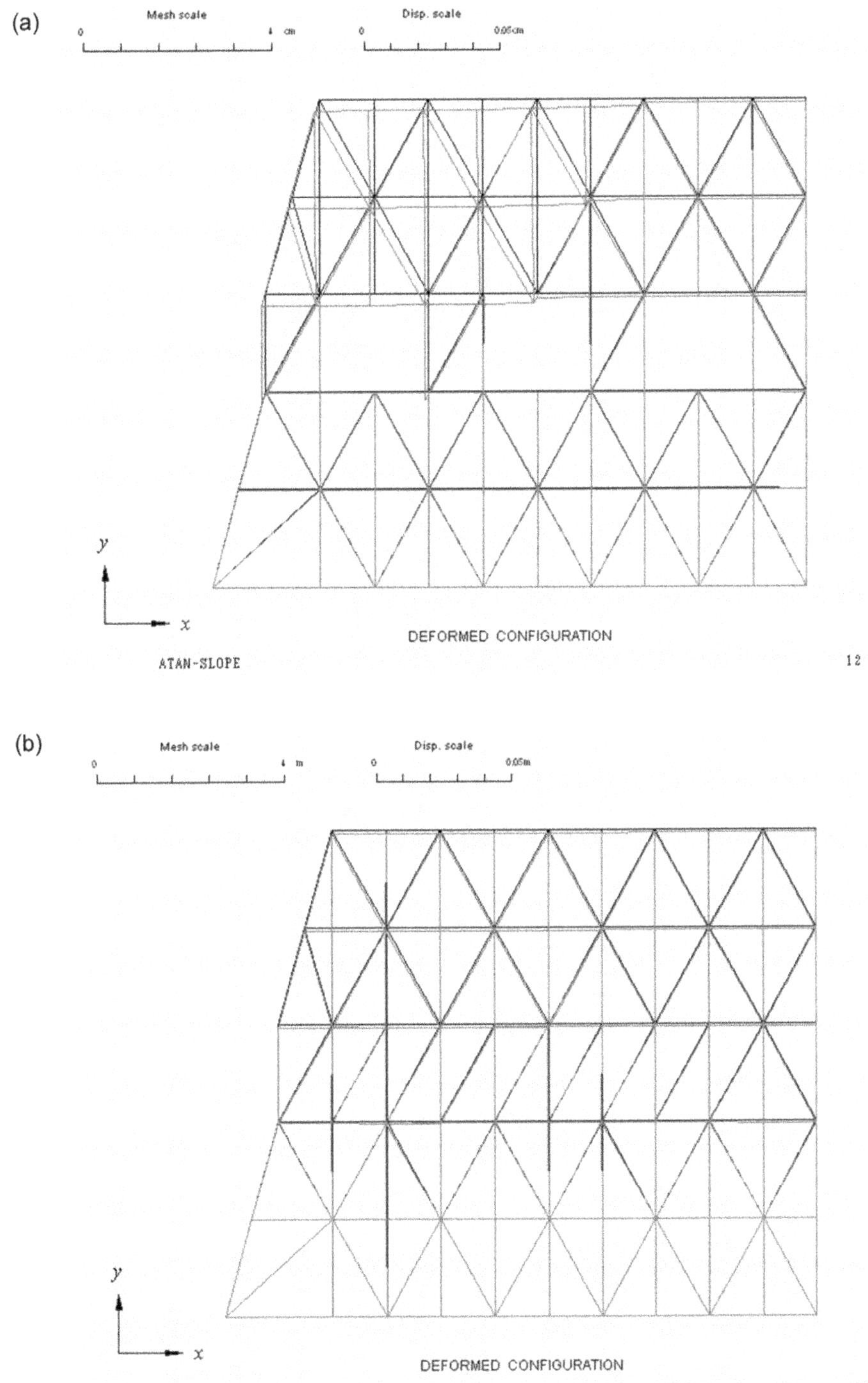

Figure 6.37 Deformed configuration of the pillar for (a) non-backfilling and (b) backfilling cases at computation step 12. (From Aydan et al., 2016b; Aydan, 2018.)

those in the previous analyses. The pillars near the cliff were assumed to have fracture planes extending to the ground surface from the second pillar. When comparing the results of the analyses, the settlement of the abandoned mine was restrained, indicating stability, while the other case demonstrates instability and continued settlement. These analyses again clearly illustrate the effect of backfilling in abandoned mines, even though its material properties are relatively soft and weak.

6.3.5 *Three-dimensional Elasto-plastic Dynamic Analyses*

6.3.5.1 *Abandoned room and pillar mine beneath Kyowa Elementary School*

It was known that there was an abandoned lignite mine beneath Kyowa Secondary School in Mitake Town, and authorities were concerned about possible damage to the school buildings and sinkhole formation due to the mine (e.g., Sugiura et al., 2019). The material properties used in the analyses are given in Table 6.8. Figure 6.38 presents a three-dimensional perspective view of the mine and a cross section. At this location, the abandoned lignite mine had a flat ground surface and an inclined lignite seam, and its dynamic response was analyzed using the strong motion anticipated from the M9 Nankai–Tonankai–Tokai Earthquake (Figure 6.39). The method is based on the finite difference technique utilizing silent boundary conditions and Rayleigh-type damping of 5%. Details of the numerical method can be found in a publication by Geniş and Aydan (2008). Figure 6.40 shows the computed ground acceleration at the surface, normalized by that of the base (Figure 6.41) at the middle section denoted in Figure 6.38. The amplification of the ground motion is about 2.2 times that of the base.

The failure mechanism of the pillars was investigated at several selected time steps. Figure 6.42 shows the failure situation at the lignite seam level at each selected time step. Time steps of 30 and 42 seconds correspond to the rupture duration and the termination of the Nankai Earthquake. Time steps of 75 and 125 seconds correspond to the termination of the Tonankai

Table 6.8 Material properties used numerical analyses

Layer	γ *kN/m*^{3}	*c kPa*	ϕ *(°)*	σ_t *kPa*	*E MPa*	ν
Soil	19	50	38	10	270	0.35
Mst-Sst-1	19	700	25	500	750	0.3
Lignite	14	656	45	500	400	0.3
Mst-Sst-2	19	1,000	45	700	1,073	0.3
Chert	19	3,000	45	2,000	3,647	0.3

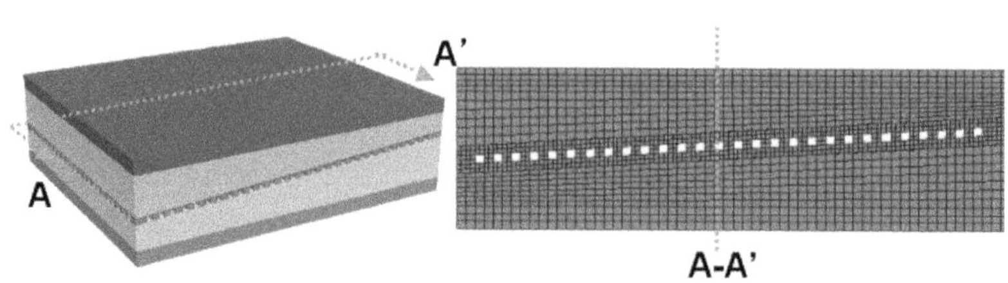

Figure 6.38 A three-dimensional view of analyzed domain and its cross section along the layer dip direction. (From Aydan & Geniş, 2015.)

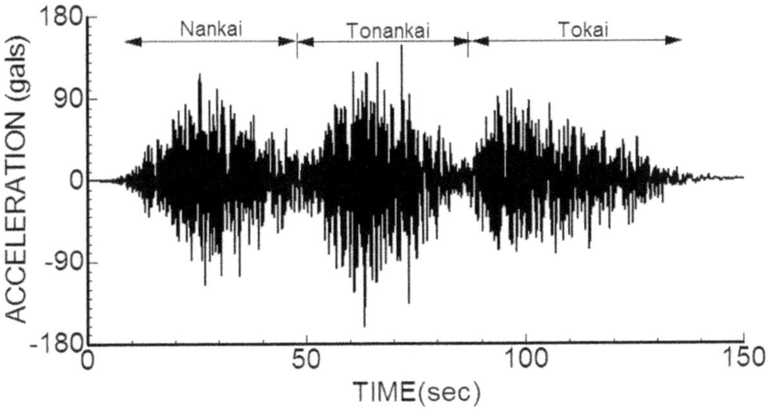

Figure 6.39 Estimated ground motions in Mitake town due to the Nankai–Tonankai–Tokai Earthquake. (From Aydan & Geniş, 2015.)

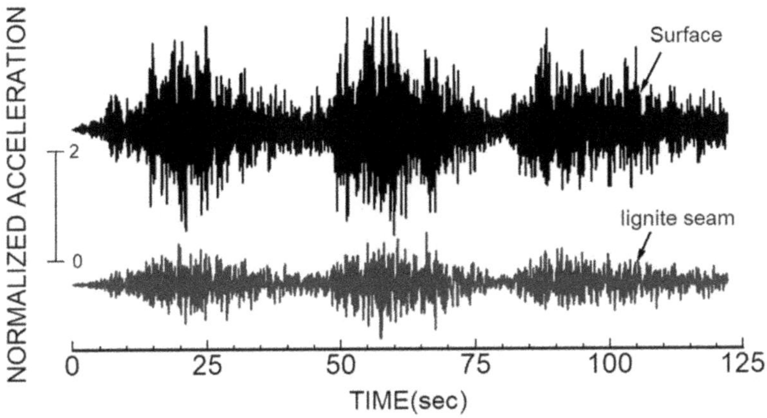

Figure 6.40 The acceleration response at the ground surface and lignite seam.

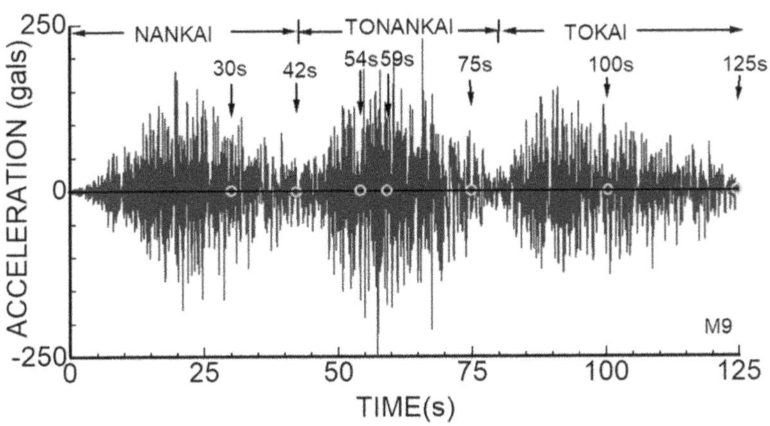

Figure 6.41 Selected time steps on the base acceleration record at the base of the numerical model. (From Sugiura et al., 2019; Geniş & Aydan, 2022.)

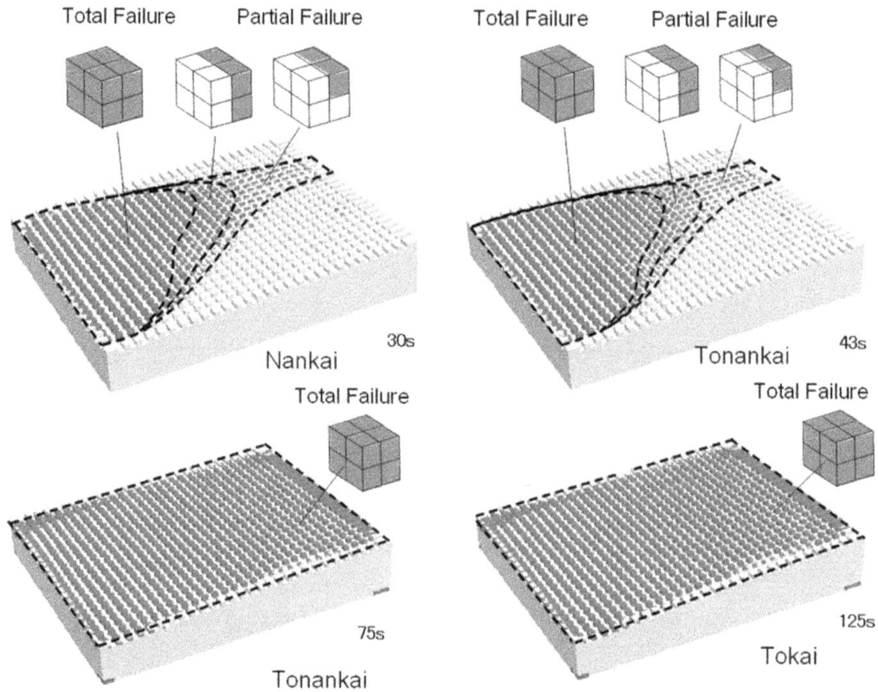

Figure 6.42 The failure situation and progression in the Kyowa abandoned mine during the successive Nankai–Tonankai and Tokai Earthquake-induced ground motions. (From Aydan, 2020b; Geniş & Aydan, 2022; Sugiura et al., 2019.)

Earthquake and Tokai Earthquake, respectively. A significant portion of the pillars fails during the rupture duration of the Nankai Earthquake. The failure starts from the deepest level and progresses toward the shallow parts. The pillars completely fail when the successive Tonankai Earthquake terminates. In other words, the total failure of the abandoned mine occurs during the Tonankai Earthquake, which was the closest earthquake to Mitake town. Since total failure had already occurred during the Tonankai Earthquake, no further yielding occurs during the Tokai Earthquake.

Figure 6.43 compares the computed acceleration responses at the ground surface due to M8.5 (old scenario) and M9.0 (new scenario) earthquakes. According to the new scenario, the maximum ground acceleration becomes 637 gals for the M9.0 earthquake, while it is 405 gals for the M8.5 earthquake. Figure 6.44 compares the maximum accelerations at the ground surface for the new and old scenarios. The largest acceleration occurs almost at the middle of the abandoned mine.

6.3.5.2 *Abandoned room and pillar analyses adjacent to Cliffs*

Geniş and Aydan (2008) and Aydan et al. (2012b) carried out a series of dynamic response analyses of the abandoned lignite mine at Mitake town in Japan. The in-situ stress state was inferred from the fault striations using Aydan's method (Aydan 2000a) (Figure 6.45). The in-situ stress condition is quite similar to those reported by Lanaro and Matsui (2008)

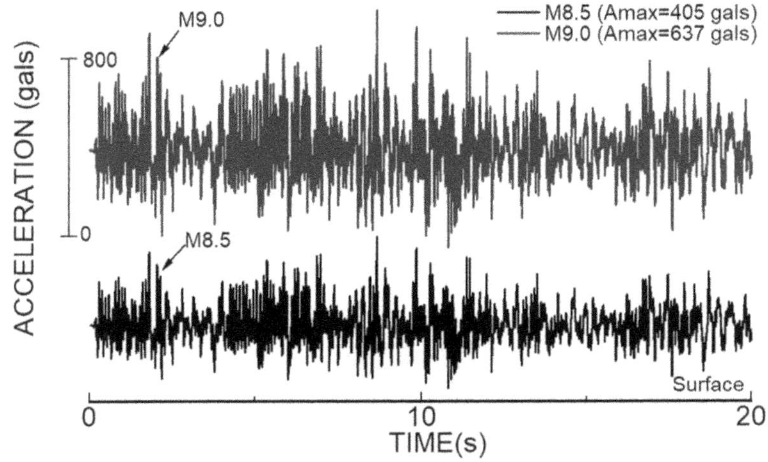

Figure 6.43 Comparison of the computed acceleration responses at the ground surface due to M8.5 and M9.0 earthquakes.

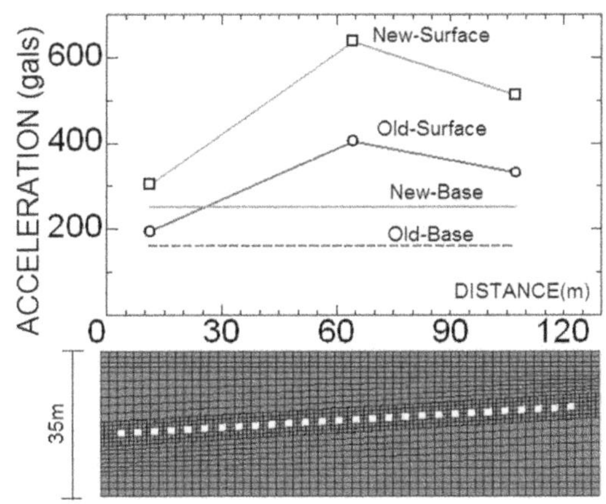

Figure 6.44 Comparison of the maximum accelerations at the ground surface for new and old scenarios.

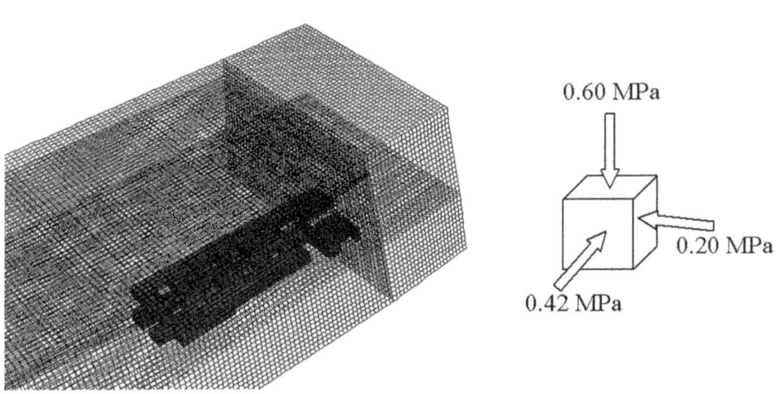

Figure 6.45 Finite difference mesh and assumed in-situ stress state. (From Geniş & Aydan, 2008; Geniş & Aydan, 2008.)

Table 6.9 Rock material properties used in numerical analyses

Unit	*Uniaxial Compressive Strength (MPa)*	*Friction Angle (°)*	*Elastic Modulus (MPa)*	*Unit Weight (kN/m³)*	*Tensile Strength (MPa)*	*Cohesion c_m (MPa)*	*Shear Modulus (MPa)*
Lignite	4	53	557	20	0.4	0.669	228.8
Sandstone	2.3	53	257	20	0.06	0.385	102.8
Mudstone	0.96	29	76	20	0.05	0.283	30.4

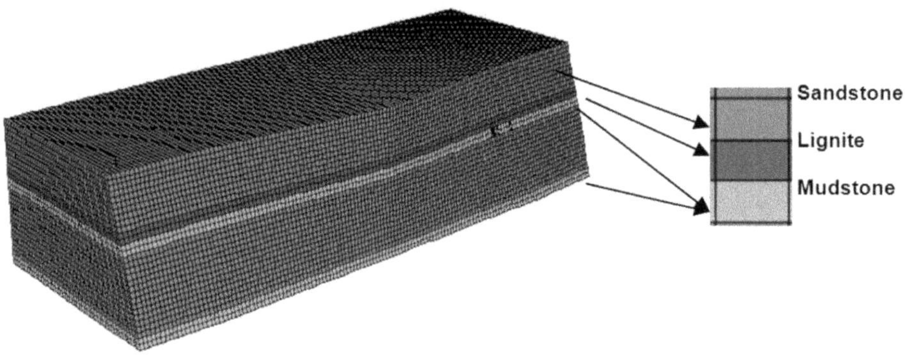

Figure 6.46 A three-dimensional view of the numerical model of the mine and geological formations. (From Geniş & Aydan, 2008.)

and Matsui and Hirano (2010) at the Tono uranium mine, which is about 12 km away from Mitake town. The abandoned mine investigated was under stable conditions, although there were some local-type failures (e.g., Aydan et al., 2005a,c, 2006). The material properties used in the analyses are given in Table 6.9. The input ground motion is assumed to be sinusoidal with a chosen period. Figure 6.46 shows the three-dimensional geological structure on the adopted numerical mesh. For these particular analyses, the FLAC3D software developed by Itasca (2005) was used. The method is based on the finite difference technique utilizing silent boundary conditions. Rayleigh-type damping was set at 5% in all analyses; however, it should be noted that Rayleigh-type damping experimentally ranges between 2% and 10% (e.g., Aydan, 2017).

Figure 6.47 shows the selected points of observation and their acceleration responses. As noted from the figure, the ground motion is amplified by three times at the seam level near the cliff and five times at the cliff crest. This example also shows the importance of ground amplification due to topography. It should be noted that the amplification would be smaller if there were no abandoned lignite mine (e.g., Geniş et al., 2009; Geniş & Aydan, 2013).

Next, sinusoidal acceleration was applied to the base of the numerical model by selecting the maximum amplitude of the acceleration as 20 and 80 gals with a frequency of 2 Hz. The acceleration is applied for a period of 1.7 seconds. Figure 6.48 shows the yield zone development for the assumed ground motions. While the yielding occurred due to tensile stresses for

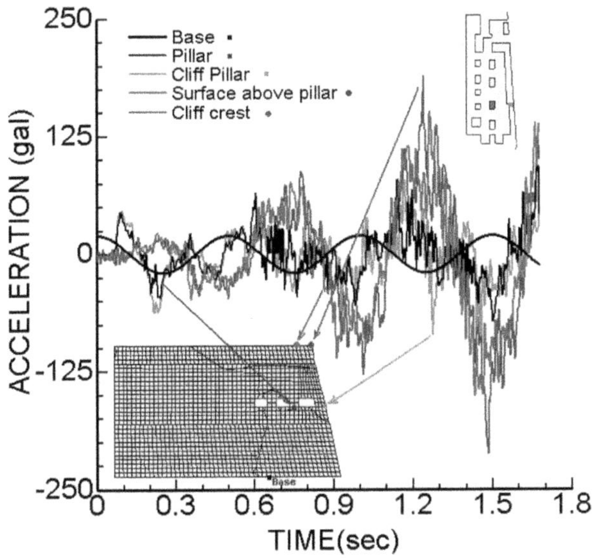

Figure 6.47 Acceleration responses of selected points in the abandoned lignite mine. (From Geniş & Aydan, 2008, 2013.)

the ground motion with an amplitude of 20 gals, yielding took place due to tensile and shearing for the ground motion with an amplitude of 80 gals. Shear yielding occurred in pillars, as shown in Figure 6.48b.

Figure 6.49 shows the acceleration responses at selected points. As noted from Figure 6.10, the maximum ground acceleration occurs at the crest of the slope. The ground acceleration at the pillar adjacent to the slope is three times higher than that at the base. This trend is quite similar for ground motions with amplitudes of 20 and 80 gals.

A final computation was carried out for the case in which the properties of the rock mass were reduced to one-eighth of those of the intact rock, while the ground acceleration amplitude was assumed to be 80 gals. This situation corresponds to a state of degradation of rock mass properties over time.Figure 6.50b shows the development of the yield zone at the model surface and at the EW and NS sections. This situation likely corresponds to a state of total collapse of the abandoned mine. As seen in Figure 6.50, the yielding at the ground surface extends over a large area, and the yielding in the EW (A–A') cross section resembles a slope failure involving the abandoned mine.

The abandoned mine was subjected to ground shaking due to the 2004 Tokaido-oki Earthquake (Mj 7.2) and the 2005 Komaki Earthquake (Mj 4.7). The accelerations recorded at the Minokamo strong motion station of K-NET were used as base accelerations, as shown in Figure 6.52. Figure 6.53 shows the acceleration responses at the same selected points, normalized by the maximum value of the base ground acceleration for the 2004 Tokaido-oki Earthquake record at the Minokamo station. Far-distant earthquakes induce larger ground motions in the vicinity of the cliff crest above the abandoned mines compared to those from near-field earthquakes.

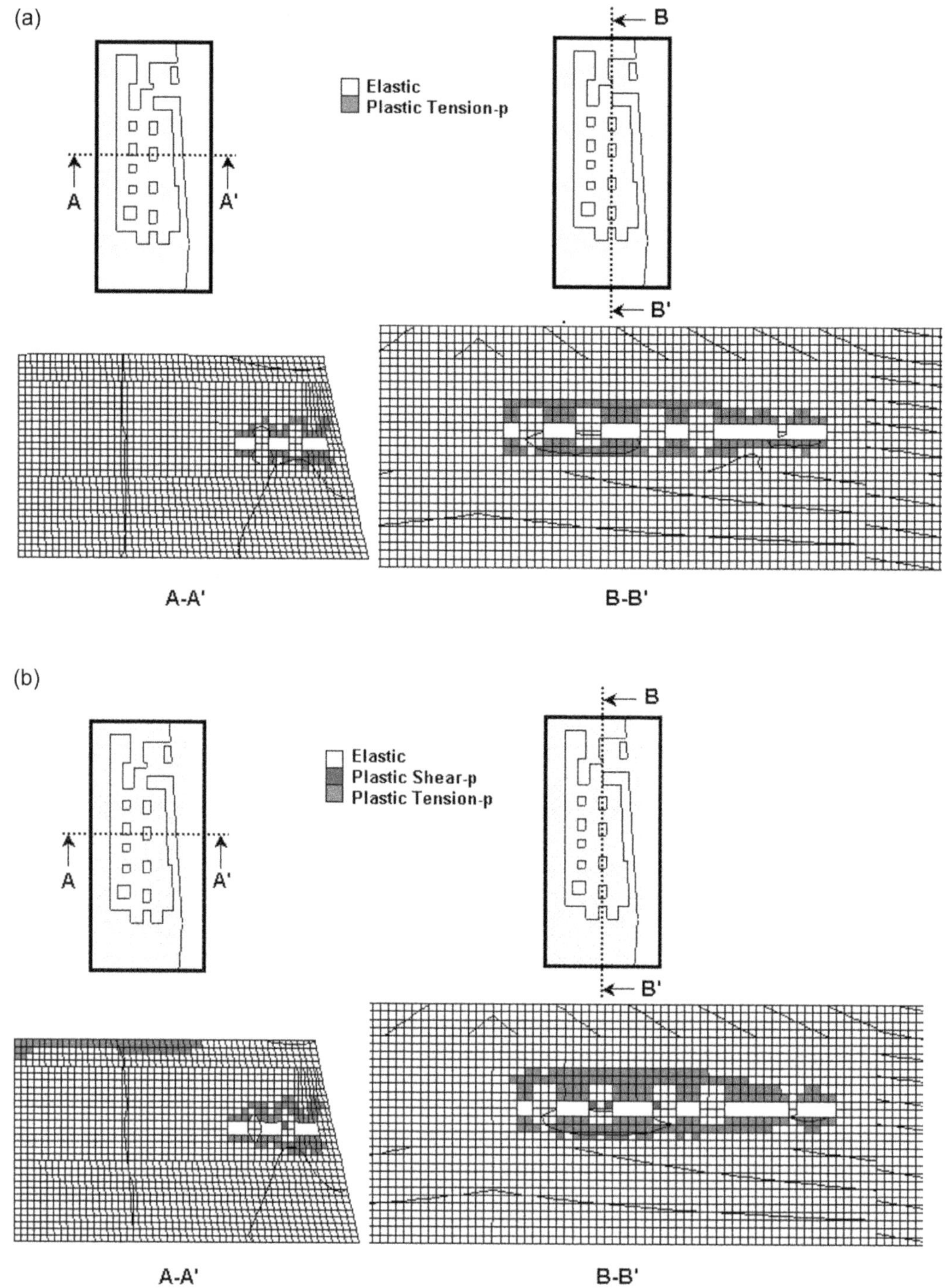

Figure 6.48 Yield zones around openings and pillars. (a) A_{max}=20 gals. (b) A_{max}=80 gals. (From Geniş & Aydan, 2008.)

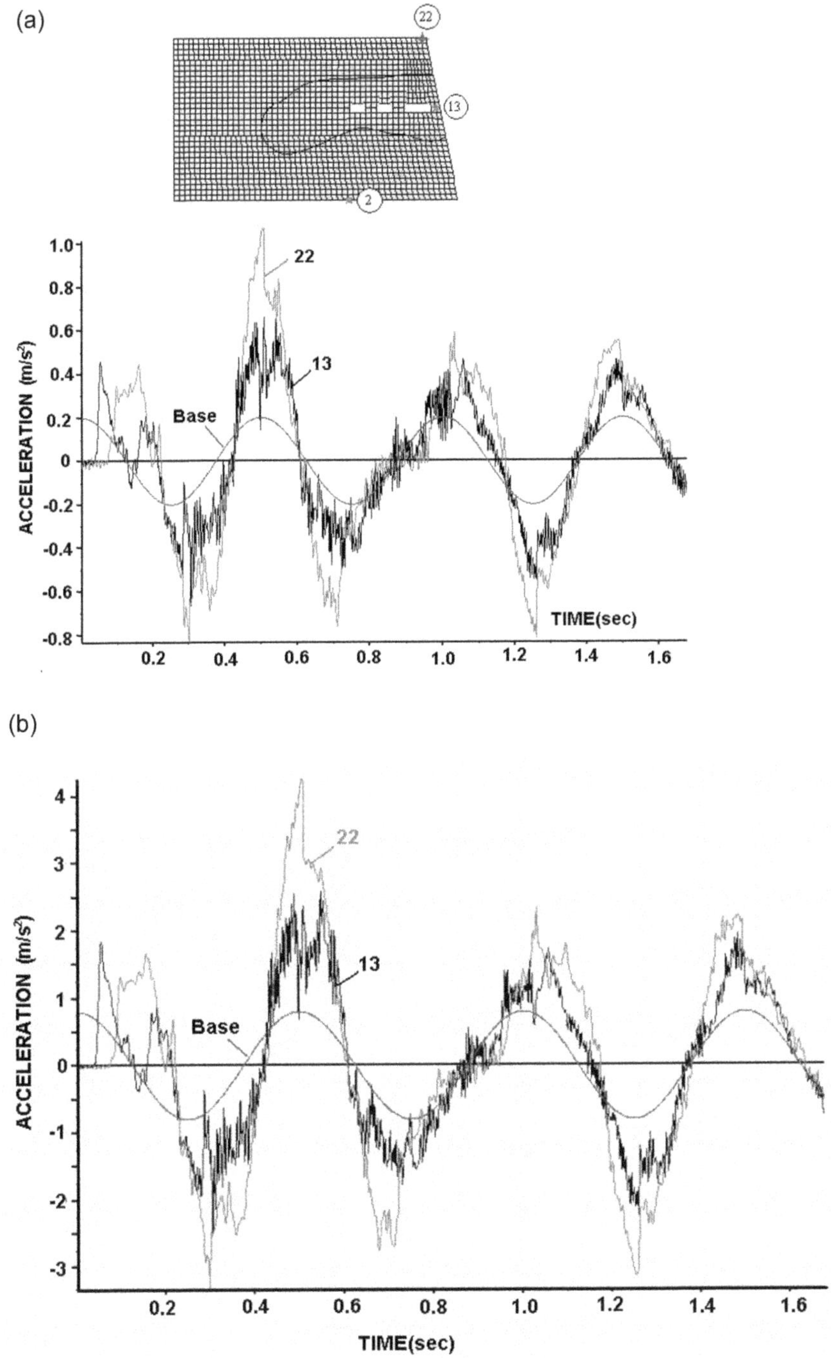

Figure 6.49 Acceleration responses of several selected points. (a) Locations of selected points. (b) A_{max} = 20 gals. (c) A_{max} = 80 gals. (From Geniş & Aydan, 2008, 2013.)

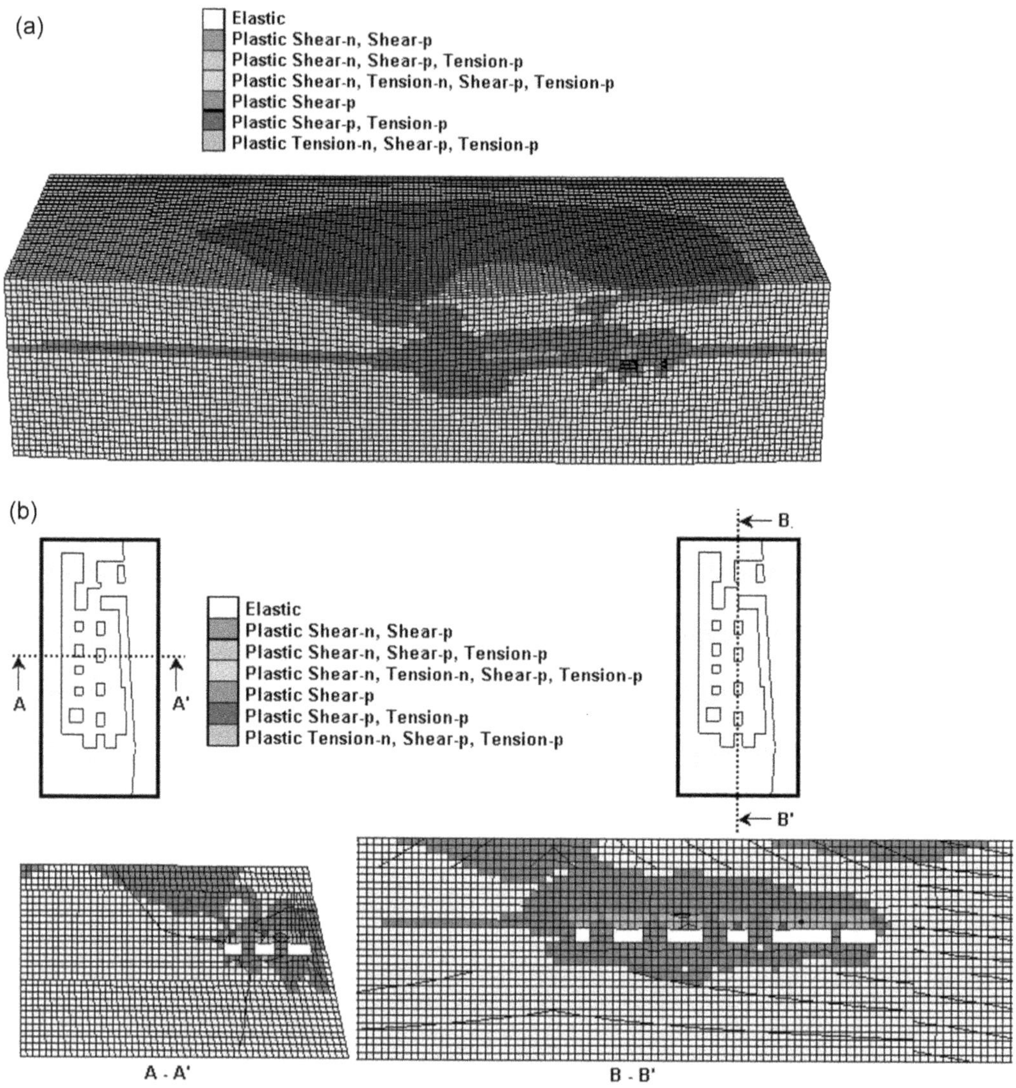

Figure 6.50 Yield zone around the mine for ground with reduced strength properties. (a) Yield zone development at model surface. (b) Yield zone development at EW and NS cross sections. (From Geniş & Aydan, 2008.)

6.3.5.3 Tunneling beneath abandoned room and pillar mine

Aydan and Genis (2014) performed three-dimensional numerical studies on the response and stability assessment of abandoned mines above a new underground excavation. Figure 6.53 shows a 3D view of the layers, abandoned mine, and circular tunnel.

Table 6.10 provides the material properties used in the numerical analyses. Under static conditions, no yielding occurred. The dynamic response of abandoned mines was investigated using

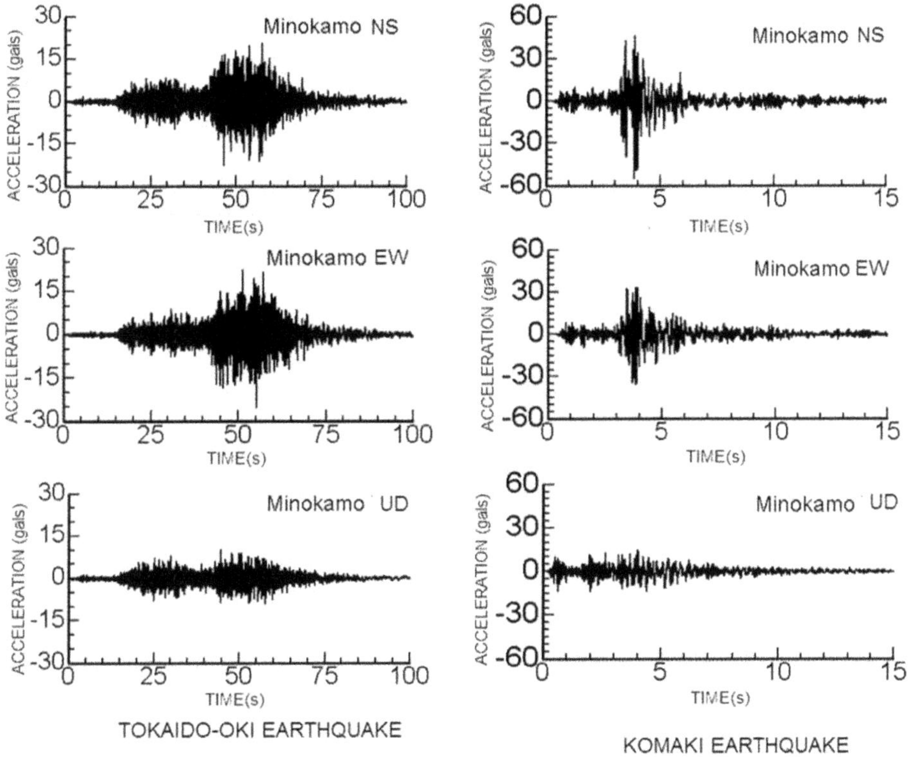

Figure 6.51 Acceleration records at the Minokamo strong motion station of K-NET. (From Aydan and Geniş; Geniş & Aydan, 2008.)

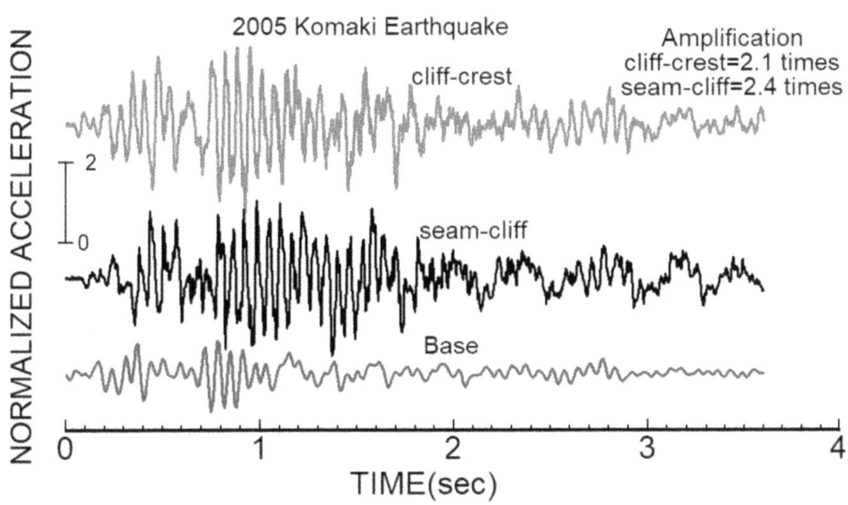

Figure 6.52 Acceleration responses of selected points during the 2004 Tokaido-oki Earthquake. (From Aydan & Geniş, 2008.)

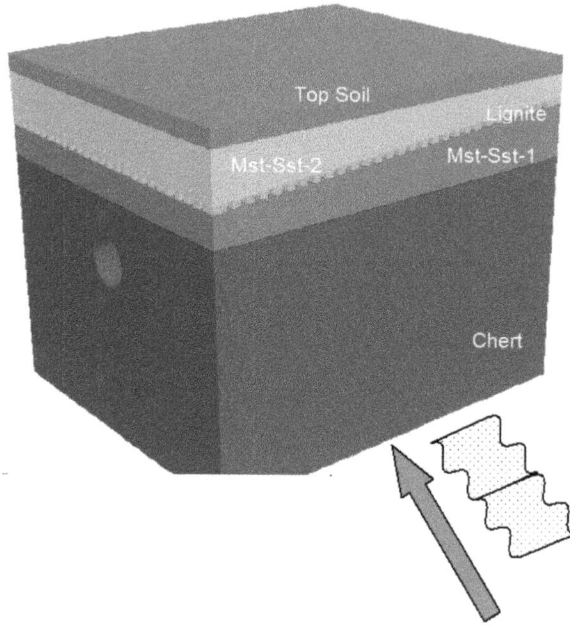

Figure 6.53 Layers of the model and direction of dynamic loading. (From Aydan & Geniş, 2015.)

Table 6.10 Parameters used numerical analyses

Layer	*Vs (m/s)*	*ρ (kN/m³)*	*E (MPa)*	*Poisson Ratio*	*Damping Ratio (%)*
Soil	300	19	270	0.35	5
Mst-Sst	500	19	750	0.30	5
Lignite	400	14	400	0.30	5
Mst-Sst	600	19	1,073	0.30	5
Chert	1,000	19	3,647	0.30	5

the ground motion due to the anticipated M9.0 class Nankai, Tonankai, and Tokai earthquakes at the site under consideration, as shown in Figure 6.54. Figure 6.55 shows the formation of the yield zone in the computational model. As noted from the figure, some yielding occurred in the pillars at the deeper section of the model. These results indicate that, under dynamic conditions, further yielding may occur in the abandoned mines.

The current seismic risk assessment in Japan and other countries does not consider the effects of abandoned mines in urbanized areas. The conclusions from this study can be summarized as follows:

1 The outcomes of the numerical studies on the dynamic characteristics of ground above abandoned mines showed that ground amplifications can be quite large in areas where extraction is carried out using the room and pillar or long-wall mining techniques (e.g., Geniş & Aydan, 2013).

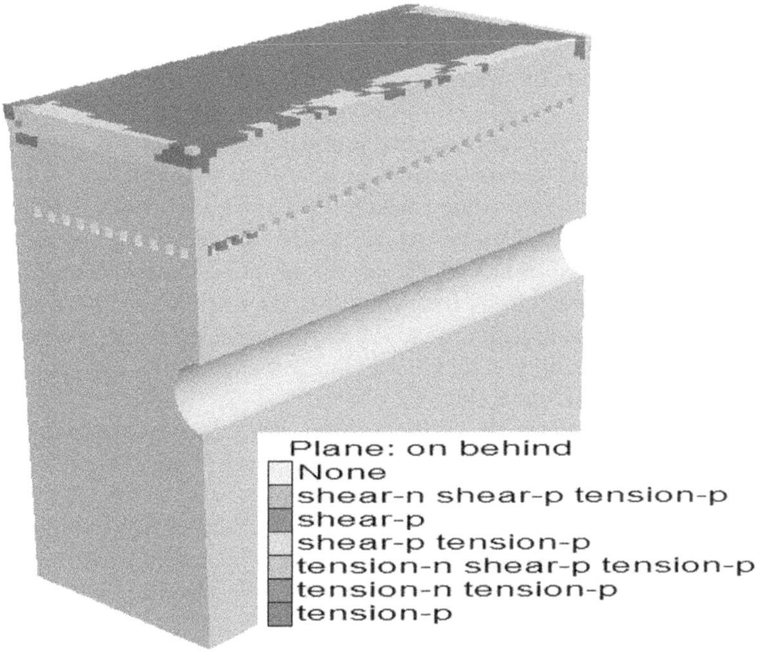

Figure 6.54 Computed yield zone in the computational model at the time step of 120 s of the record shown in Figure 6.14. (From Aydan & Geniş, 2015.)

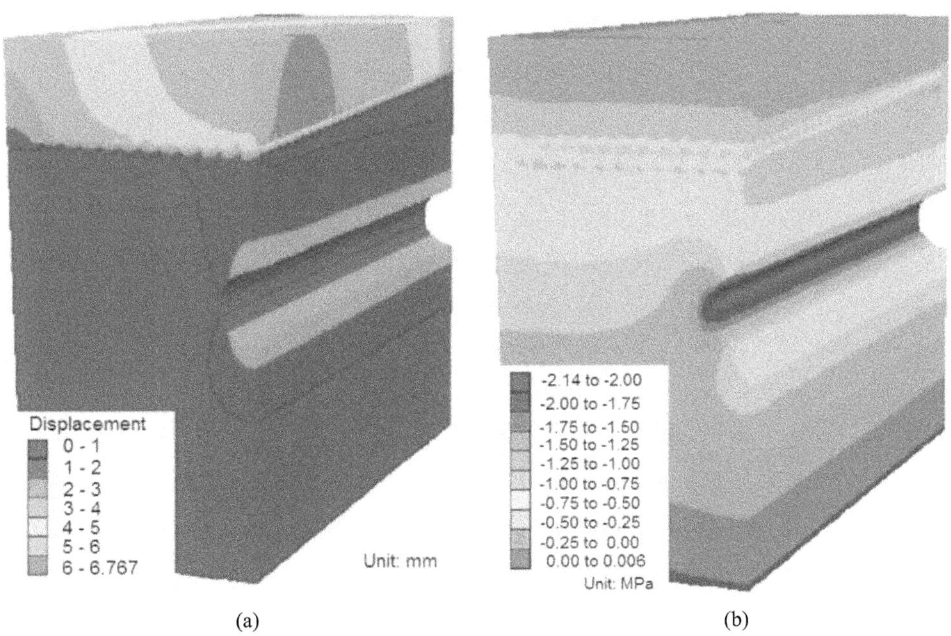

Figure 6.55 Displacement and maximum principal stress contours for static excavation analysis. (a) Displacement. (b) Maximum principal stress. (From Aydan & Geniş, 2015.)

2 Ground amplifications may be 3–5 times greater than those at the base, particularly near steep cliffs. Furthermore, the frequency content of ground accelerations can have pronounced effects on ground amplifications when comparing responses induced by nearby and distant earthquakes.
3 Far-distant earthquakes induce larger ground motions in the vicinity of the cliff crest above abandoned mines compared to those induced by near-field earthquakes.
4 Regarding the abandoned lignite mine beneath the Kyowa Secondary School, a significant portion of the pillars failed during the rupture duration of the Nankai Earthquake, with failure starting from the deepest level and progressing upwards. The pillars totally failed during the successive Tonankai earthquake. In other words, the total failure of the abandoned mine occurred during the Tonankai earthquake, which is the closest earthquake to Mitake town. Since total failure had already occurred during the Tonankai Earthquake, no further yielding occurred during the Tokai Earthquake.
5 Further studies on the effects of multiple seam extraction or new large underground excavations beneath abandoned mines on ground motions and their amplifications should be carried out.
6 The construction of a new tunnel may have profound effects on the stability of abandoned mines under both static and dynamic conditions. In particular, the distance between the abandoned mine level and the crown of the tunnel is a major factor affecting the development of the yielding zone.
7 From numerical analyses, it was found that if the distance between the crown of the tunnel and the abandoned mine level is less than 1D, the ground above the tunnel may yield. Additionally, the distance should generally be greater than 2D. Under dynamic conditions, further yielding may occur in the abandoned mine.

Chapter 7

In-situ Stresses and Their Evaluations

The stress state of the Earth is crucial in geomechanics and geophysics. The virgin stress state in the Earth's crust is especially relevant in mining and civil engineering, as it greatly affects excavation stability. Geophysicists are also concerned with the stress state of the crust, as it relates to understanding earthquake mechanisms and predicting earthquakes. This chapter is fundamentally based on the chapter on in-situ stress chapter with the title of "1In-situ Stress Estimation, Measurement and Inference Methods" in the book by Aydan (2020) with some attention to those of abandoned room and pillar mines in order to make this book complete regarding the whole aspects of the abandoned room and pillar mines.

There are many in-situ stress inference techniques can be broadly classified into direct and indirect methods (Amadei and Stephanson, 1997). Direct techniques are generally costly and are typically employed for significant structures. These techniques utilize boreholes and assume that the surrounding rock behaves elastically. However, the acoustic emission (AE) method can serve as a direct stress measurement technique, as it is less expensive and can be performed under well-controlled laboratory conditions once sampling is completed.

Indirect stress inference techniques, which utilize borehole breakouts, fault striations, and earthquake focal mechanism solutions, have also been proposed and employed (Angellier, 1984; Aydan, 2000a; Aydan & Kim, 2002; Aydan et al. 2002, 2013b; Zoback & Healy, 1992). Recently, a new stress inference technique utilizing the damage zone around blast holes was proposed by Aydan (2012) and has been applied at several sites in Japan and Turkey.

7.1 In-situ Stress Estimation Methods

7.1.1 *Empirical Approaches*

Empirical approaches to estimate crustal stresses are based on empirical formulas derived from stress measurements in various engineering projects and earthquake prediction studies. The initial measurements of in-situ stresses were conducted in South Africa (Hast, 1969) and later expanded to other countries. These measurements indicated that horizontal stresses could be several times greater than vertical stresses in shields such as the Canadian or Scandinavian shields (Herget, 1986, Stephanson et al., 1986). Furthermore, many measurements associated with excavations in mining, civil engineering, and earthquake prediction projects have been undertaken over the last 50 years to assess stress measurements in the Earth's crust. However, most of these measurements are restricted to depths below 5,000 m. Brown and Hoek (1978) proposed some empirical relations, given below, to estimate the ratio of horizontal stress to normalized vertical stress.

$$\lambda = 1.0 + \frac{2}{h} \tag{7.1}$$

DOI: 10.1201/9781003517214-7

7.1.2 *Analytical Approaches*

Many proposals for the stress state of the Earth have been made in this century, and these can be classified based on their main characteristics as follows:

7.1.2.1 *Approaches assuming the Earth is in a liquid state*

Jeffreys and Bullen (1940) considered that the stress state of the Earth was hydrostatic, calculating the pressure by considering variations in density and gravitational acceleration. Anderson and Hart (1976) modified this approach based on recent findings.

Nadai (1950) also derived the following formula under the assumption that the density of the Earth is constant and gravitational acceleration varies linearly with depth:

$$\sigma_r = \sigma_\theta, \quad \sigma_r = \frac{\rho_m g_o R_o}{2}\left(1 - \left(\frac{r}{R_o}\right)^2\right) \tag{7.2}$$

where σ_r is the radial stress, σ_θ is the tangential stress, R_o is the radius of the Earth, r is the radial distance, ρ_m is the mean density of the Earth, and g_o is the gravitational acceleration at the Earth's surface.

7.1.2.2 *Approaches assuming the Earth is in a solid state*

Terzaghi (Terzaghi & Richart, 1950) considered a vertical column and formulated the stress state of the Earth by assuming that lateral strains are zero and the medium is elastic as follows:

$$\sigma_\theta = \frac{\nu}{1-\nu}\sigma_r, \quad \sigma_r = \rho g H \tag{7.3}$$

where ν is Poisson's ratio, H is the depth from the ground surface, ρ is the density, and g is the gravitational acceleration.

Salustowicz (1968) derived the following formula by assuming that the density of the Earth is constant and that it consists of a homogeneous elastic material, while also considering linearly decreasing gravitational acceleration:

$$\sigma_r = \rho_m g_o R_o \alpha\left(1 - \left(\frac{r}{R_o}\right)^2\right), \tag{7.4}$$

$$\sigma_\theta = \rho_m g_o R_o \alpha\left(1 - \beta\left(\frac{r}{R_o}\right)^2\right), \tag{7.5}$$

where

$$\alpha = \frac{3-\nu}{10(1-\nu)}, \quad \beta = \frac{1+3\nu}{3-\nu}$$

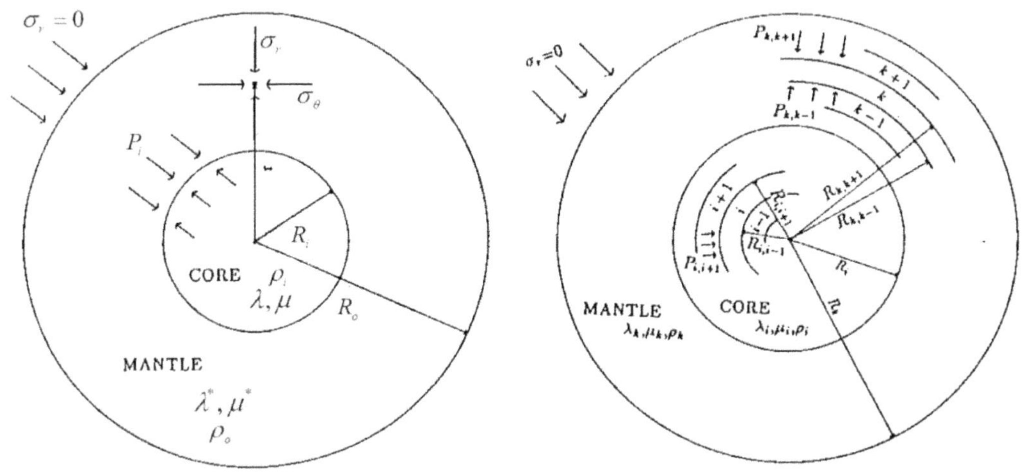

Figure 7.1 (a) Double-layer model and (b) multi-layer model. (From Aydan, 1993.)

7.1.2.3 *Approaches based on the assumption that the crust and mantle are in a solid state and the core is in a liquid state*

Aydan (Aydan, 1993; Aydan & Kawamoto, 1994) proposed two models, namely:

- Two-layered model (TLM)
- Multi-layer model (MLM)

TLM (Figure 7.1a: By considering that gravitational acceleration remains constant up to the interface between the lower mantle and the outer core, and thereafter decreases linearly (Fowler, 1990), the Earth was modeled as a simple spherical body consisting of a core and a mantle. By introducing averaged values of physical and Lame's constants for each zone, the following formulas were developed for radial and tangential stresses in the mantle and the core:

Mantle

$$\sigma_r^* = \frac{R_i^3 R_o^3}{R_o^3 - R_i^3}\left[P_i\left(\frac{1}{r^3} - \frac{1}{R_o^3}\right) + \frac{\lambda^* + \mu^*}{\lambda^* + 2\mu^*}\rho_o g_o\left(\frac{R_o^4 - R_i^4}{R_o^3 R_i^3} - \frac{R_o - R_i}{r^3}\right)\right] - \frac{\lambda^* + \mu^*}{\lambda^* + 2\mu^*}\rho_o g_o r \quad (7.6)$$

$$\sigma_\theta^* = \frac{R_i^3 R_o^3}{R_o^3 - R_i^3}\left[-P_i\left(\frac{1}{2r^3} + \frac{1}{R_o^3}\right) + \frac{\lambda^* + \mu^*}{\lambda^* + 2\mu^*}\rho_o g_o\left(\frac{R_o^4 - R_i^4}{R_o^3 R_i^3} + \frac{R_o - R_i}{2r^3}\right)\right] - \frac{2\lambda^* + \mu^*}{\lambda^* + 2\mu^*}\rho_o g_o \frac{r}{2} \quad (7.7)$$

Core

$$\sigma_r = P_i + \frac{\rho_i g_o R_i}{2}\left(1 - \frac{r^2}{R_i^2}\right), \quad \sigma_\theta = \sigma_r \quad (7.8)$$

where $\lambda, \mu, \lambda^*, \mu^*$: average Lame's constants for the outer and inner zones, respectively, R_i: radius of the inner zone,

$$P_i = \frac{\frac{R_i^3 R_o^3}{R_o^3 - R_i^3} \frac{\lambda^* + \mu^*}{\lambda^* + 2\mu^*} \rho_o g_o \left[\frac{1}{4\mu^*} \frac{R_o - R_i}{R_i^2} + \frac{R_i}{3\lambda^* + 2\mu^*} \frac{R_o^4 - R_i^4}{R_o^3 R_i^3} \right] - \frac{\rho_o g_o}{\lambda^* + 2\mu^*} \frac{R_i^2}{4} - \frac{\rho_i g_o}{\lambda} \frac{R_i^2}{15}}{\frac{R_i^3 R_o^3}{R_o^3 - R_i^3} \left[\frac{1}{4\mu^*} \frac{1}{R_i^2} + \frac{1}{3\lambda^* + 2\mu^*} \frac{R_i}{R_o^3} \right] + \frac{R_i}{3\lambda}} \tag{7.9}$$

MLM (Figure 7.1b: Since physical and Lame's constants vary with depth, the preceding approach was extended to simulate a multi-layered structure of the Earth to provide a better solution for its stress state. The solutions for each layer were essentially similar to those for the two-layered model. The displacement for a layer numbered i in the core is obtained by assuming the Earth consists of n layers with constant physical and mechanical parameters in each respective layer as follows:

$$u^i = A_1^i r + A_2^i \frac{1}{r^2} - \frac{\rho_i g_o}{10\lambda_i} \frac{r^3}{R_{mci}} \tag{7.10}$$

By introducing the following conditions at the interface between layer i and layer $i+1$,

$$\sigma_r = P_{i,i+1} \text{ at } r = R_{i,i+1}$$

integration constants A_1^i and A_2^i are obtained as

$$A_1^i = \frac{1}{3\lambda_i} \left[P_{i,i+1} + \rho_i g_o \frac{R_{i,i+1}^2}{2R_{mc}} \right], \quad A_2^i = 0$$

where R_{mc} is the radius of the interface between the mantle and the core.

Similarly, for a typical layer numbered k in the mantle, displacement u^k can be obtained in the following form:

$$u^k = A_1^k r + A_2^k \frac{1}{r^2} - \frac{\rho_k g_o}{\lambda^k + 2\mu^k} \frac{r^2}{4} \tag{7.11}$$

By introducing the following conditions at the interface between layer k and layer $k+1$ and the interface between layer k and layer $k-1$,

$$\sigma_r = P_{k,k-1} \text{ at } r = R_{k,k-1}$$

$$\sigma_r = P_{k,k+1} \text{ at } r = R_{k,k+1}$$

integration constants A_1^k and A_2^k can be obtained as

$$A_1^k = \frac{1}{3\lambda^k + 2\mu^k} \frac{R_{k,k-1}^3 R_{k,k+1}^3}{R_{k,k+1}^3 - R_{k,k-1}^3} \left[\frac{\lambda^k + \mu^k}{\lambda^k + 2\mu^k} \rho_k g_o \left(\frac{R_{k,k+1}^4 - R_{k,k-1}^4}{R_{k,k-1}^3 R_{k,k-1}^3} \right) + \frac{P_{k,k+1} R_{k,k+1}^3 - P_{k,k-1} R_{k,k-1}^3}{R_{k,k-1}^3 R_{k,k+1}^3} \right]$$

$$A_2^k = \frac{1}{4\mu^k} \frac{R_{k,k-1}^3 R_{k,k+1}^3}{R_{k,k+1}^3 - R_{k,k-1}^3} \left[\frac{\lambda^k + \mu^k}{\lambda^k + 2\mu^k} \rho_k g_o \left(\frac{R_{k,k+1}^4 - R_{k,k-1}^4}{R_{k,k-1}^3 R_{k,k-1}^3} \right) + (P_{k,k+1} - P_{k,k-1}) \right]$$

Introducing the continuity condition of displacement, that was, $u_i - u_{i+1} = 0$ at each interface $r = R_{i,i+1}$ yielded a linear equation system for interlayer pressures $P_{1,2}, \dots P_{i,i-1}, P_{i,i+1}, P_{n-1,n}$. The solution of this equation system together with the condition of $\sigma_r = 0$ at the Earth's surface, provided the pressures on both sides of layers, from which the displacement and stresses of each layer could be easily obtained.

7.1.2.4 *Approaches assuming that the crust is in solid state and plastic*

Possible stress states in the crust associated with normal and thrust faulting were first discussed by Hubbert (1951) and Anderson (1951). This approach assumes that the crust is in a plastic state and that the materials obey a Mohr–Coulomb (MC) type yield criterion. The stress states associated with various faulting regimes are summarized and discussed in a textbook by Jaeger and Cook (1976) as follows:

Normal faulting regime

$$\sigma_v = q\sigma_h + \sigma_c, \quad q = \frac{1 + \sin\phi}{1 - \sin\phi} \tag{7.12}$$

Thrust faulting regime

$$\sigma_H = q\sigma_v + \sigma_c \tag{7.13}$$

Strike–slip faulting regime

$$\sigma_H = q\sigma_h + \sigma_c \tag{7.14}$$

where $\sigma_v = \rho g H$, ϕ is the friction angle, σ_c is the uniaxial strength of crust, σ_H is the maximum horizontal stress, and σ_h is the minimum horizontal stress.

As noted from the above equations, one of the stress components is always indeterminate. Parameters q and σ_c may be temperature and/or confining pressure dependent. This fact is also considered in some publications (i.e., Brace & Kohlstedt, 1980; McCarr, 1980; Shimada, 1993).

7.1.2.5 *Comparisons*

The approaches summarized in this subsection are compared with each other, and their main characteristics are discussed herein. Figure 7.2 shows plots of the distributions of radial and tangential stresses, as well as the ratio of tangential stress to radial stress in the Earth obtained from

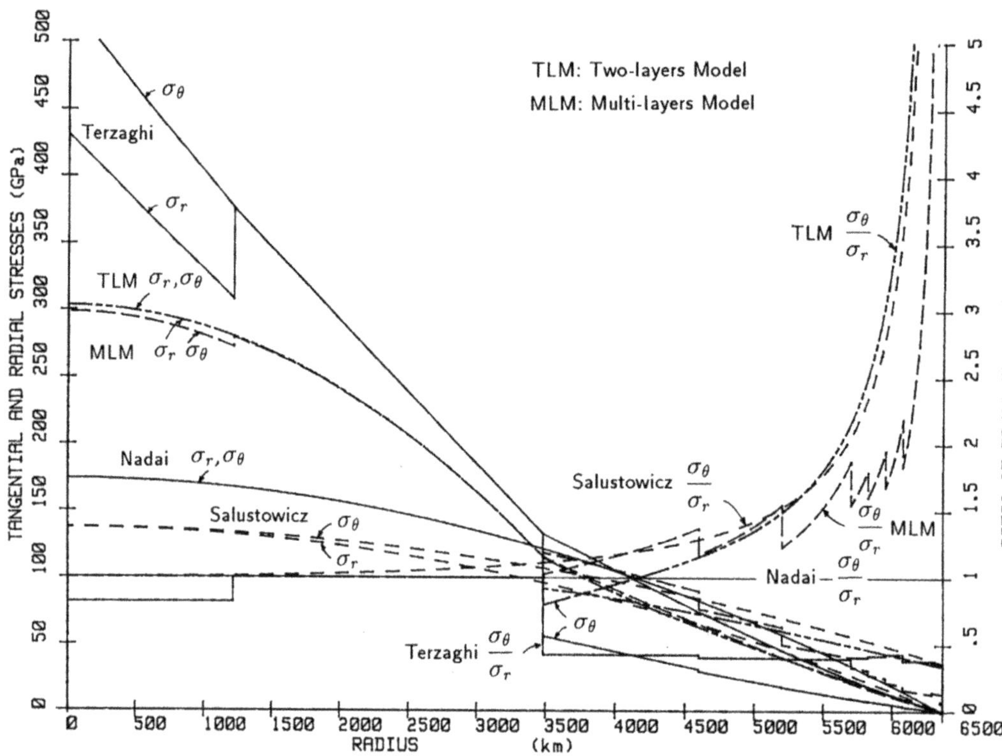

Figure 7.2 Comparison of various approaches. (From Aydan, 1993,1995a.)

various models. Nadai's and Salustowicz's models underestimate the radial stress distributions. While the lateral stress coefficient remains 1 for Nadai's model, it is remarkable to note that the coefficients predicted by the TLM, MLM, and Salustowicz's model increase exponentially near the Earth's surface and approach infinity.

The TLM predicts the radial stress distribution, which is very close to the pressure distribution reported by Anderson and Hart (1976). While the tangential stress predicted by the TLM is equal to the radial stress throughout the core, its distribution varies within the mantle. At the core–mantle interface, there is a discontinuity between the tangential stresses of the core and the mantle; the tangential stress is less than the radial stress in the mantle. However, it becomes greater than the radial stress and reaches a finite value at the Earth's surface, while the radial stress is nil there. This implies that the ratio of the horizontal stress to the vertical stress should be infinite at the Earth's surface.

The vertical stress distribution obtained from Terzaghi's proposal is close to that of the radial stress, but it becomes very large in the core, which is quite different from those predicted by other solutions. The lateral stress distribution is also different from the tangential stress distribution obtained from other solutions, and it cannot explain why lateral stresses should be large at the Earth's surface. Amadei and Savage (1985) proposed a model to explain large horizontal stresses near the Earth's surface by introducing the anisotropy of elastic constants of rock mass while using a one-dimensional column model. However, it must be noted that large horizontal stresses near the Earth's surface can also be obtained by taking into account the sphericity of the Earth, even if the rock mass remains isotropic.

A better solution for the stress state in the Earth is provided by the multi-layered model (Figure 7.2). In the calculations, the Earth is assumed to consist of nine co-centric spherical layers. While the radial stress distribution almost coincides with that of the two-layered model and pressure model, the tangential stress distribution is remarkably different from that of the two-layered model. Since the mechanical and physical properties used in the analysis differ in each layer, the tangential stress at each interface becomes discontinuous. Furthermore, the value of the tangential stress is much less than that predicted by the two-layered model or Salustowicz's model.

7.1.3 Numerical Approaches

7.1.3.1 Approach assuming that the Earth is in a fluid state

The original approach is based on the assumption that the Earth is in a fluid state. This approach was proposed by Jeffrey and Bullen (1940), and Anderson and Hart (1976) used a better model for the density distribution of the Earth. They utilized the finite difference technique and spherical symmetry to obtain the pressure distribution in the Earth.

7.1.3.2 Approach assuming that the Earth is a spherically symmetric thermo-elasto-plastic body

Triaxial tests on rocks were undertaken by many experimenters in the fields of geomechanics and geophysics under high confining pressure and high-temperature regimes (i.e., Byerlee, 1978; Edmonton & Paterson, 1972, Hirth & Tullis, 1994; Shimada, 1993; von Karman, 1911, etc.). These experiments showed that the mechanical behavior of rocks changes from brittle to ductile as the confining pressure increases (Figure 7.3a). Furthermore, the strength decreases as temperature increases, as shown in Figure 7.3b. Taking into account these facts, Aydan (1995a) proposed a thermo-plastic yield criterion:

$$\sigma_1 - \sigma_3 = \left[S_\infty - (S_\infty - S_0)e^{-b_1\sigma_3} \right] e^{-b_2 T} \tag{7.15}$$

where S_0 is the uniaxial compressive strength at room temperature, S_∞ is the ultimate deviatoric compressive strength at room temperature, T is the temperature, b_1, b_2 are physical constants,

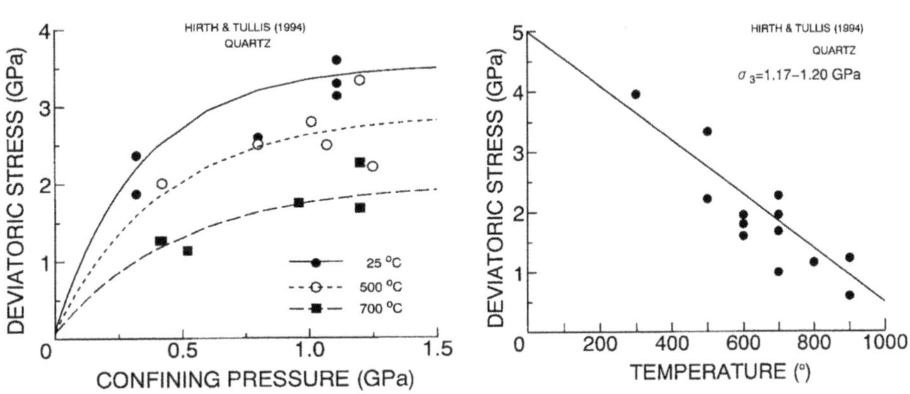

Figure 7.3 Triaxial compression tests on quartz. (Data from Hirth & Tullis, 1994.)

σ_1 is the maximum principal stress, and σ_3 is the minimum principal stress (confining pressure). Figure 7.5 shows a plot of the above yield criterion in the space of temperature T and confining pressure σ_3.

The governing equation for a spherically symmetric body can be written as follows:

$$\frac{\partial \sigma_r}{\partial r} + 2\frac{\sigma_r - \sigma_\theta}{r} = \rho g \tag{7.16}$$

where ρ and g are density and gravitational acceleration which may vary with depth. The constitutive law for elastic behavior is written as follows:

$$\begin{Bmatrix} \sigma_r \\ \sigma_\theta \end{Bmatrix} = \begin{bmatrix} 2\mu+\lambda & 2\lambda \\ \lambda & 2(\mu+\lambda) \end{bmatrix} \begin{Bmatrix} \varepsilon_r \\ \varepsilon_\theta \end{Bmatrix} \tag{7.17}$$

It should be noted that Lame's constants in the above equation can vary with depth.

Taking variations on δu, the following integral form of the above expression is obtained:

$$\int_\Omega \left(\frac{\partial \sigma_r}{\partial r} + 2\frac{\sigma_r - \sigma_\theta}{r} \right) \delta u d\Omega = \int_\Omega \rho g \delta u d\Omega \tag{7.18}$$

where $d\Omega = 4\pi r^2 dr$. With the use of the Gauss divergence theorem and the following conditions:

$$\delta u = 0 \text{ on } \Gamma_u, \text{ and } \hat{t} = 0 \text{ on } \Gamma_t,$$

the weak form of Eq. (7.17) takes the following form:

$$\int_\Omega \left(\sigma_r \delta\varepsilon_r + 2\sigma_\theta \delta\varepsilon_\theta \right) d\Omega = \int_\Omega \rho g \delta u d\Omega \tag{7.19}$$

Let us assume that the displacement in a given element is approximated by the following expression:

$$u = [N]\{U\} \tag{7.20}$$

where $[N]$ is the shape function and $\{U\}$ is the nodal displacement vector. Using the above approximate form and the constitutive law (7.17), the following finite element form is obtained for a typical finite element:

$$[K]\{U\} = \{F\} \tag{7.21}$$

where

$$[K] = \int_{\Omega_e} [B]^T [D][B] \, d\Omega, \quad \{F\} = \int_{\Omega_e} \rho g \, [N]^T \, d\Omega$$

If a linear type shape function is chosen as given below:

$$N^i = \frac{r_j - r}{L}, \quad N^j = \frac{r - r_i}{L}, \quad L = r_j - r_i \tag{7.22}$$

The stiffness matrix $[K]$ and load vector $\{F\}$ given in Eq. (7.21) are specifically obtained as follows:

$$[K] = \frac{4\pi}{L^2}\begin{bmatrix} K_{11} & K_{12} \\ K_{12} & K_{22} \end{bmatrix}, \quad \{F\} = \frac{\rho g \pi}{3}\begin{Bmatrix} r_j^3 + r_j^2 r_i + r_j r_i^2 - 3r_i^3 \\ 3r_j^3 - r_j^2 r_i - r_j r_i^2 - r_i^3 \end{Bmatrix} \tag{7.23}$$

where

$$K_{11} = \frac{D_1 + 2D_2 + D_1^*}{3}(r_j^3 - r_i^3) - (D_2 + D_1^*)r_j(r_j^2 - r_i^2) + D_1^* r_j^2 (r_j - r_i)$$

$$K_{22} = \frac{D_1 + 2D_2 + D_1^*}{3}(r_j^3 - r_i^3) - (D_2 + D_1^*)r_i(r_j^2 - r_i^2) + D_1^* r_i^2 (r_j - r_i)$$

$$K_{12} = K_{21} = -\frac{D_1 + 2D_2 + D_1^*}{3}(r_j^3 - r_i^3) + (D_2 + D_1^*)\frac{r_j + r_i}{2}(r_j^2 - r_i^2) - D_1^* r_i r_j (r_j - r_i)$$

$$D_1 = 2\mu + \lambda, \quad D_2 = 2\lambda \quad D_1^* = 4(\mu + \lambda)$$

In the analyses, it was assumed that the stress–strain response of the rocks constituting the Earth exhibits elastic-perfectly plastic behavior. The elastic constants of the rocks were taken from a report by Anderson and Hart (1976). The initial stiffness technique was chosen as the iteration method to address the plastic behavior of the rocks in finite element analyses (Owen & Hinton, 1980). The temperature distribution of the Earth was input as known, using a distribution reported in a textbook (p. 248) by Fowler (1990) (Figure 7.4).

A series of case studies conducted by Aydan (1995a) (Figure 7.5) are as follows:

CASE 1: The Earth was in a liquid state (hydrostatic).
CASE 2: The crust and mantle were elastic solids, and the core was in a liquid state (non-hydrostatic).
CASE 3: The crust and mantle were elasto-plastic solids, and the core was in a liquid state (isothermic: room temperature)
CASE 4: The crust and mantle were thermo-elasto-plastic solids and the core was in a liquid state (non-isothermic).
Parameters b_1, b_2, S_o, S_∞ of the yield criterion were chosen as 0.2 GPa^{-1}, 0.0014 °C^{-1}, 0 MPa, and 5 GPa, respectively.

Figure 7.6 shows the distributions of radial and tangential stresses, as well as the ratio of tangential stress to radial stress in the lithosphere for each case. The elasto-plastic analyses showed that the entire crust and mantle became plastic. Large tangential stresses observed in CASE 2

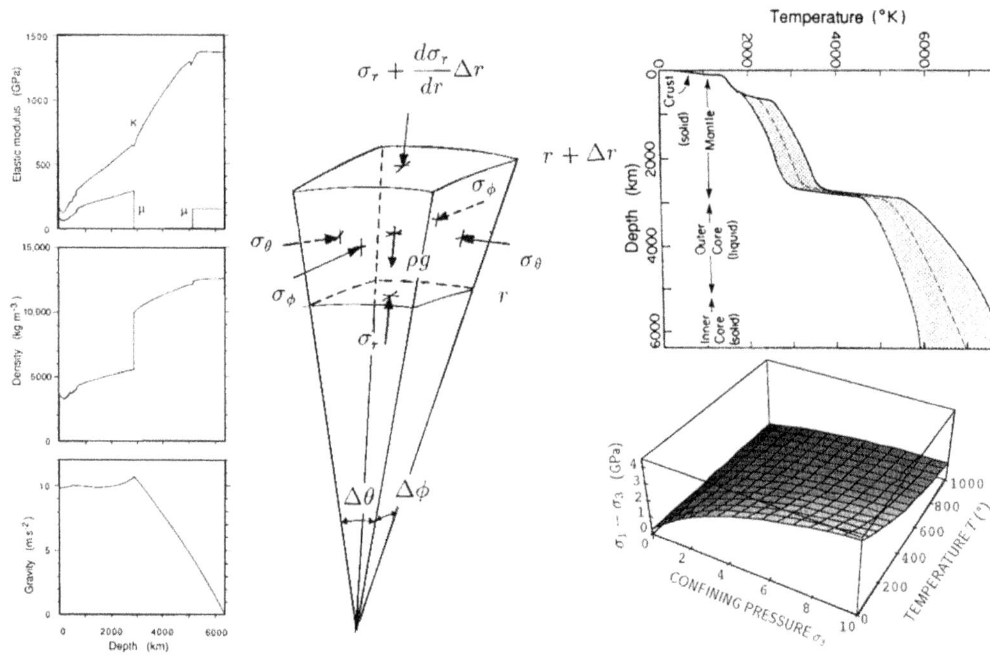

Figure 7.4 Illustration of the spherical model of the Earth and material properties. (Arranged from Aydan, 1993 and Fowler, 1990.)

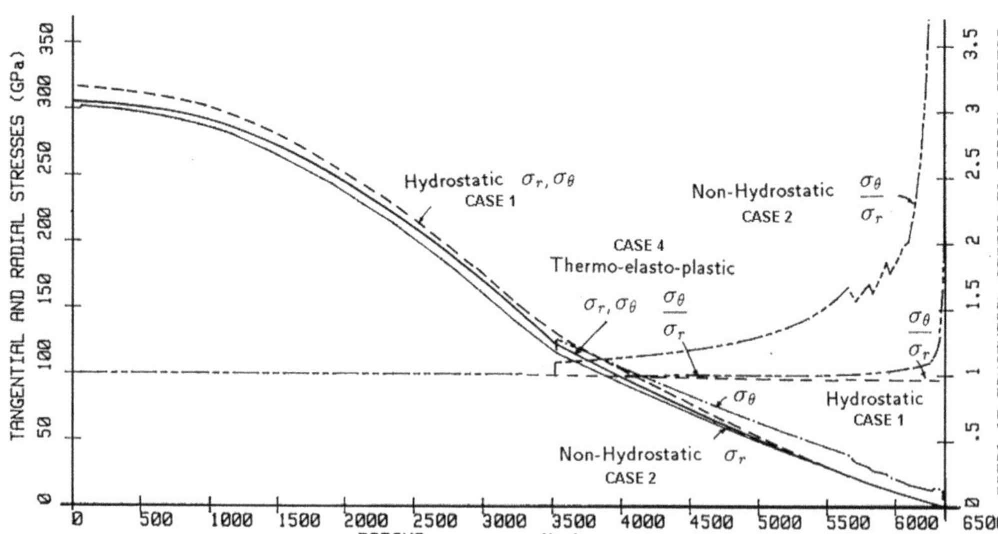

Figure 7.5 Stress state of the Earth. (From Aydan, 1995a.)

were dissipated when the plastic behavior of the crust and mantle was considered. The tangential stresses dissipated in the mantle increased both radial and tangential stresses in the core. Considering the thermo-elasto-plastic behavior of the mantle and the crust of the Earth further decreased the deviatoric stresses in both the mantle and the crust.

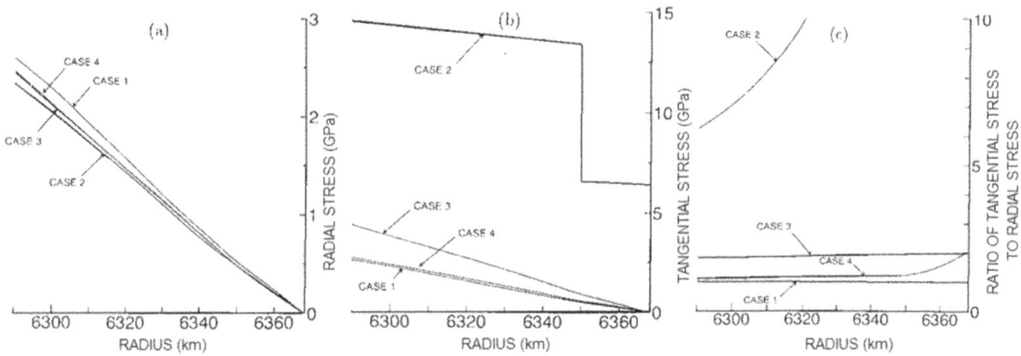

Figure 7.6 Computed radial and tangential stresses in the lithosphere. (From Aydan, 1995a.)

The lateral stress coefficient, which has a value of infinity at the ground surface for CASE 2, also decreased in magnitude when plastic behavior was considered.

In the next series of parametric studies, the uniaxial compressive strength parameter S_o was varied from 0 MPa to 40 MPa in increments of 20 MPa, while the other parameters were kept the same as those in the previous thermo-elasto-plastic analysis. The computed distributions of radial and tangential stresses, along with the ratio of tangential stress to radial stress, were plotted together with in-situ observations (Aydan & Paşamehmetoğlu, 1994) and predictions by Shimada (1993) for a depth of 4 km in the Earth's crust in Figure 7.7. The radial stress (vertical stress) almost coincided with each other (Figure 7.7a). They are almost equal to the vertical stress calculated from ρgH (Shimada, 1993), and it is a good fit to in-situ measurements.

The tangential stresses (horizontal stress) for each value of S_o fit fairly well to in-situ measurements for a depth of 1 km from the ground surface (Figure 7.7b). However, finite element computations tend to overestimate the tangential stresses at depths greater than 1 km.

The ratio of tangential stress to radial stress for each value of S_o provides upper bounds for in-situ measurements (Figure 7.7c). The discrepancy between computed and measured values may be eliminated if a thermo-elasto-visco-plastic constitutive law is employed in finite element analysis.

7.2 In-situ Stress Measurement Methods

7.2.1 *Stress Relief (Overcoring) Method*

The overcoring method has been used for many decades. This method is based on the principle of relieving the stress state of instrumented rock core inside a borehole (Figure 7.8) by overcoring. It is a widely used technique around the world. There are two variations of this technique:

- Leeman's method (or doorstopper method), which utilizes a flat-ended borehole.
- Sugawara-Obara's, which utilizes a semi-spherical-ended borehole. Another variation of this approach is proposed by Kobayashi and Mizuta, who used a conical-ended borehole (conical-ended borehole overcoring).

It involves creating a circular hole, placing a device for measuring strains or displacements in the hole, and then drilling over the top of that hole to relieve the stress, causing a deformation

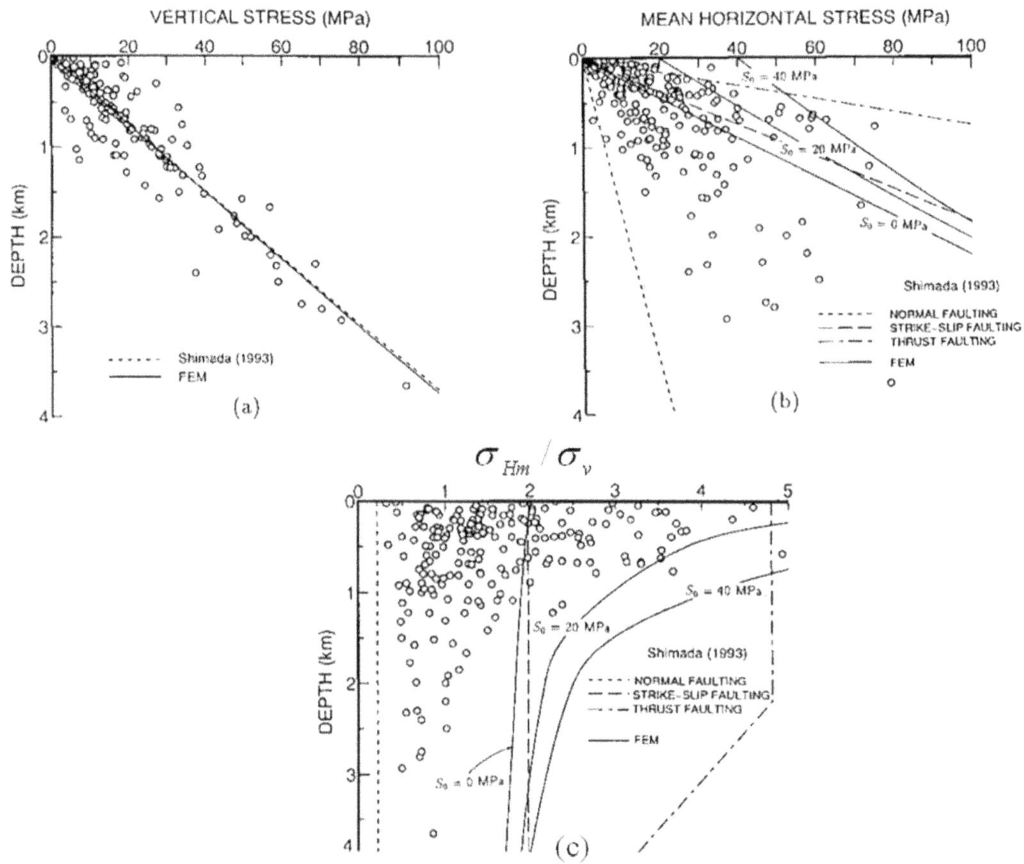

Figure 7.7 Comparison of computed and measured in-situ stresses. (a) Depth vs. vertical stress; (b) depth vs. mean horizontal stress, (c) depth vs. horizontal stress coefficient. (From Aydan & Kawamoto, 1994.)

change. The measured deformation, along with the rock modulus and Poisson's ratio measured in the laboratory, is used to calculate the magnitude and direction of the stresses existing in the rock by considering the geometry and elasticity of the surrounding rock.

7.2.2 *Flat Jack Method*

This method is based on the principle of relieving the stress state of a cavern first and then restoring it by pressurizing the flat jack (Figure 7.9). It is also a widely used technique around the world.

7.2.3 *Hydro-Fracturing and Sleeve Fracturing Method*

The hydro-fracturing method of stress measurement involves pressurizing a fluid into a section of borehole bounded by two packers, sufficient to fracture the rock around the hole. The fracture is intended to occur parallel to the axis of the borehole. The method assumes that one of the principal stresses (generally vertical stress) is parallel to the borehole axis, and the maximum

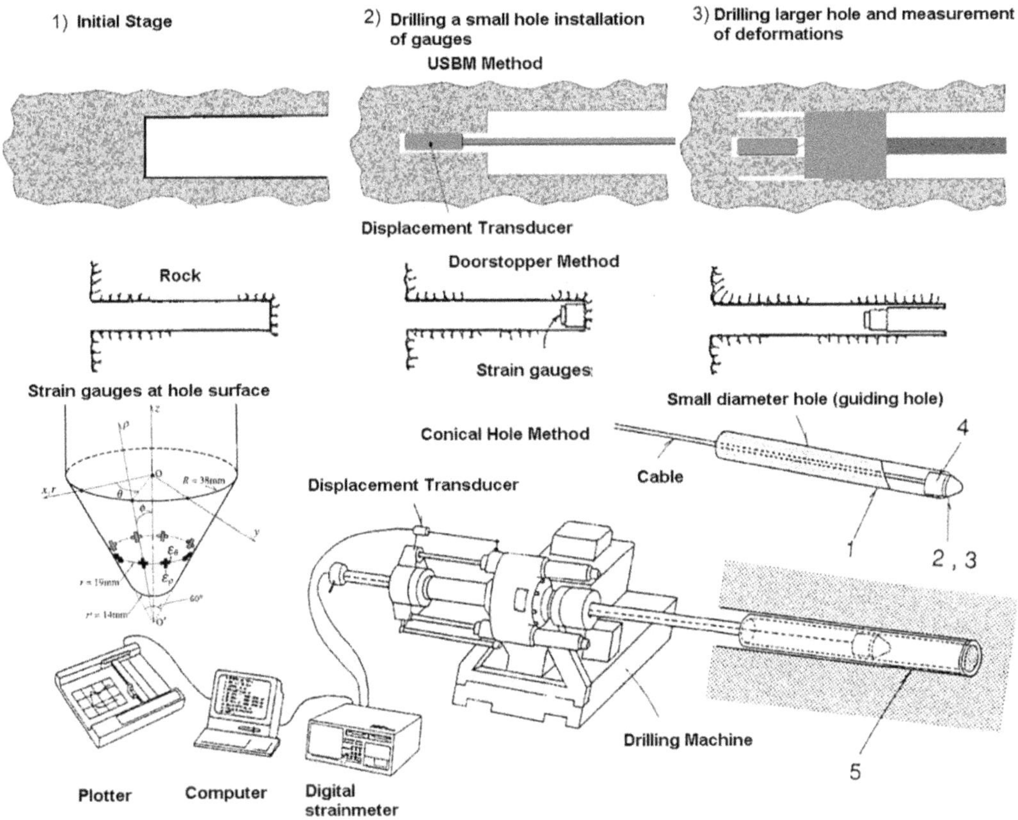

Figure 7.8 Various stress measurement techniques. (Arranged from Sugawara & Obara, 1993; Obert et al., 1962; Leeman, 1964).

and minimum stresses in the plane perpendicular to the borehole axis are obtained from pressure readings (Figure 7.10). In addition, the tensile strength of the surrounding rock is necessary for computing stress components. The sleeve-fracturing method applies internal pressure in the borehole through mechanical jacks instead of pressurized fluid.

7.2.4 AE Method

The AE method utilizes the Kaiser effect to infer the stress state. Since the stress tensor is a symmetric second-order tensor, it has six independent components. As a result, it is necessary to perform uniaxial or triaxial tests in six different directions. According to the Kaiser effect (Kaiser, 1953), it is expected that the AE response will differ when the (deviatoric) stress level exceeds the level to which the material was previously subjected, as illustrated in Figure 7.11. The fundamental complexity in rock mechanics is that the Earth's crust has a stress history. In actual experiments, one may find one or several stress levels. The question is how to select or define the one that reflects the current stress level to which the rock was subjected before unloading.

The AE method (AEM) was first suggested by Kanagawa et al. (1976, 1981) for inferring in-situ stresses. Since then, many attempts have been made to measure the stress state using this method.

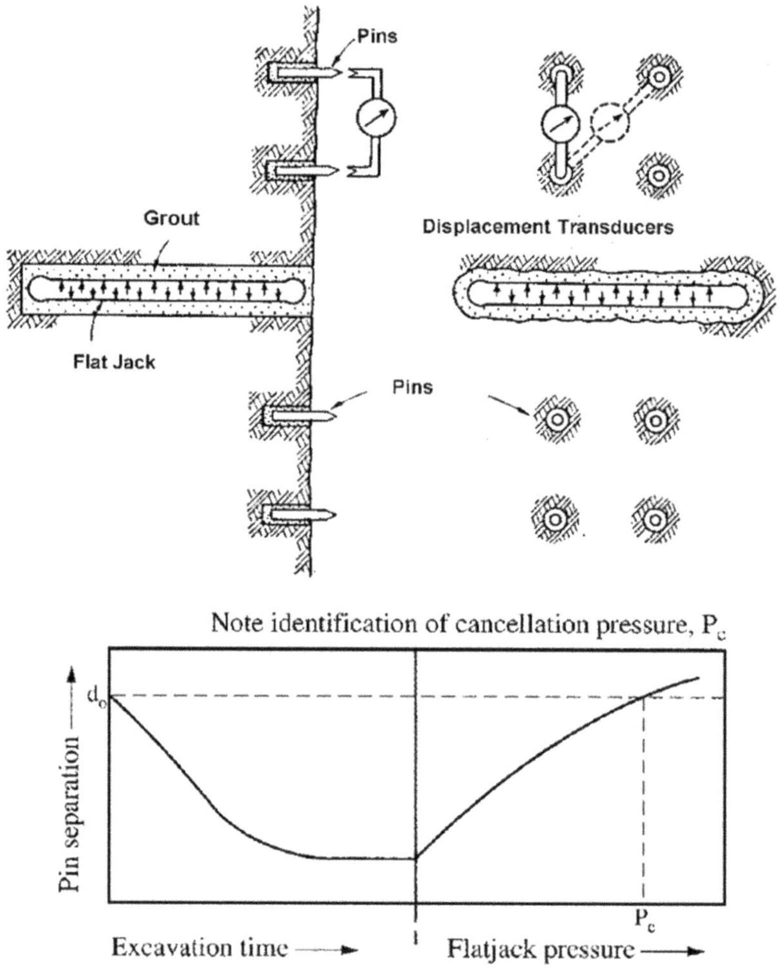

Figure 7.9 Illustration of stress measurement by the flat jack method. (Arranged from ISRM, 1986.)

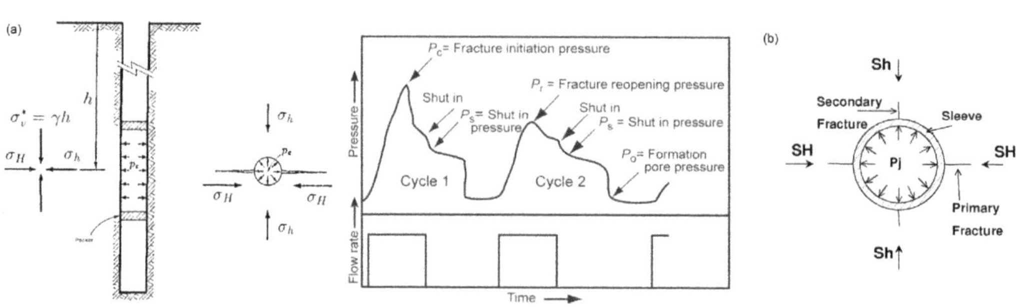

Figure 7.10 Illustration of stress measurement by hydro-fracturing (a) and sleeve fracturing method (b). (Arranged from Haimson, 1987; Stephanson et al., 1986.)

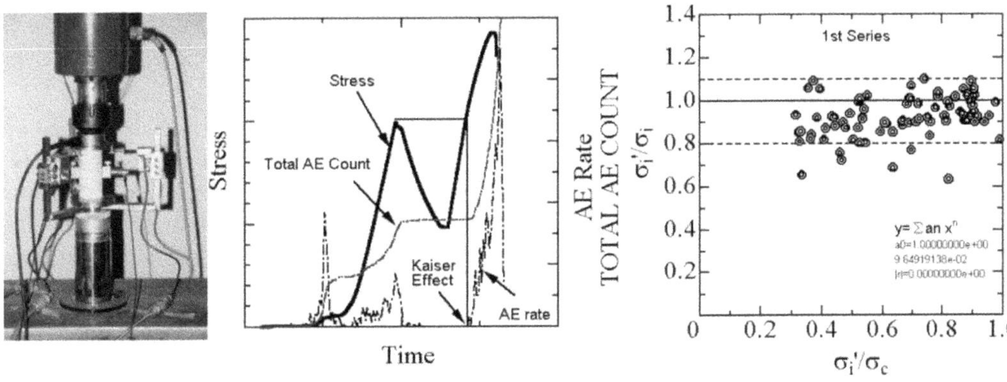

Figure 7.11 Illustration of stress measurement by the AE method and Kaiser effect. (Arranged from Daido et al., 2003.)

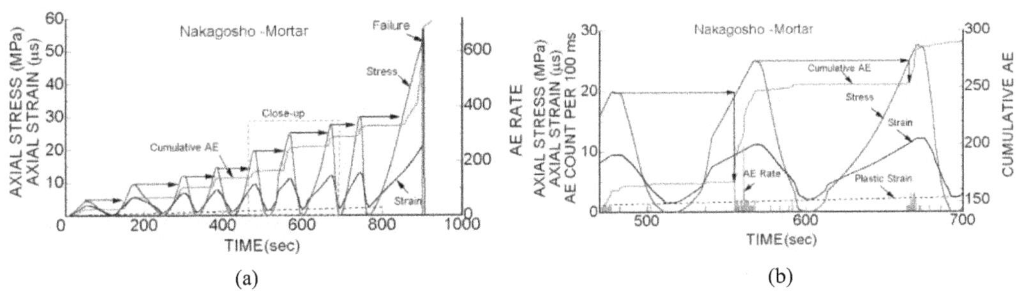

Figure 7.12 Acoustic emission response of mortar during cyclic loading. (a) Total response. (b) Close-up plot of selected time span.

Nevertheless, the cost of AE measurement equipment was quite high in the past, preventing it from becoming a widely accepted method. Recently, the cost of equipment has decreased, allowing experiments to be easily performed under laboratory conditions. As a result, there is renewed interest in stress measurements using this method (Holcomb, 1993; Hughson & Crawford, 1987; Seto et al., 1999; Wang et al., 2000; Watanabe & Tano, 1999; Watanabe et al., 1994, 1999).

The validity of the AE method is often questioned since other methods of in-situ stress measurements have already been conducted at various sites. Daido et al. (2003) performed fundamental experiments and found that the Kaiser effect is a sound concept for inferring the stress state to which the rock was previously subjected. However, there may be some deviations depending on the stress level in relation to the strength of the rock (Figure 7.12). Particularly, if the stress level exceeds the unstable crack threshold value defined by Bieniawski (1967), the inferred stresses may be less than the actual level to which the rock was subjected. Conversely, if the stress level is below the unstable crack propagation threshold stress level, the inferred stress levels may be greater than the actual ones. Although many researchers associate AE events in rocks with micro-cracking, it would be better to utilize the concept of permanent straining associated with the plastic behavior of materials, macroscopically, as also seen in Figures 7.11 and 7.12. If such a concept is employed, the fundamentals of the AEM would be based on firm principles of mechanics.

7.3 In-situ Stress Inference Method

7.3.1 Borehole Breakout Method

When induced stresses exceed the strength of the rock around the borehole, methods based on the theory of elasticity cannot be used. Borehole breakouts are formed by the spalling of fragments of the wellbore in a direction parallel to the minimum stress. Borehole spalling occurs during drilling and progresses over time. The identification and analysis of borehole breakouts as a technique for in-situ measurement of stress orientation and magnitude, as well as for identifying the orientation of both naturally occurring and induced fractures, have received a great deal of attention. The inference of the dimensions of borehole breakouts is based on the Kirsch solution for circular holes, using the MC yield criterion. In practice, the dimensions of borehole breakouts can be measured using a borehole camera, laser scanning, or a mechanical caliper. This method is based on the principle of determining the stress state by observing the borehole breakouts (Figure 7.13). Zoback applied the Kirsch solution along with the MC yield criterion to estimate the shape, extent, and orientation of breakouts to assess the stress state. Although this technique is attributed to Zoback, the main principle was already proposed by Kastner (1962) for circular tunnels and has been used by many (Talobre, 1964).

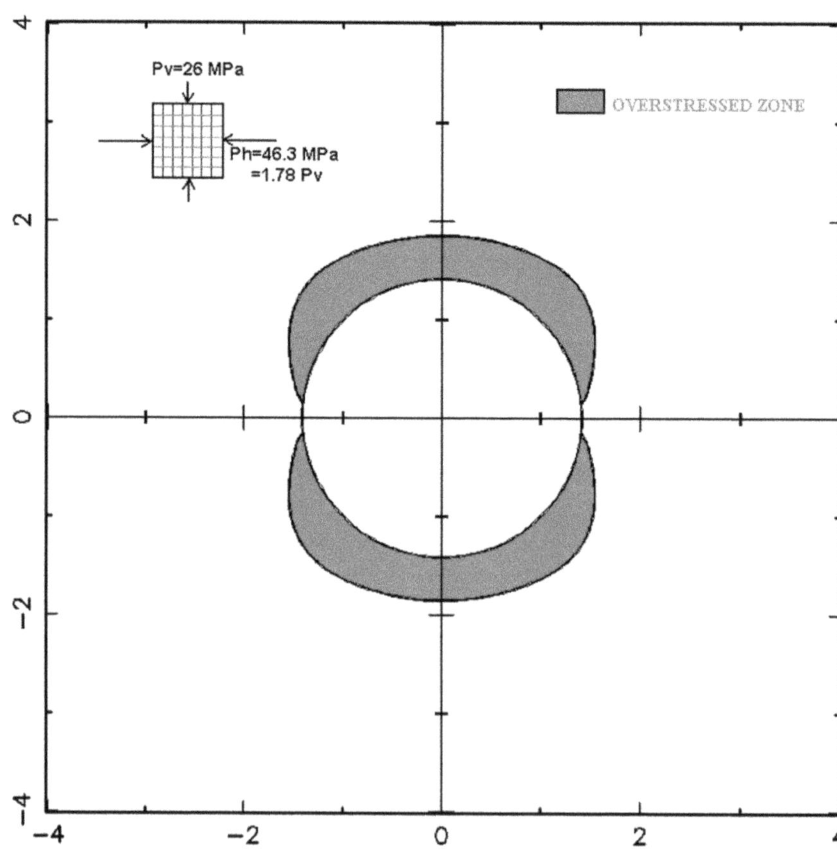

Figure 7.13 Borehole breakout method for stress estimation.

7.3.2 *Fault Type Method*

Possible stress states in the crust associated with faulting were first discussed by Hubbert (1951) and Anderson (1951). This approach assumes that the crust is in a plastic state and obeys a MC type yield criterion. The stress states associated with various faulting regimes are outlined and discussed in a textbook by Jaeger and Cook (1976). When this technique is used, one of the stress components is always indeterminate. However, the vertical stress σ_z for the strike–slip faulting regime is assumed to be equal to $(\sigma_H + \sigma_h)/2$ (Sibson, 1974). Parameters of the MC yield criterion may be temperature and/or confining pressure-dependent. This fact is also considered in some publications (i.e., Brace & Kohlstedt, 1980; Shimada, 1993).

7.3.3 *Fault Striation Method*

Aydan (2000a) proposed a new method to infer the crustal stresses from the striations of the faults or other structural geological features, which may be quite useful for studying the stress state associated with past and current earthquakes. Figure 7.14 illustrates the notation used in this method, and the final expression to infer the stress state is

$$\begin{bmatrix} -q & l_1^2 & m_1^2 & 2l_1m_1 & 2l_1n_1 & 2m_1n_1 \\ -\beta & l_2^2 & m_2^2 & 2l_2m_2 & 2l_2n_2 & 2m_2n_2 \\ -1 & l_3^2 & m_3^2 & 2l_3m_3 & 2l_3n_3 & 2m_3n_3 \\ 0 & l_1l_2 & m_1m_2 & (l_1m_2+l_2m_1) & (l_1n_2+l_2n_1) & (m_1n_2+m_2n_1) \\ 0 & l_1l_3 & m_1m_3 & (l_1m_3+l_3m_1) & (l_1n_3+l_3n_1) & (m_1n_3+m_3n_1) \\ 0 & l_2l_3 & m_2m_3 & (l_2m_3+l_3m_2) & (l_2n_3+l_3n_2) & (m_2n_3+m_3n_2) \end{bmatrix} \begin{Bmatrix} N_{III} \\ N_{xx} \\ N_{yy} \\ N_{xy} \\ N_{xz} \\ N_{yz} \end{Bmatrix}$$

$$= \begin{Bmatrix} -n_1^2 \\ -n_2^2 \\ -n_3^2 \\ -n_1n_2 \\ -n_1n_3 \\ -n_2n_3 \end{Bmatrix} \tag{7.24}$$

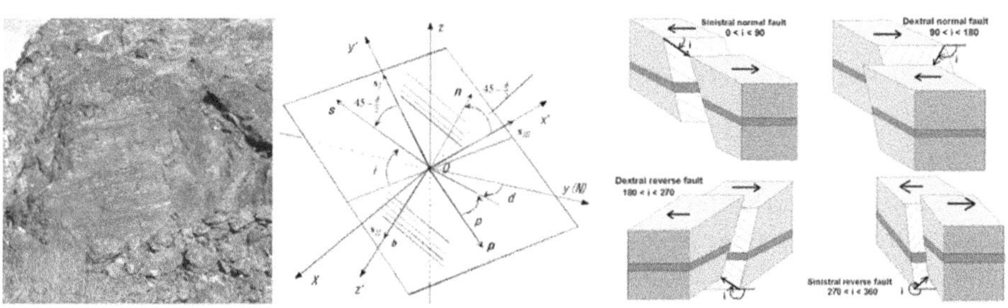

Figure 7.14 (a) A view of fault striation at Tanyeri near Erzincan and illustrations for notation for fault striation method (b,c). (Arranged from Aydan, 2000a.)

where

$$\mathbf{s}_I = \begin{Bmatrix} l_1 \\ m_1 \\ n_1 \end{Bmatrix} = \begin{bmatrix} s_x & s_y & s_z \\ n_x & n_y & n_z \\ b_x & b_y & b_z \end{bmatrix}^{-1} \begin{Bmatrix} \cos\left(45 - \frac{\phi}{2}\right) \\ \cos\left(45 + \frac{\phi}{2}\right) \\ 0 \end{Bmatrix};$$

$$\mathbf{s}_{III} = \begin{Bmatrix} l_3 \\ m_3 \\ n_3 \end{Bmatrix} = \begin{bmatrix} s_x & s_y & s_z \\ n_x & n_y & n_z \\ b_x & b_y & b_z \end{bmatrix}^{-1} \begin{Bmatrix} \cos\left(135 - \frac{\phi}{2}\right) \\ \cos\left(45 - \frac{\phi}{2}\right) \\ 0 \end{Bmatrix}$$

$$\mathbf{n} = \{ n_x \quad n_y \quad n_z \}; \mathbf{s} = \{ s_x \quad s_y \quad s_z \}; n_x = \sin p \sin d;$$

$$n_y = \sin p \cos d; n_z = \cos p \sin i$$

$$s_x = -\cos p \sin d \sin i; \; s_y = -\cos p \cos d \sin i; \; s_z = \cos p \sin i$$

$$\mathbf{b} = \mathbf{s} \times \mathbf{n}; \; \mathbf{b} = \{ b_x \quad b_y \quad b_z \}; \; b_x = s_y n_z - s_z n_y; \; b_y = s_z n_x - s_x n_z; \; b_x = s_x n_y - s_y n_x$$

$$\beta = \frac{\sigma_{II}}{\sigma_{III}} = \frac{1}{2}\left(b - \sqrt{b^2 - 4c}\right); \; N_{III} = \frac{\sigma_{III}}{\sigma_z}; \; N_{xx} = \frac{\sigma_{xx}}{\sigma_z}; \; N_{yy} = \frac{\sigma_{yy}}{\sigma_z}; \; N_{zz} = 1;$$

$$N_{xy} = \frac{\sigma_{xy}}{\sigma_z}; \; N_{yz} = \frac{\sigma_{zy}}{\sigma_z}; \; N_{zx} = \frac{\sigma_{zx}}{\sigma_z}; \; \sigma_z = \gamma h;$$

$$b = \frac{(1+6\alpha^2)(q+1)}{1-3\alpha^2}; \; c = \frac{(1-3\alpha^2)(q^2+1) - q(1+6\alpha^2)}{1-3\alpha^2}; \; \alpha = \frac{2\sin\phi}{\sqrt{3}(3+\sin\phi)}; \; q = \frac{1+\sin\phi}{1-\sin\phi};$$

and p is plunge, d is the dip direction, i is the striation angle, h is overburden, γ is unit weight, ϕ is the friction angle, and c is cohesion.

7.3.4 *Focal Plane Solution Method*

The author of this book (Aydan & Kim, 2002) recently advanced this method to infer the stress state of the Earth's crust from focal plane solutions. The focal plane solutions used in geoscience are derived by assuming that pure-shear conditions hold As a result of this assumption, one of the principal stresses is compressive while the other is tensile in focal plane solutions. This condition may also imply that the friction angle of the fault is nil. Therefore, the principal stresses are inclined at an angle of 45° with respect to the normal of the slip direction. This condition is used to determine the P-axis and T-axis in focal plane solutions. Each focal plane solution

(a) (b)

Figure 7.15 Core disking at the Kaore Underground Power House site. (a) Core disking. (b) Surface of fracture plane.

involves the fault plane on which sliding takes place and the auxiliary plane. The normal of the auxiliary plane corresponds to the slip vector, and it is orthogonal to the neutral plane on which the P-axis and T-axis exist.

7.3.5 Core Disking Method

When drilling is done in brittle rock under high stress, it is often reported that core disking occurs. Jaeger (1963) was the first to investigate the stress state resulting in core disking (Figure 7.15a). The stress state is very complicated during the drilling operation, as the compressive and torsional tractions are imposed on the rock stubs by the drill bit in the close vicinity of the borehole end. Numerical analyses clearly indicate local high tensile stresses in the close vicinity of the borehole end. Nevertheless, the ridges on the surface of disking fractures are indicative of the maximum in-situ stress perpendicular to the borehole axis (Figure 7.15b).

7.3.6 Blast Hole Damage Method

Aydan (2013) proposed a method to infer the in-situ stresses from the damage zones around the blast holes, and it is named the blast hole damage method. Several applications of the method to various sites, where in-situ stress states are obtained using direct or indirect techniques, are discussed, and its validity is evaluated in view of the measurements from other methods.

The stress state around a circular cavity in an elastic medium under biaxial far-field stresses was first obtained by Kirsch (1898). These solutions were modified to incorporate the effect of uniform internal pressures (Jaeger & Cook, 1979). In a polar coordinate system, the radial, tangential, and shear stresses around the circular cavity can be expressed in the following forms:

$$\sigma_r = \frac{\sigma_{10}+\sigma_{30}}{2}\left(1-\left(\frac{a}{r}\right)^2\right)-\frac{\sigma_{10}-\sigma_{30}}{2}\left(1-4\left(\frac{a}{r}\right)^2+3\left(\frac{a}{r}\right)^4\right)\cos 2(\theta-\beta)+p_i\left(\frac{a}{r}\right)^2 \quad (7.25a)$$

$$\sigma_\theta = \frac{\sigma_{10}+\sigma_{30}}{2}\left(1+\left(\frac{a}{r}\right)^2\right)+\frac{\sigma_{10}-\sigma_{30}}{2}\left(1+3\left(\frac{a}{r}\right)^4\right)\cos 2(\theta-\beta)-p_i\left(\frac{a}{r}\right)^2 \quad (7.25b)$$

$$\tau_{r\theta} = \frac{\sigma_{10} - \sigma_{30}}{2}\left(1 - 4\left(\frac{a}{r}\right)^2 + 3\left(\frac{a}{r}\right)^4\right)\sin 2(\theta - \beta) \tag{7.25c}$$

where

σ_{10}, σ_{30} are far-field principal stresses, a is the radius of the hole, r is the radial distance, β is the inclination of σ_{10} far-field stress from horizontal, θ is the angle of the point from horizontal, and p_i is the internal pressure applied onto the hole perimetry.

The yield criteria available in rock mechanics are follows:

MC

$$\sigma_1 = \sigma_c + q\sigma_3 \tag{7.26a}$$

Drucker–Prager

$$\alpha I_1 + \sqrt{J_2} = k \tag{7.26b}$$

Hoek–Brown (HB) (1980)

$$\sigma_1 = \sigma_3 + \sqrt{m\sigma_c\sigma_3 + s\sigma_c^2} \tag{7.26c}$$

Aydan (1995)

$$\sigma_1 = \sigma_3 + [S_\infty - (S_\infty - \sigma_c)e^{-b_1\sigma_3}]e^{-b_2 T} \tag{7.26d}$$

where

$$I_1 = \sigma_I + \sigma_{II} + \sigma_{III};\ J_2 = \frac{1}{6}((\sigma_I - \sigma_{II})^2 + (\sigma_{II} - \sigma_{III})^2 + (\sigma_{III} - \sigma_I)^2$$

$$\alpha = \frac{2\sin\phi}{\sqrt{3}(3+\sin\phi)};\ k = \frac{6c\cos\phi}{\sqrt{3}(3+\sin\phi)};\ c: \text{cohesion};\ \phi: \text{friction angle};\ q = \frac{1+\sin\phi}{1-\sin\phi};$$

σ_∞ is the ultimate deviatoric strength, T is the temperature, and m, s, b_1, b_2 are empirical constants. MC and Drucker–Prager yield criteria are linear functions of confining or mean stress, while the criteria of HB and Aydan are of a non-linear type (Figure 7.16). Furthermore, Aydan's criterion also accounts for the effect of temperature. However, the effect of temperature is omitted in Figure 7.16 for the sake of comparison. If the yield criterion is chosen to be a function of minimum and maximum principal stresses, they can be expressed in the following form in terms of stress components given by Eq. (7.25):

$$\sigma_1 = \frac{\sigma_\theta + \sigma_r}{2} + \sqrt{\left(\frac{\sigma_\theta - \sigma_r}{2}\right)^2 + \tau_{r\theta}^2} \tag{7.27a}$$

$$\sigma_3 = \frac{\sigma_\theta + \sigma_r}{2} - \sqrt{\left(\frac{\sigma_\theta - \sigma_r}{2}\right)^2 + \tau_{r\theta}^2} \tag{7.27b}$$

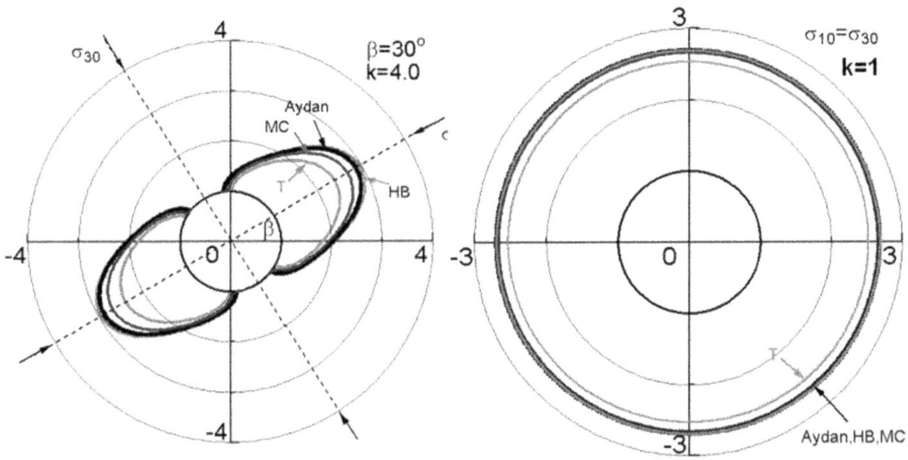

Figure 7.16 Estimated yield zones around the blast hole. (From Aydan, 2013.)

Table 7.1 Values of in-situ stress parameters and properties of yield functions

σ_{10} (MPa)	k	σ_c (MPa)	σ_t (MPa)	ϕ (o)	M	σ_∞ (MPa)	b
50.0	4.0	100.0	5.0	60	20	400	31.39

The damage zone around the blast hole under high internal pressure can be estimated using one of the yield criteria listed above. It should be noted that yielding is induced by the high internal pressure in the blast hole, which is fundamentally different from in-situ stress-induced borehole breakout. In other words, there will always be a damage zone around the blast hole perimeter when the blasting technique is employed. The blast hole pressure depends on the characteristics of the surrounding medium, the amount, layout, and type of explosive, blasting velocity, and the geometry of the blast hole. The blast hole pressure ranges from (i.e., Jaeger & Cook, 1979; Brady & Brown, 1985).

First, we assume that the properties of the surrounding rock have the values given in Table 7.1, and the blast hole (internal) pressure has a value of 400 MPa. The rock chosen roughly corresponds to an igneous rock such as granitic rocks. The maximum far-field stress is inclined at an angle of 30° from horizontal, and the lateral stress coefficient has a value of 4. Figure 7.16 shows the example of computation for the given conditions. The largest yield zone is obtained for the HB criterion, while the tension cut-off criterion (T) results in a smaller yield zone. The Aydan criterion estimates a slightly larger yield zone than the MC criterion. It is very interesting to note that yielding propagates in the direction of maximum far-field stress. In other words, the elongation direction of the yield zone would be the best indicator of the maximum far-field stress in the plane of the blast hole.

7.3.7 Global Positioning System Method

Although it is difficult to obtain total stresses from GPS measurements, it may be possible to determine principal stress variations and their orientations in the tangential plane to the Earth'

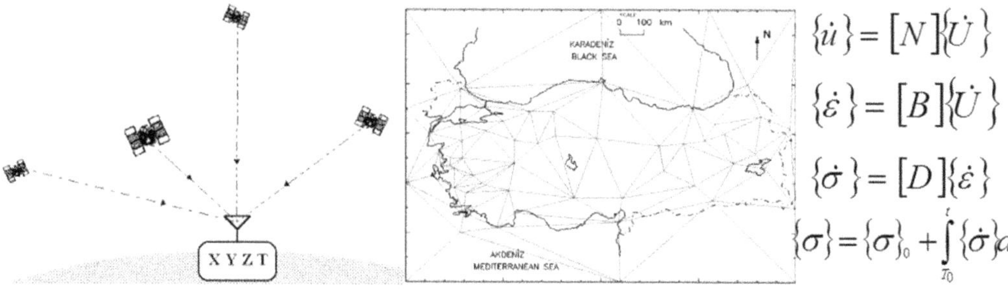

Figure 7.17 Illustration of computation of strain and stress increments using the crustal deformations. (Arranged from Aydan, 2000b.)

surface from the processing of GPS measurements as proposed by Aydan (2000b, 2004) and Aydan et al. (2011). However, if the co-axiality condition between stress rate and total stress in the crust holds, it may also be possible to infer the orientation of maximum and minimum horizontal stresses in the crust. It should be noted that it is geometrically possible to compute strain rate components in a plane tangential to the Earth's surface from the variation of positions of stations at a given time interval, although it is unlikely to obtain the components of strain rate on other planes. For this purpose, a mesh and interpolation technique similar to those used in the finite element method is utilized, and the nodes are selected to correspond to the locations of GPS antennas (Figure 7.17). The element type can be selected as desired with consideration of the GPS network. Displacement increments of each node for a given time interval are used to compute the strain increments first, and then stress increments are computed with the incorporation of constitutive models of the Earth's crust. Aydan (2000b) used Hooke's law as a preliminary approach to compute the strain and stress increments.

7.4 Comparisons

In this section, the applications of the methods to the evaluation of the stress state of several abandoned room and pillar mine sites are discussed.

7.4.1 Abandoned Room and Pillar Mines in Mitake Town

There are no stress measurements at the location of the abandoned mines in Mitake. The nearest site where in-situ stress measurements were carried out is Tono mine. This mine serves as an underground laboratory for nuclear waste disposal and is located 30 km south of Mitake Town. The stress measurements were performed using different techniques at a depth of 130 m in relatively weak sandstone. The results are summarized in Table 7.2 (Maeda et al., 1999; Sugihara, 2002).

There are several normal faults in the abandoned mine (Figure 7.18). The major ones have a strike parallel to the flow direction of the Kiso River. Using Aydan's method (2000), the stress state is inferred. Results are given in Tables 7.3 and 7.4, and stereo projections are shown in Figure 7.19. As noted from the inferred results, the maximum horizontal stress is almost aligned in the E–W direction.

Figure 7.20 shows the directions and magnitude of principal strains measured and computed over about 100 years through the triangulation method. It is of great interest that the maximum compressive strain direction is very close to that inferred from fault striations. From these results, it may be said that the maximum principal stress is the vertical stress, which should be

Table 7.2 Stress measurement results at 130 m in Tono URL

Method	σ_H*(MPa)*	σ_h *(MPa)*	σ_v*(MPa)*
Overcoring techniques	2.85(NS)	1.30(EW)	1.97
Hydro-fracturing	4.44	1.89	2.33

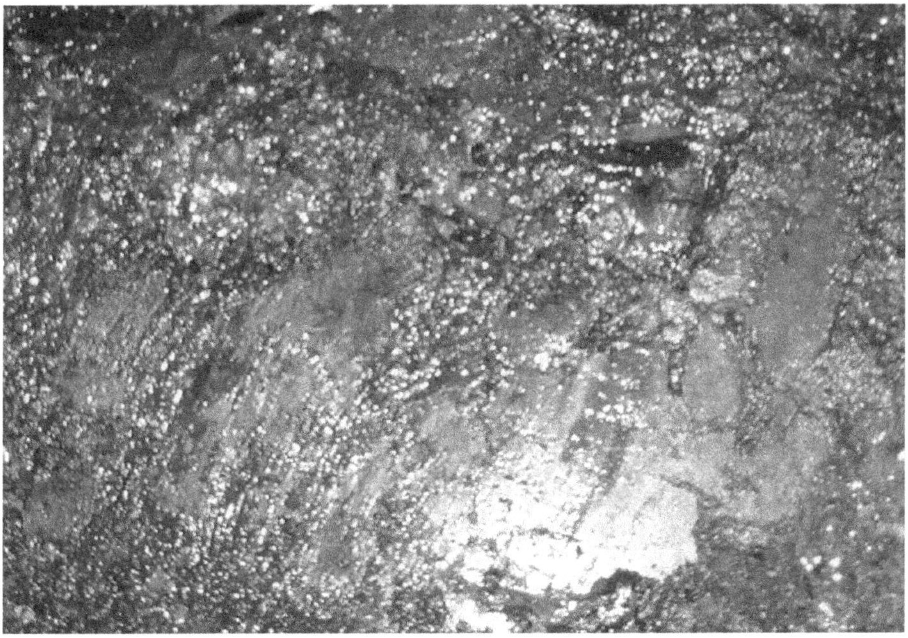

Figure 7.18 A view of a normal fault in an abandoned lignite mine. (Mitake–Fushimi region).

Table 7.3 Inferred stress state from fault striation for Mitake abandoned mine

σ_1			σ_2			σ_3					
$\frac{\sigma_1}{\sigma_v}$	θ_1	ϕ_1	$\frac{\sigma_2}{\sigma_v}$	θ_2	ϕ_2	$\frac{\sigma_3}{\sigma_v}$	θ_3	ϕ_3	$\frac{\sigma_h}{\sigma_v}$	$\frac{\sigma_H}{\sigma_v}$	α_{σ_H}
1.007	82.5	81.3	0.694	265	8.7	0.336	175.0	0.4	0.336	0.701	84.93
1.01	327	80	0.696	124	9	0.337	215	4	0.339	0.704	305

Table 7.4 Inferred stress state from the focal mechanism of 2011 Tajimi Earthquake

σ_1			σ_2			σ_3					
$\frac{\sigma_1}{\sigma_v}$	θ_1	ϕ_1	$\frac{\sigma_2}{\sigma_v}$	θ_2	ϕ_2	$\frac{\sigma_3}{\sigma_v}$	θ_3	ϕ_3	$\frac{\sigma_h}{\sigma_v}$	$\frac{\sigma_H}{\sigma_v}$	α_{σ_H}
1.225	298	55	0.89	179	19	0.408	70	28	0.568	0.909	337

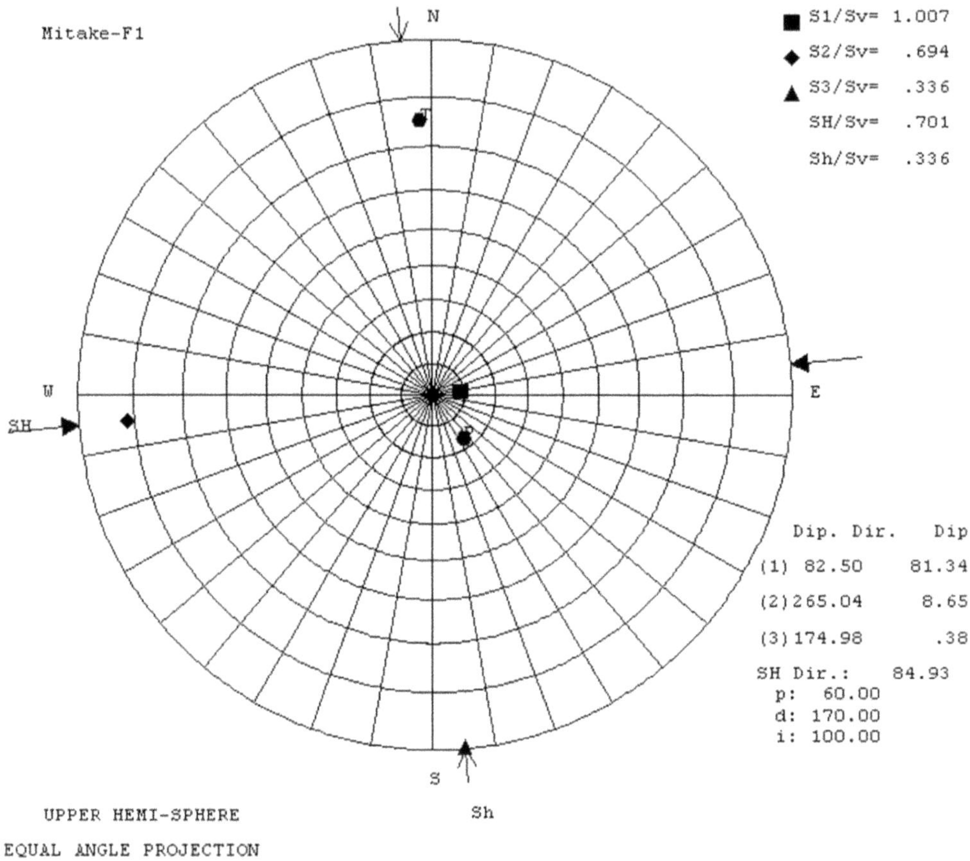

Figure 7.19 Stereo projections of inferred stress state and possible faulting mechanisms.

equivalent to overburden. The intermediate principal stress is almost horizontal and parallel to the flow direction of the Kiso River. The least principal stress is perpendicular to the cliff and the flow direction of the Kiso River. For this direction, the horizontal stress coefficient is about 0.34.

7.4.2 Abandoned Room and Pillar Mines in Oya Town

The stress state obtained from the AEM is given in Table 7.5. The results indicate that the horizontal stress ratio is about 0.8–0.9. There are two underground powerhouses in Tochigi Prefecture. The stress measurements for Imaichi and Sabigawa Power Houses are also given in Table 7.5. The stress state in the powerhouses located in the mountainous region is somewhat different from the AEM method (Watanabe et al., 2003). Table 7.6 presents the inferred stress states for some earthquakes that occurred in Tochigi Prefecture. The distance between the earthquake epicenters and the Oya underground quarry is less than 50 km, and the hypocenter depth is less than 10km. The results imply that the maximum horizontal stress direction remains the same (almost EW), while the horizontal stress ratio varies with time.

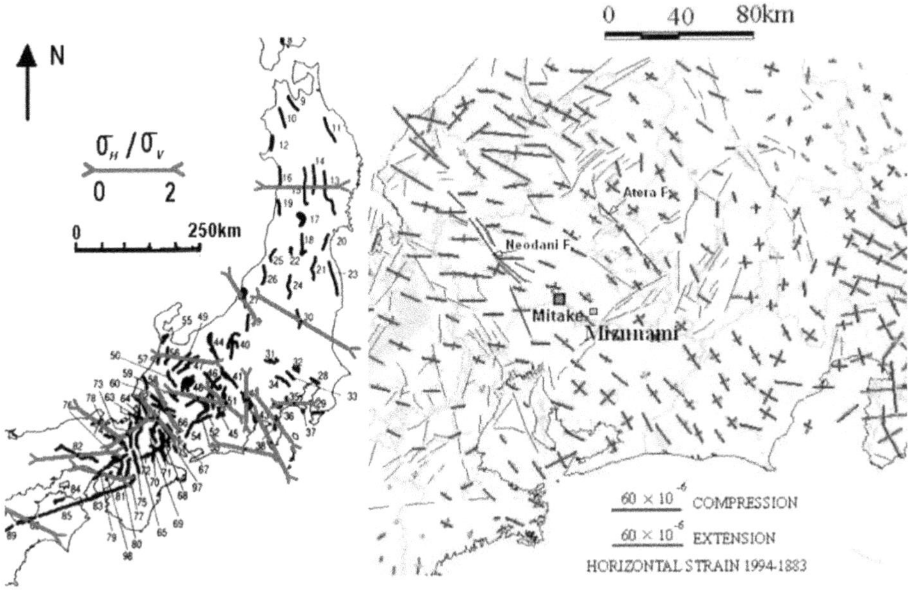

Figure 7.20 Principal horizontal stresses in Central Japan for over the past 100 years.

Table 7.5 Comparison of inferred stresses from the AEM method, and the striations of faults

Method	σ_1			σ_2			σ_3			$\frac{\sigma_h}{\sigma_v}$	$\frac{\sigma_H}{\sigma_v}$	$p\sigma_H$
	$\frac{\sigma_1}{\sigma_v}$	d_1	p_1	$\frac{\sigma_2}{\sigma_v}$	d_2	p_2	$\frac{\sigma_3}{\sigma_v}$	d_3	p_3			
AEM	1.16	253.2	417	0.896	22.5	32.9	0.802	1310	28.0			
Imaichi PH	1.41	313.0	25	1.000	109.0	63.0	0.897	210.0	10.0	0.89	1.42	313
Sabikawa PH										0.58	0.80	
Ashio GPP										0.93	1.16	315
Sekiya Fault	2.913	118.0	7.0	2.007	208.0	0.0	0.971	298.0	83.0	2.007	2.885	298
Futaba Fault	1.085	85.0	70.0	0.747	355.0	0.0	0.362	265.0	20.0	0.446	0.747	355

7.4.3 *Abandoned Room and Pillar Mines at Bazda Antique Underground Limestone Quarry*

There are no in-situ stress measurements at this site yet. However, preliminary experiments on the samples from the site indicate that the Kaiser effect, which is used in in-situ stress estimations, is clearly observed. Therefore, oriented sampling at this site should provide some information on the possible stress state. Preliminary stress state inferences for the site and its close vicinity have been made using the method of Aydan (Aydan, 2000a, 2024; Aydan & Kim, 2002; Aydan et al., 2002) from the striations of the faults at Bazda quarries and focal mechanism

Table 7.6 Comparison of inferred stresses from focal plane solutions of earthquakes in close vicinity

Earthquake Year	σ_1			σ_2			σ_3					
	$\frac{\sigma_1}{\sigma_v}$	d_1	p_1	$\frac{\sigma_2}{\sigma_v}$	d_2	p_2	$\frac{\sigma_3}{\sigma_v}$	d_3	p_3	$\frac{\sigma_h}{\sigma_v}$	$\frac{\sigma_H}{\sigma_v}$	$p\sigma_H$
1998.Tune.11 M3.8	1.47	278.7	28.9	1.013	45.1	47.0	0.49	171.1	28.7	0.628	1.346	89.2
2001.Sep.28 M3.3	1.45	262.7	5.9	0.996	104.3	83.6	0.48	353.0	2.3	0.438	1.441	82.8
2002.Mar.10 M3.8	2.41	116.5	15.2	1.662	211.7	18.4	0.81	349.5	65.8	1.564	2.316	291.2

Table 7.7 Inferred crustal stresses for striation and large earthquakes for Bazda quarries.

		Fault 1	*Fault 2*	*2102 Bozova E.*	*2010 Hilvan E.*
σ_1	σ_1/σ_v	1.05	1.06	1.45	1.43
	d_1	178	187	206	17
	p_1	69	67	8	12
σ_2	σ_2/σ_v	0.72	0.73	1.00	0.98
	d_2	32	33	72	218
	p_2	19	21	77	77
σ_3	σ_3/σ_v	0.35	0.35	0.48	0.48
	d_3	65	299	297	108
	p_3	8	9	8	5
	σ_h/σ_v	0.36	0.37	0.50	0.48
	σ_H/σ_v	0.76	0.77	1.45	1.41
	θ	337	27	27	17

solutions of some recent earthquakes in the vicinity (Figure 7.21). The inferred stress states for faults and some recent earthquakes are given in Table 7.7. As noted from the table, the maximum horizontal stress is almost north–south, and the vertical stress is close to the maximum principal stress inferred from fault striations. The horizontal stress ratio ranges between 0.36 and 0.5. Therefore, it is expected that the overburden stress would be the dominant factor governing the stability of the quarries. The overburden at underground quarries ranges between 3 and 35 m.

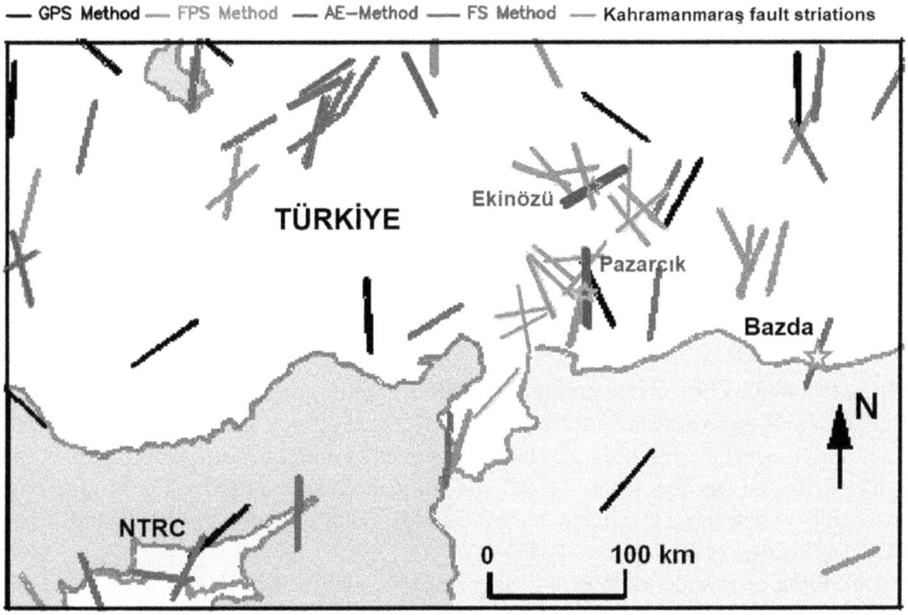

Figure 7.21 Horizontal stress directions in the vicinity of Bazda underground quarries. (Modified from Aydan, 2024.)

Chapter 8

Remediation Measures for Abandoned Room and Pillar Mines

Backfilling techniques have been commonly used in mining engineering for a very long time, as mining methods were devised 9,000 years ago (Kaptan, 1990). Particularly, tailing materials resulting from mineral processing, together with a small amount of cement as a binder, are used to backfill excavated cavities when the cut and fill technique is employed in mining engineering for stability concerns (e.g., Belem & Benzaazoua, 2008; Doherty et al., 2015; Grice, 1998; Stones, 2014; Whittaker & Reddish, 1989).

The backfilling of abandoned mines is also employed when there is a possibility of sinkhole hazards or occurrences observed in urbanized areas and/or transportation structures (e.g., Bell & Genske, 2001; Hawkins & McDonald, 1994; Karfakis & Topuz, 1991; Poulsen & Shen, 2013). This type of backfilling is not systematically carried out, and the procedures vary depending on the geological and geometrical features of the sites.

The extraction of lignite and building stone using the room and pillar mining technique was extensive in various parts of Japan until the 1960s. The lignite mines have since been abandoned, and some of these areas have become urbanized. The Tokai region in Central Japan is a well-known example of such a case. Authorities are now concerned about how to fill these abandoned mines in urbanized areas due to significant concerns regarding the risk of sinkhole occurrences and/or subsidence of the ground above these abandoned mines. The authors have been involved with cavity filling for some time in the Tokai region (Sakamoto, 2007; Sakamoto et al., 2005a, 2005b; Sugiura et al., 2003, 2005, 2019) and have developed their own integrated cavity filling technique.

The selection and supply of filling materials are critical aspects of the filling operations. Given the high costs associated with filling materials in Japan, the utilization of non-hazardous industrial by-products (recycling materials) may help reduce costs as well as ease the disposal problem of such by-products.

There are many abandoned lignite mines as well as underground stone quarries in various parts of Japan, and recent earthquakes have shown that these mines are quite vulnerable to causing sinkholes and subsidence (Aydan & Tano, 2012a, b, c). Therefore, there is growing concern about the safety of areas above such abandoned mines in relation to the anticipated Nankai–Tonankai–Tokai Earthquake. The backfilling technique is presumed to be the most effective and reliable technique for addressing potential damage due to the failure of abandoned mines during earthquakes and over the long term.

In this chapter, an integrated cavity filling technique developed for abandoned lignite mines in the Tokai region, as well as for abandoned underground quarries, is explained. Emphasis is

DOI: 10.1201/9781003517214-8

also placed on the development of new cavity filling materials utilizing non-hazardous industrial by-products. Furthermore, some experimental studies on the effectiveness of backfilling operations are presented.

The major issue is how to determine the critical depth of the abandoned lignite mines for the implementation of backfilling. When the mines are deep, the surface effects of failed abandoned mines are considered to be minimal due to the bulking of the failed overburden rock. It is also claimed that if the mines are submerged in groundwater, the degradation of surrounding rock becomes negligible, eliminating the necessity for backfilling. Although this claim may hold true under static conditions, model tests and observations indicate that the formation of sinkholes or surface ground subsidence may occur in areas of abandoned room and pillar lignite mines due to earthquakes (Aydan & Kawamoto, 2004; Aydan & Tano, 2012a,b).

8.1 Backfilling of Abandoned Room and Pillar Mines

The cavity filling or backfilling technique involves several fundamental steps, such as exploring the dimensions of cavities to be filled, establishing filling plant facilities, backfilling, and conducting constructional and environmental monitoring and surveying. All these steps must be performed with due consideration of environmental restrictions and regulations. Figure 8.1 shows the schematic flowchart of the integrated cavity filling technique. The details of the filling technique are explained in the following sub-sections.

8.1.1 Explorations and Surveying

Explorations and surveys involve determining the layouts and spatial distributions of cavities to be filled, as well as assessing underground water conditions using past documents, maps, and geological explorations. Although some mining plans submitted to the authorities at the time of mining operations are available, these documents generally do not coincide with actual mining layouts for various reasons. Furthermore, the geometry of the cavities may change over time due to partial degradation and the collapse of pillars and chimneying roof layers in abandoned mines.

To perform underground filling works effectively, the following explorations and surveys are specifically conducted in urbanized areas.

1 Collection and analysis of

- existing data on topography, strata, and geology,
- treatises and documents issued by public offices, and
- site checks and observations.

2 Site investigations, which involve geological and geophysical explorations, including boring, sonar, and radar-type soundings for cavities to be filled, as well as selecting sites for plant yards, offices, and backfilling material sources while considering environmental effects,
3 Investigations of existing cracks in exterior walls, ground floors, and elevations of nearby buildings, as well as the presence of wells and their conditions,
4 Examination of groundwater and the waters of rivers and wells,
5 Investigation of buried and/or underground structures,

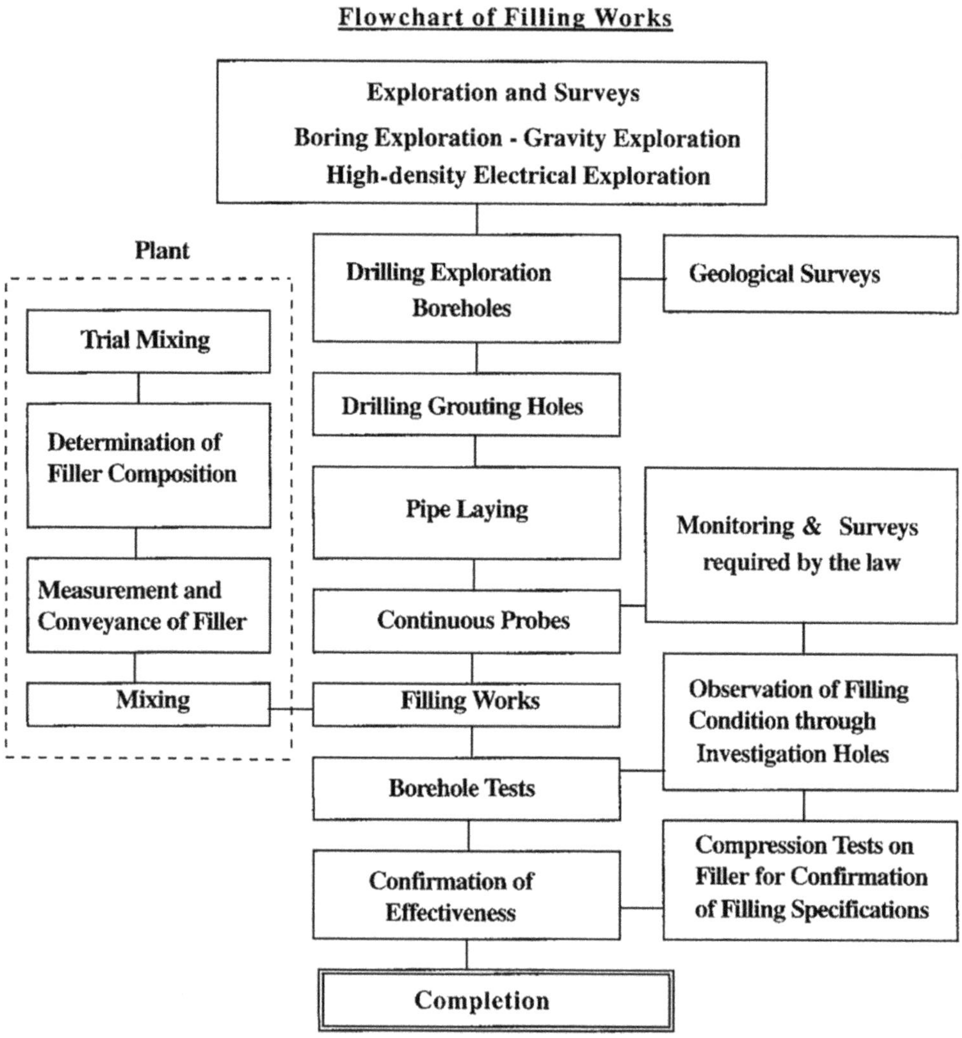

Figure 8.1 A schematic illustration of the integrated cavity filling technique. (From Sakamoto et al., 2005.)

Several exploration techniques, such as borings, gravity exploration, high-density electrical resistivity, and sonar exploration, are utilized depending on the geological and environmental conditions of the sites. Among all these exploration techniques, the most efficient method is the boring technique. This technique is carried out to check the available exploitation maps of abandoned mines, assess the conditions of roofs and pillars, determine the size of openings to be backfilled, and enable real-time monitoring of the backfilling operation, along with sonar exploration utilizing drillholes. Figure 8.2 illustrates the sonar exploration technique utilizing a drillhole and its typical output. This system can be used in immersed underground environments to determine the geometry of cavities along project routes.

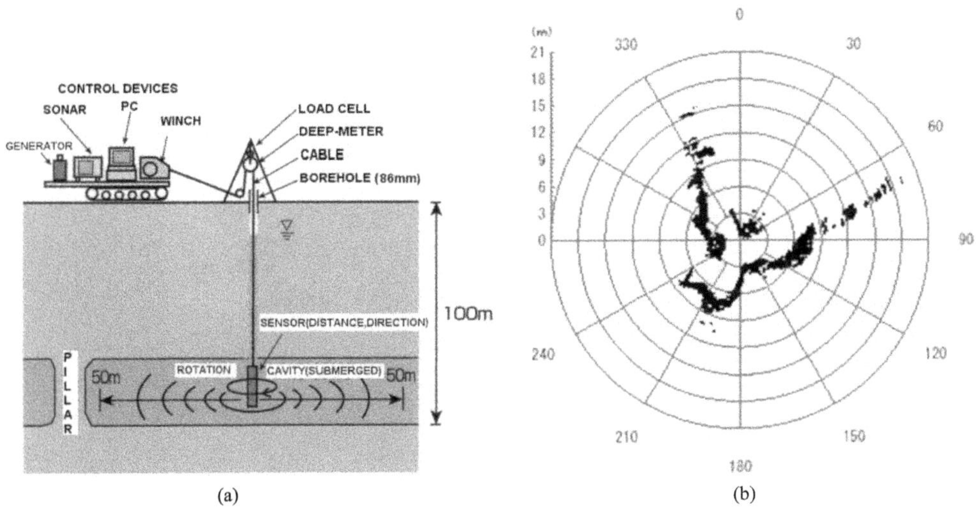

Figure 8.2 (a) Sonar exploration technique and (b) its application to an abandoned lignite mine. (Arranged from Sugiura et al., 2019.)

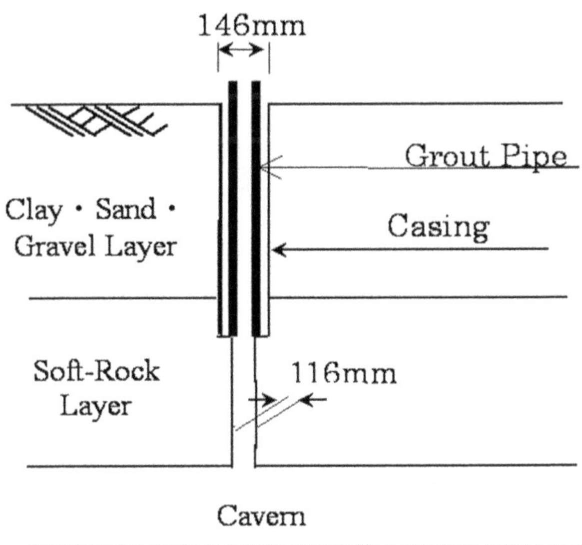

Figure 8.3 An illustration of dimensions of drillholes. (From Sugiura et al., 2019.)

8.1.2 Design of Grouting and Monitoring Drillholes

The next step is to design the layout of drillholes for grouting and piping. Grouting drillholes (Figure 8.3) should be drilled at points about 0.5 m away from exploration drillholes where the existence of caverns is confirmed. Although the spacing between drillholes depends on the characteristics of the backfilling material, an area with a radius of about 20 m can be backfilled through a single grouting drillhole using the backfilling materials employed by the authors.

Figure 8.4 A monitoring probe for grouting pressure. (From Sugiura et al., 2019.)

Therefore, the spacing distance between the grouting boreholes can be chosen as approximately 40 m. However, if there are significant deposits in the caverns or the area is densely urbanized, the spacing of grouting boreholes could be less than 40 m in order to control grouting pressure (Figure 8.4).

The water level in other grouting drillholes is continuously monitored through probes after infusing water into some grouting boreholes. By monitoring the water level, it is possible to confirm whether the drillholes are connected to each other. These probes are also used to assess whether the cavities are connected or independent, and to determine the filling order. Ground settlement or uplift during and after the backfilling operation is also continuously monitored through probes, as shown in Figure 8.4. The data collected is used to check the safety of the construction and to prevent any undesirable effects on the environment and nearby structures. The monitoring of the grouting holes is conducted for essentially the duration of the backfilling period. Nevertheless, some holes are used to check the effectiveness of backfilling through the acoustic emission (AE) technique for about two months following the completion of backfilling and during earthquakes. Additional monitoring of seismic responses of the ground before and after backfilling operations, as well as micro-tremor measurements, is also implemented. These additional monitoring operations are carried out after about one or two months.

8.1.3 Filling Plant

A typical filling plant consists of suppliers of scums, agitators, mixers, and slurry pumps, as illustrated in Figures 8.5–8.8. The capacity of the plant depends on the size of the site, the duration of backfilling, and the available space. For example, a plant with a maximum mixing capacity of 300 m^3 per day should have an area of about 2,000 m^2, considering the filling materials transported to the plant by 10–20 dump trucks with an 11-ton capacity per day, as well as the materials that should be stored for a few days in the plant yard.

Mixers are used to combine the various scums before pumping filling material into cavities. They can dissolve even viscous materials like clay scum in a short time and mix them easily. The maximum mixing capacity ranges between 20 and 550 m^3/day, depending on the plant size. However, the usual mixing capacity ranges between 10 and 300 m^3/day. The amount of backfilling per day also depends on cavity conditions.

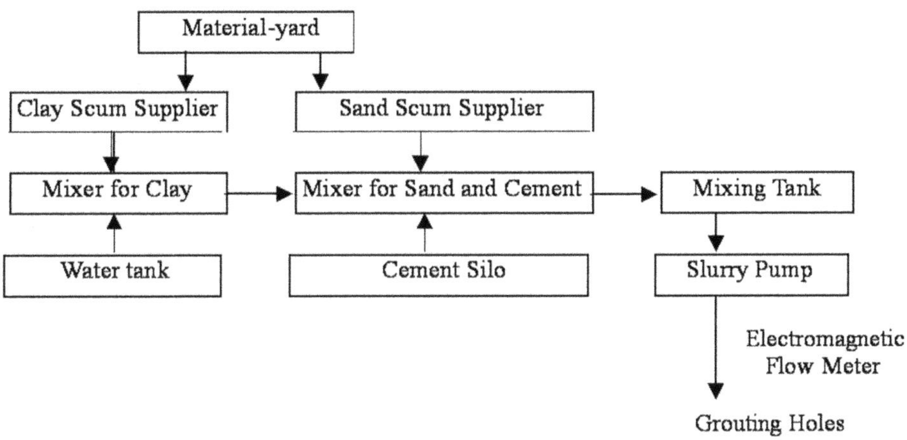

Figure 8.5 Flowchart of the backfilling plant. (From Sugiura et al., 2019.)

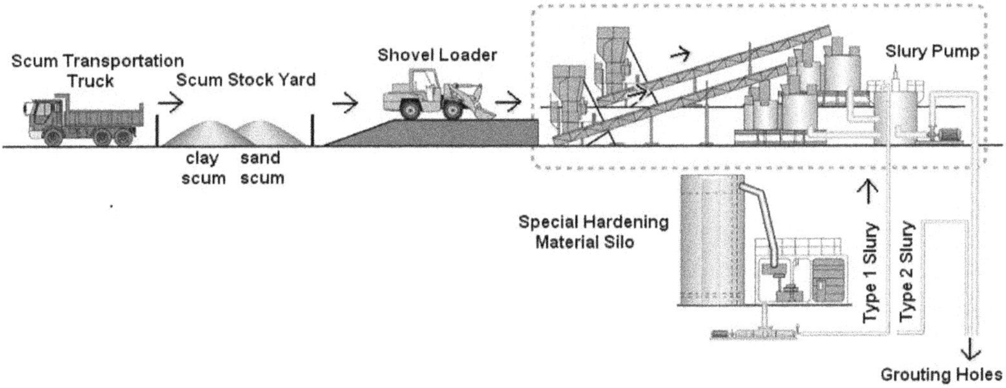

Figure 8.6 An illustration of the backfilling plant and backfilling operation. (From Sugiura et al., 2019.)

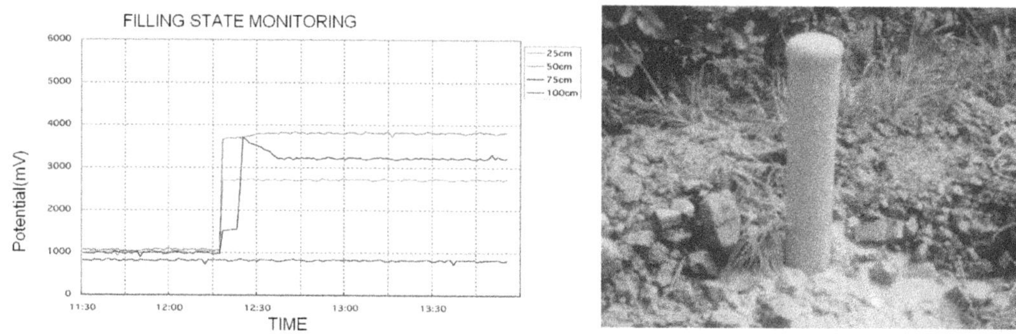

Figure 8.7 Monitoring result for cavity filling state and the inference from the overflow of wells. (From Sugiura et al., 2019.)

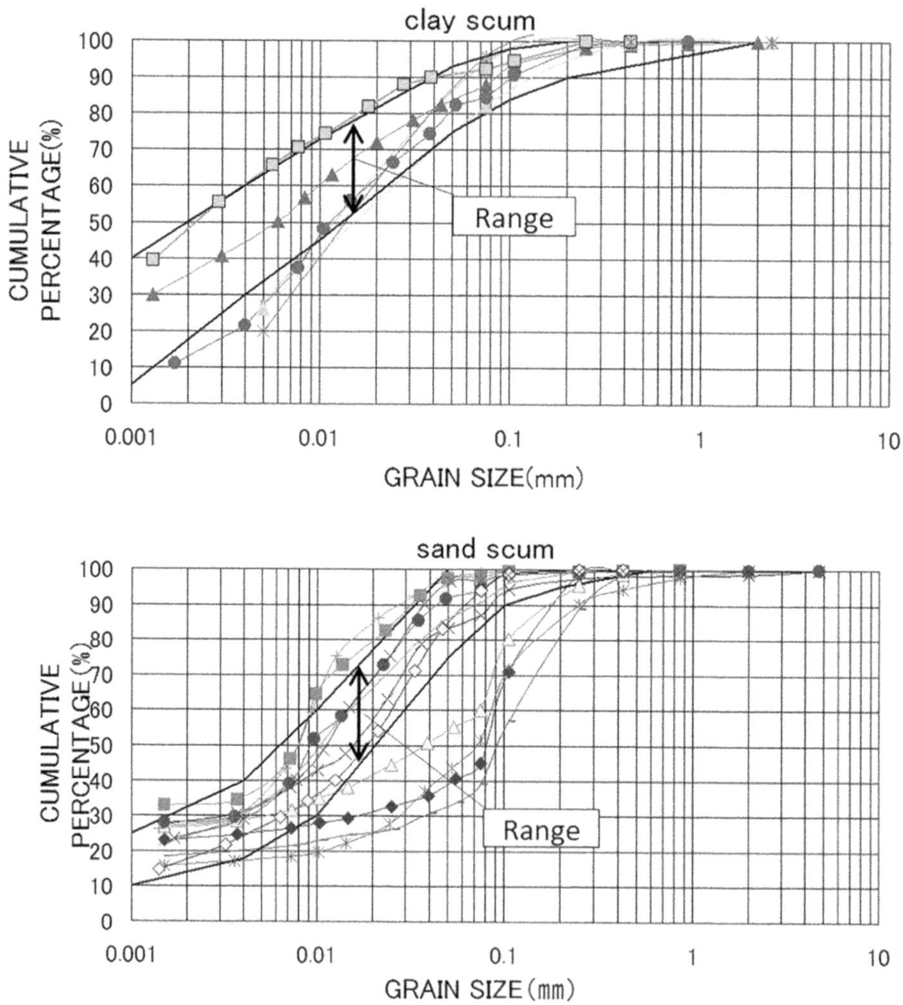

Figure 8.8 Grain size distributions of scums.

Slurry pumps are used to pump the mixed slurry of filling material and have the capability of conveying slurry with a density ranging between 1.3 and 1.5 kg/cm^3, even when grouting drillholes may be about 1,000 m away from the location of the pumps.

8.1.4 Cavity Filling Procedure

Filling materials are pumped into cavities, with priority given to those at a lower elevation. During the filling operation, utmost attention is paid to the pressure rise in the grouting holes to prevent accidents such as spurts of filler or protuberances caused by overfilling. The following monitoring items are used to infer complete filling (Figure 8.7).

a To confirm the slurry flow by filler sensors or tape measurement in connected cavities
b The overflow from grouting holes changes from water to slurry
c The water pressure rise in cavities

8.1.5 Limited Area Backfilling Method

Basic concept of the limited area backfilling method is illustrated in Figures 8.9 and 8.10.

Environmental monitoring involves assessing ground deformations due to filling operations, the migration of solvents in underground water, and groundwater levels. Protuberances caused by overfilling are prevented by monitoring pore water pressure, tape measurements of

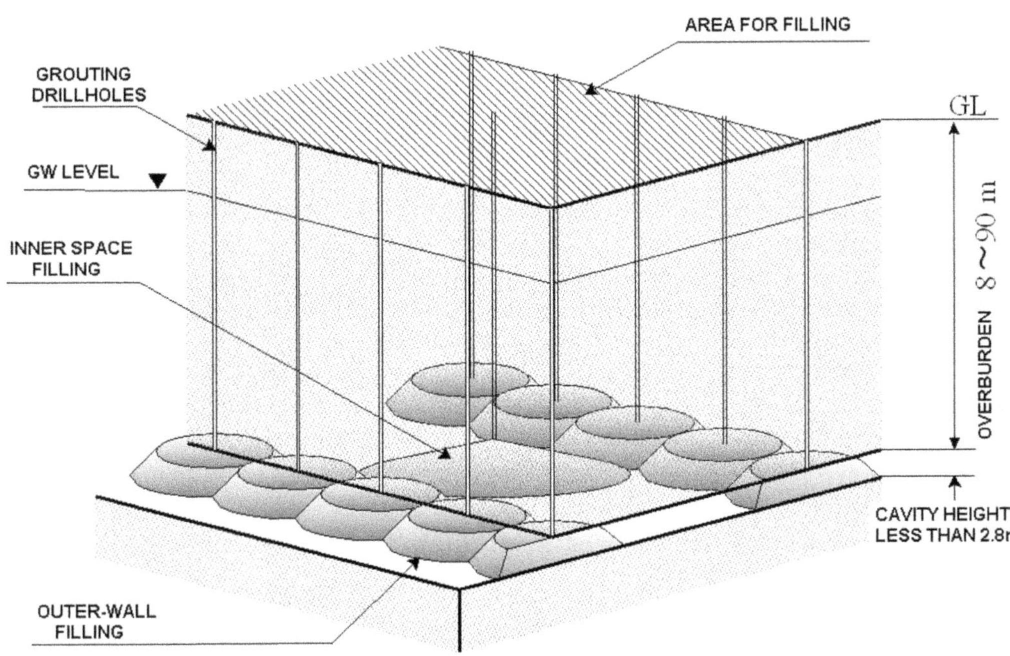

Figure 8.9 Illustration of limited area filling technique (translated and arranged from Sakamoto et al., 2005).

Figure 8.10 Illustration of actual cavity filling operations of limited areas. (From Sakamoto et al., 2005a.)

Table 8.1 Criteria of quality control

A List of Tests	*Criteria of Quality Control*
Flow value	9–14 seconds
Bleeding rate	3% or less than
Uniaxial compressive strength after 28 days	Uniaxial strength cured samples in water should be 50 kPa or more Uniaxial strength from the boring cores should be 20 kPa or more
Analysis of toxic materials	Less than the statutory criteria

the slurry level in grouting boreholes, and observed changes through tilt-meters and crack-meters on the ground.

The spurts of water from grouting boreholes after neutralizing are drained. The spurts of slurry are pumped up with sand pumps and vacuum systems, and the wastes are discarded after the required processing. The pH value of groundwater and solvents are monitored to assess the effects of grouting on the environmental conditions of groundwater. To ensure the quality of grouting and the filling state, several tests and measurements are conducted, as listed in Table 8.1. Besides groundwater monitoring, the uplift of the ground due to backfilling pressure is also monitored using probes, such as those shown in Figure 8.4.

8.2 Examples of Applications of Backfilling Technique

This integrated backfilling technique was first applied in the Tokai Region and has since been extended to other parts of Japan, as shown in Figure 8.11. The locations of applications cover the three major islands of Japan: Honshu, Kyushu, and Hokkaido. However, there are more than 60 implementation sites of this technique in the Tokai region, where it was initiated (Figure 8.12). Among all locations, the applications at two sites will be described in detail.

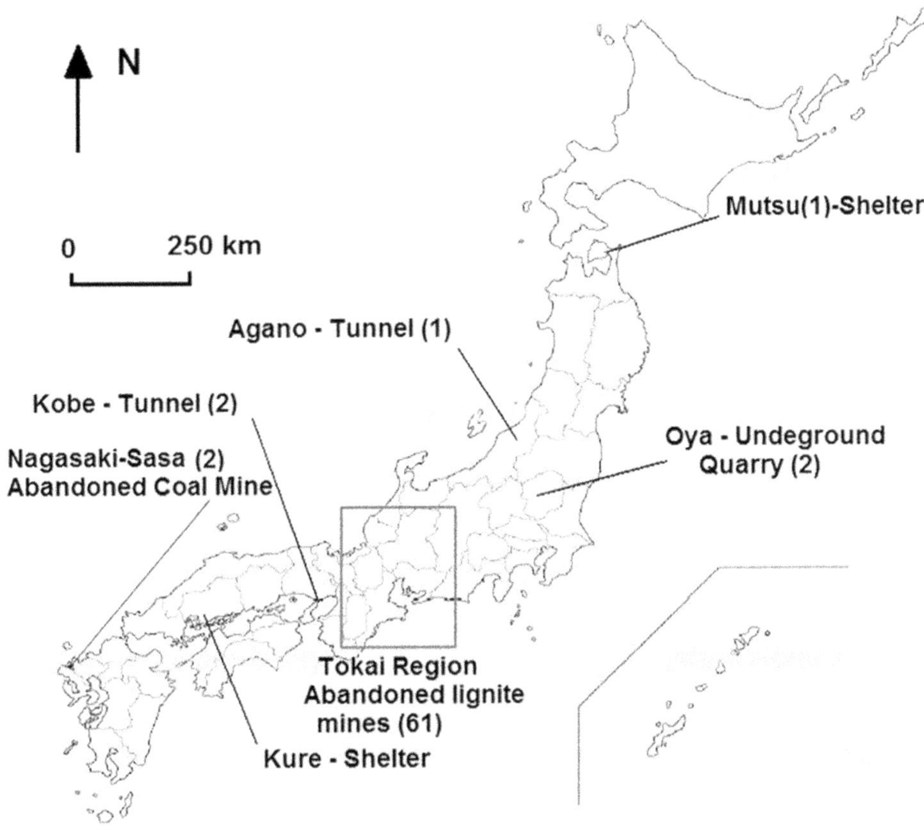

Figure 8.11 Implementation of backfilling in Japan.

8.2.1 *Application to Tokai By-Pass Expressway*

The first example is associated with the cavity filling of abandoned lignite mines in Mitake Town and Kani City at an interchange on the route of the Tokai By-Pass Expressway. A plan view and geological cross sections are shown in Figure 8.13. The N-S section coincides with the alignment of the Nagoya-bound expressway, while the E-W section is aligned with the local roadway. Figure 8.14a shows an artist's illustration of the interchange and backfilling works. In the same figure, a part of the uncovered abandoned lignite mine is also shown (Figure 8.14b). One of the challenging problems associated with cavity filling operations is determining the exact geometry of the cavities to be filled, as pointed out previously. Although some mining plans submitted to the authorities at the time of mining operations are generally available, these documents do not always coincide with actual mining layouts. Furthermore, the geometry may change due to partial degradation and collapses of pillars and roof layers. Figure 8.15 shows the

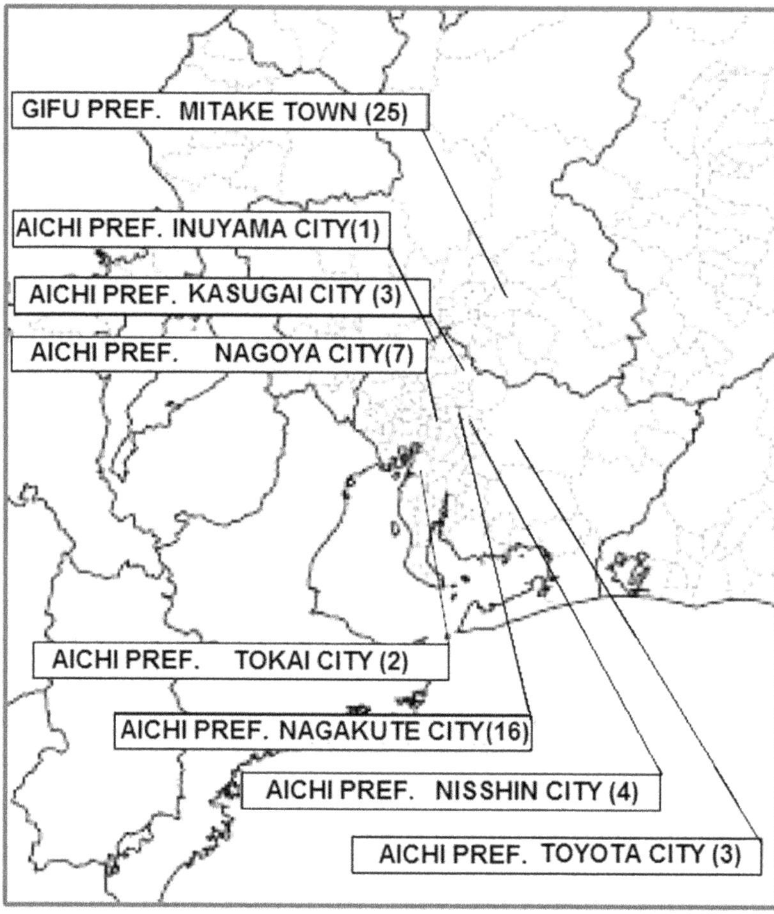

Figure 8.12 A close-up view of backfilling operations in the Tokai region (numbers in parentheses are the number of backfilling at each location.

distribution of cavities and pillars beneath the interchange area of the Tokai By-Pass Expressway, obtained from sonar exploration methods through drillholes, which were carried out as part of an experimental backfilling operation.

As only a limited area needs to be backfilled, the limited area filling technique is employed. The composition of the backfilling materials for outer walls and inner space differs in the short term, but they are almost the same in the long term. Thus, their strengths also differ, as given in Table 8.2. Furthermore, the composition of the filling materials varies depending on the characteristics of the superstructure. Higher strength filling materials are used beneath the foundations of the viaducts of the expressway. In addition to clay and sand scums, the amount and type of hardening materials differ according to the required strength characteristics of the filled space

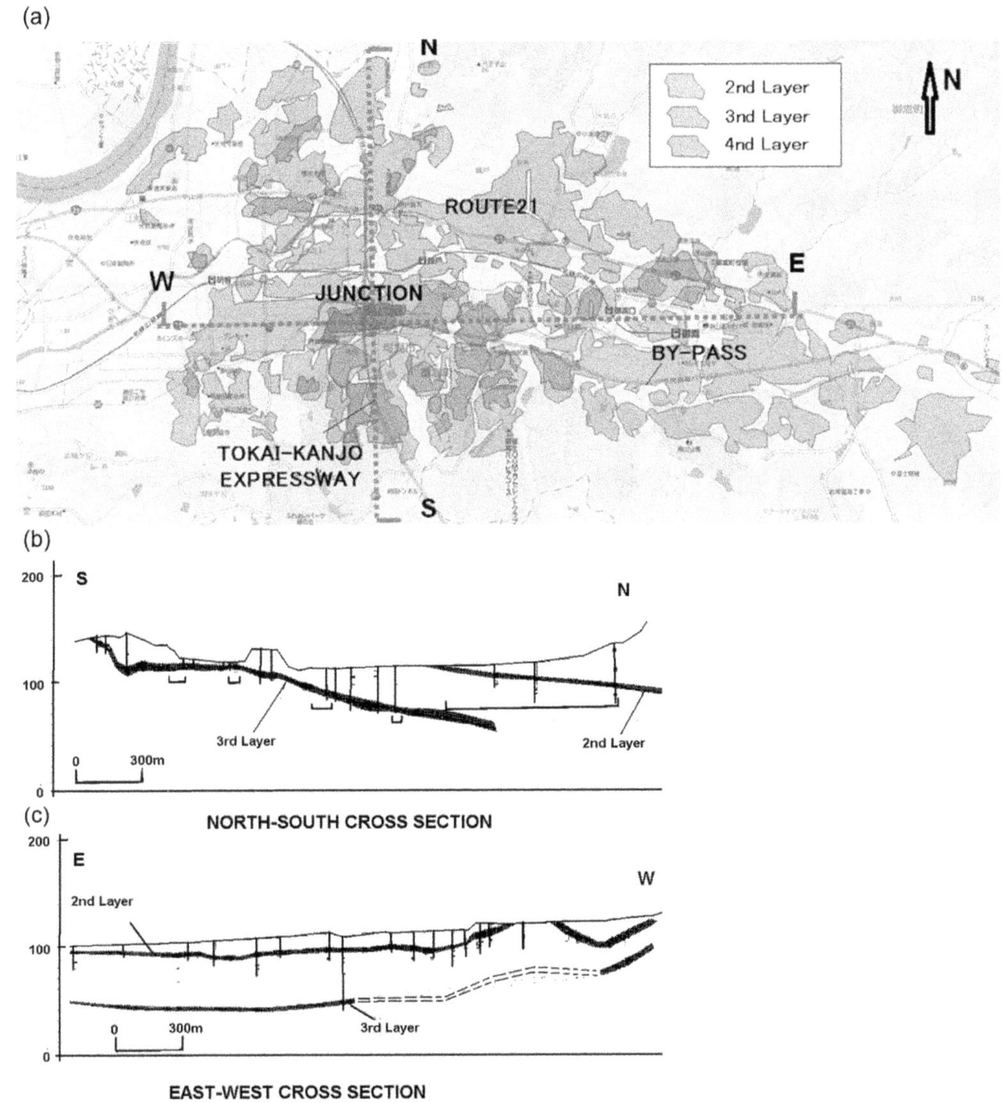

Figure 8.13 (a) Plan view, (b) and (c) geological cross sections, and lignite seams in the vicinity of the interchange of Tokai By-Pass Expressway. (From Sakamoto et al., 2005a.)

in relation to the type of superstructures of the expressway. The diameter of the drillholes for grouting was 100 mm, and the average spacing of the drillholes was 15 m. The volume of the experimental filling operation was 2,155 m^3. However, the total volume of the experimental filling operation was 81,000 m^3. Figure 8.16 shows the areal distribution of outerwall filling and inner space filling at this particular location.

Various monitoring was carried out to measure ground tilting, outerwall gradient, filling state, quality control of filling materials, and groundwater characteristics during the filling work.

(a)

(b)

Figure 8.14 Tokai Expressway passing through abandoned mines in the Mitake–Kani area. (a) Artist's illustration of cavity filling and constructed interchange at Kani City. (b) Exposed abandoned lignite mine. (Arranged from Sugiura and Yamada, 2003; Sakamoto et al., 2005a.)

Figure 8.15 The distribution of pillars and cavities obtained from sonar exploration technique.

Table 8.2 Examples of mix proportions of filling materials (limited area type)

	Low Strength Type		*High Strength Type*	
	Outside	*Inside*	*Outside*	*Inside*
Clay scum (kg)	360	340	360	360
Sand scum (kg)	180	170	180	180
Cement (kg)	90	100	120	150
Silicate grout (kg)	48.8	12.2	48.8	24.4
Water (L)	724	763	715	725

8.2.2 Application to Aichi Exposition Linear Motor Car Railway Project

The second example is associated with the route of the Aichi Exposition linear motor car transportation system in Nagakute Town, Aichi Prefecture, Japan. The railway passes over Nagakute lignite deposits, which were extracted until the 1950s (Figure 8.17). The railway is elevated

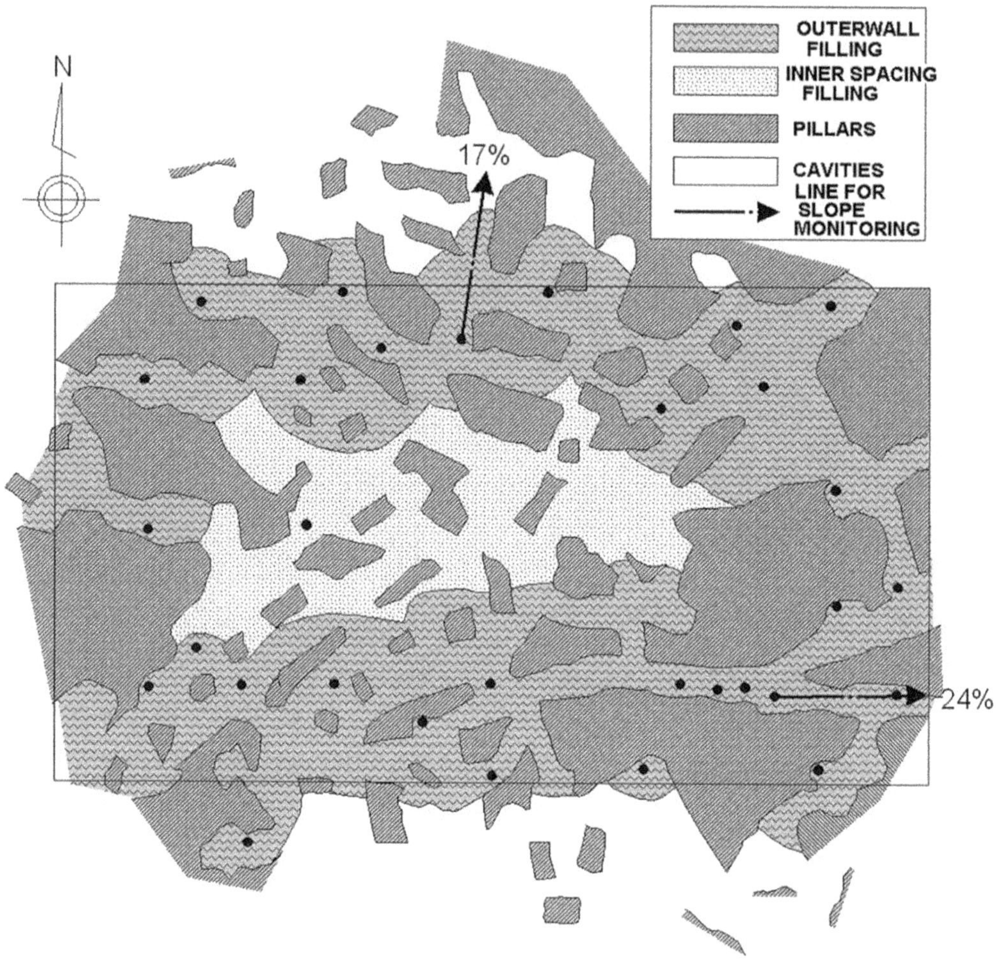

Figure 8.16 Distribution of outerwall and inner space backfilling.

through viaducts, and the lignite seams dip west. The length of the line with cavities beneath it is about 1 km, and the depth of the cavities ranges between 6 and 57 m. The average height of the extracted lignite seam was about 1 m, and the tributary area ratio was 65%.

As the Aichi Exposition linear motor car route passes through a highly urbanized area, the construction space is very limited. The total volume of grouting was 19,000 m^3, and 990 drillholes (with a total length of 35.7 km) were bored. The drillholes for the outer wall and inner space had to be bored from the central pavement of the existing roadway (Figures 8.18 and 8.19). The backfilling materials, denoted as type A and type B, were prepared separately at two plants and mixed just before grouting into the drillholes. Their compositions are given in Table 8.3. The distance between the plants and grouting locations was up to 630 m.

Deviation of drillholes, grouting performance, ground tilting, outerwall gradient, filling state, quality control of filling materials, and groundwater characteristics were monitored. Figure 8.20 shows some of the results.

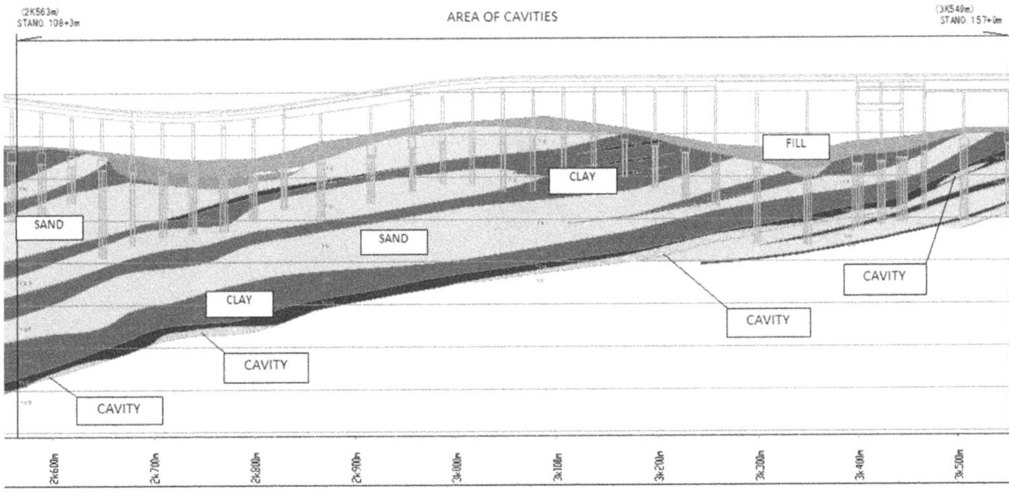

Figure 8.17 Geological cross section and locations of viaducts of the monorail and cavities. (From Sakamoto et al., 2005a.)

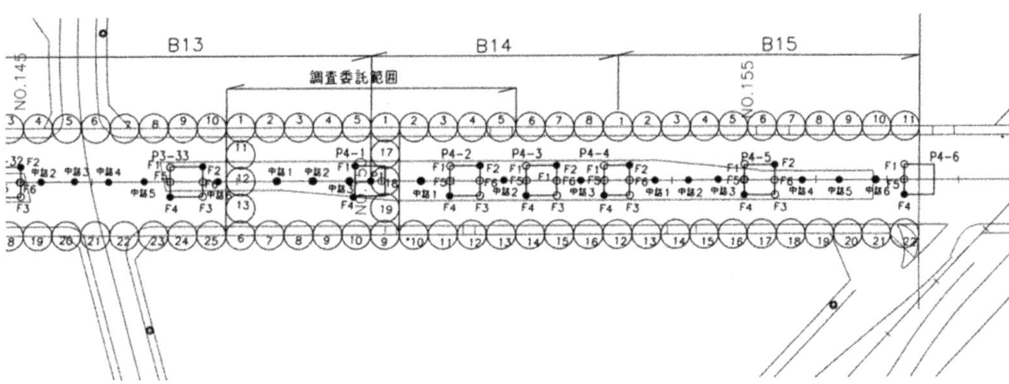

Figure 8.18 Planned layout of drillholes for outerwall and inner space filling works. (From Sakamoto et al., 2005a.)

8.3 Restriction of Urbanization over Abandoned Room and Pillar Mines

One of the remediation measures could be the restriction of urbanization over the abandoned room and pillar mines. However, it is always difficult to implement such measures due to land ownership rights. Furthermore, the old exploitation maps should be preserved and made available for public use. Many times, such maps are either unavailable or inaccurate. It is particularly well known that old exploitation maps do not reflect the actual exploitation geometry. Additionally, it is always risky to check those geometries.

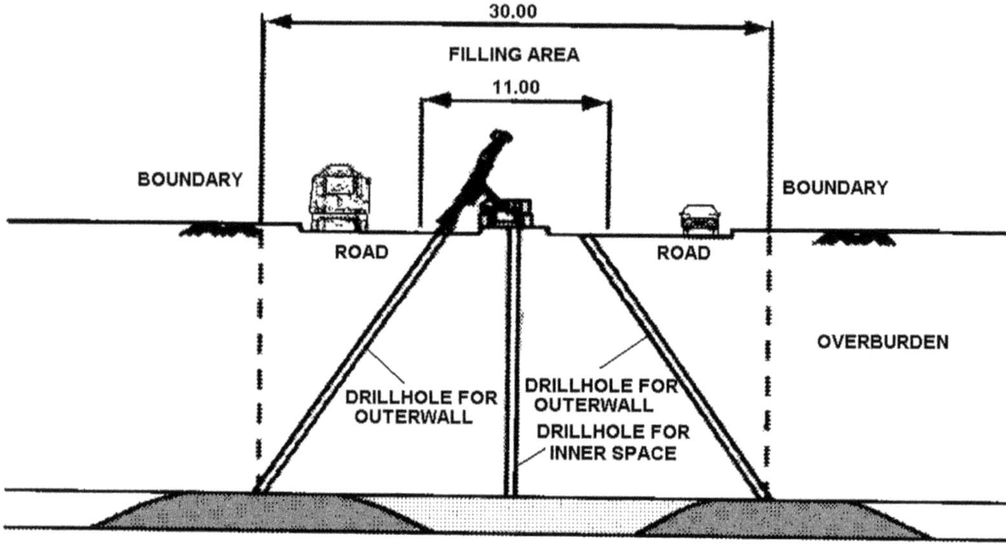

Figure 8.19 An illustration of how drillholes for outerwall and inner space backfilling works are bored. (From Sakamoto et al., 2005a.)

Table 8.3 Composition of filling materials (unit kgf/m^3)

Filling Materials	*Required Strength (kPa)*	*Type A Filling Material*				*Type B Filling Material*		
		Clay Scum	*Sand Scum*	*Special Sodium Silica*	*Water*	*Special Hardening Agent*	*Water*	*Retarder*
Outer walls	400	360	180	48.8	595	120	120	$C \times 0.5\%$
Inner space	400	360	180	24.4	575	150	150	$C \times 0.5\%$

8.4 Inundation (flooding) of Abandoned Room and Pillar Mines

Abandoned room and pillar mines would naturally become inundated when they are located below the groundwater level. However, they may not be fully inundated if they are shallow, as seen in Figures 2.16 and 2.29. Furthermore, seasonal groundwater level variations may occur. It is generally assumed that the degradation of surrounding rocks would be halted when abandoned mines are inundated. However, as surrounding rocks are always vulnerable to degradation, some degree of degradation would still occur. Additionally, most rocks have lower strength when saturated. Therefore, inundation would not be a permanent remediation measure. Groundwater may exert tremendous pressures on the perimeters of underground openings during earthquakes, as seen in Figures 2.8, 2.15–2.18, potentially inducing sinkholes. This issue is particularly important in seismic countries such as Japan and Türkiye and must be taken into account.

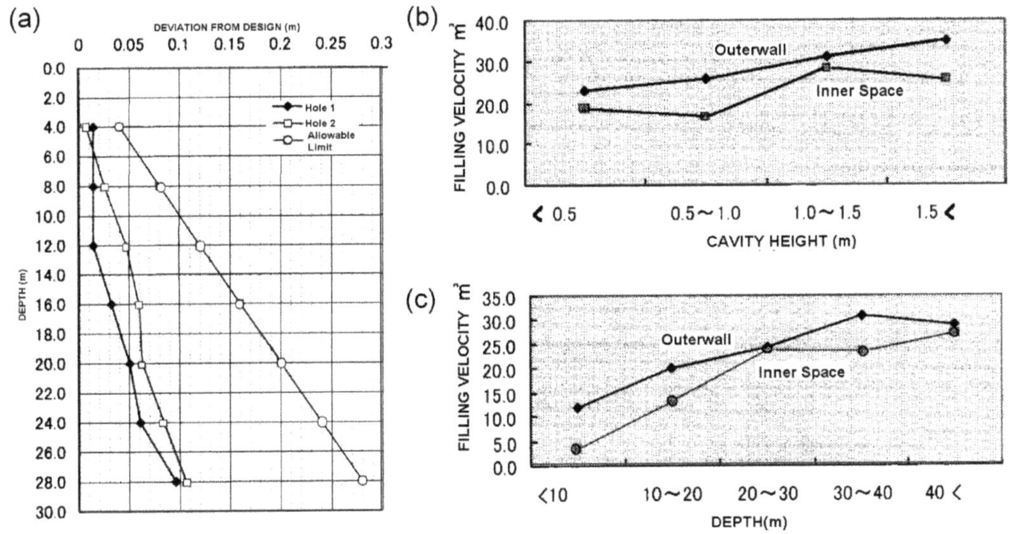

Figure 8.20 (a) Depth vs. deviation from design. (b) Filling velocity vs. cavity height. (c) Filling velocity vs. depth.

8.5 Verification of Backfilling

8.5.1 Backfilling Project of Takenoyama Abandoned Lignite Mine Area

Nagakute and Nissin cities near Nagoya in Aichi Prefecture have been developing residential areas to accommodate newcomers. As there are many abandoned lignite mines in these two cities, the stability of these abandoned mines is a serious concern for developers. Therefore, backfilling is now commonly adopted to ensure the safety of areas against sinkhole formations and subsidence. The authors were asked how to verify the effect of backfilling of abandoned mines in-situ. On the basis of the outcomes of experiments under well-controlled laboratory conditions, it is expected that if the backfilling is effective, it should stop the further degradation of the rocks surrounding the abandoned mines, reducing the velocity of groundwater seepage and the deviatoric stress due to the confinement effect in view of Kaiser's effect. Therefore, the AE occurrences before and after backfilling would be entirely different. If there are no technical problems with the devices for monitoring AEs, the AE counts would drastically decrease following the completion of backfilling, as illustrated in Figure 8.21. Three boreholes were drilled in the area of abandoned lignite mines in the Takenoyama region of Nisshin City before the commencement of the backfilling operation. Wave-guides reaching the cavities were equipped with AE sensors, and measurements were carried out for a period of six months. Figure 8.22 shows the monitoring results for borehole No. 3. As noted from the figure, the monitoring results are in accordance with the anticipated AE responses before, during, and after filling. AE occurrences drastically disappear following the completion of the backfilling operation.

8.5.2 Backfilling Project of Karstic Caves below Gushikawa Castle Remains

Two karstic caves exist beneath the remains of Gushikawa Castle in Itoman City on Okinawa Island, Japan (Figure 8.23). An extensive monitoring program was undertaken to observe the

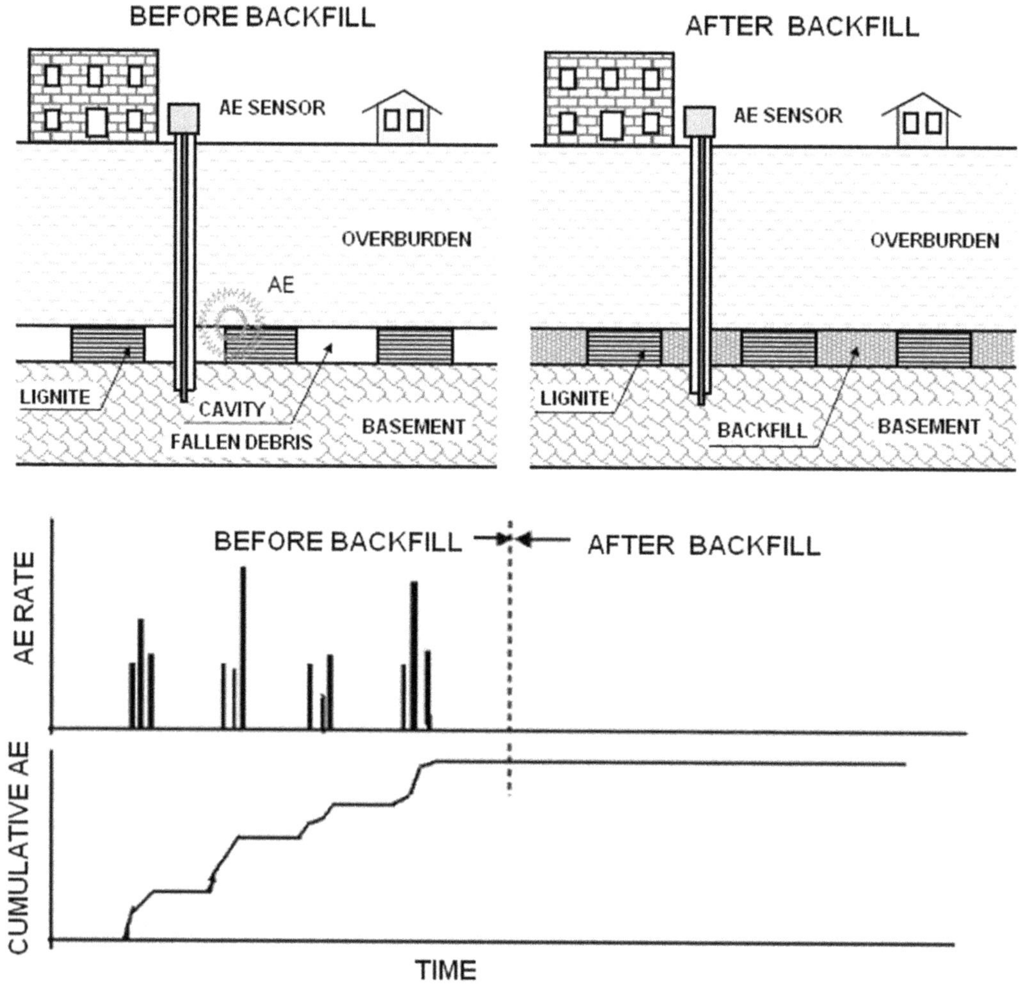

Figure 8.21 Conceptual model for the verification of the effect of backfilling from acoustic emission response before, during, and after backfilling. (From Aydan et al., 2013.)

performance of the caves before and after filling. The monitoring program involved measuring the displacement response of a continuous crack passing through Cave A using six gap gauges, the tilting of pillars and the ground surface using inclinometers, longitudinal and transverse strains, and AEs from the roofs of Caves A and B, as well as temperature, humidity, and rainfall at the ground surface in the Gushikawa Castle remains.

The instruments were installed about one month before the backfilling construction began. Particularly, the AE response is of great significance for observing the effect of filling on the caves before and after filling. As cavities are subjected to daily variations in temperature, humidity, tidal loading, and rainfall, they experience gradual degradation. As a result, some AE events associated with this gradual degradation take place. Filling the cavities should terminate this degradation process, leading to a subsiding of AE events over time. Figure 8.24 shows the acoustic emission responses of Caves A and B. As expected, the number of acoustic emission

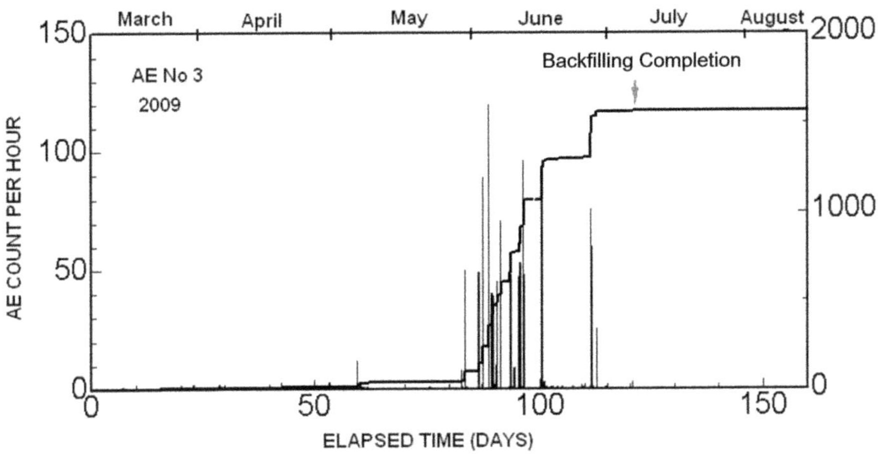

Figure 8.22 Acoustic emission responses of AE No.3 station during the period between March 2009 and August 2009. (From Aydan et al., 2013.)

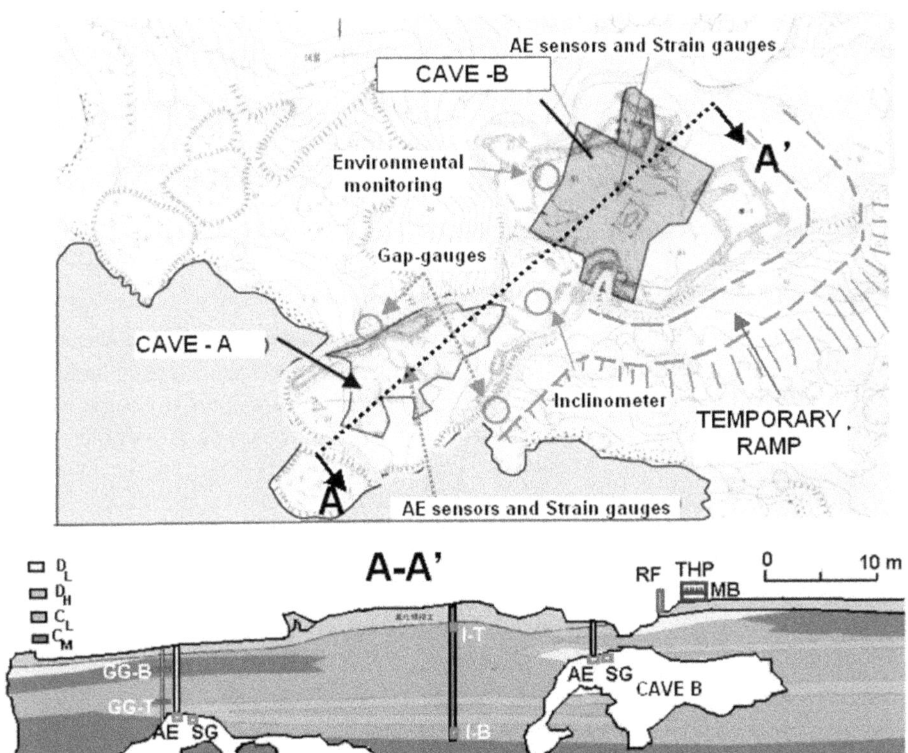

Figure 8.23 Layout of instrumentation. (From Aydan et al., 2013.)

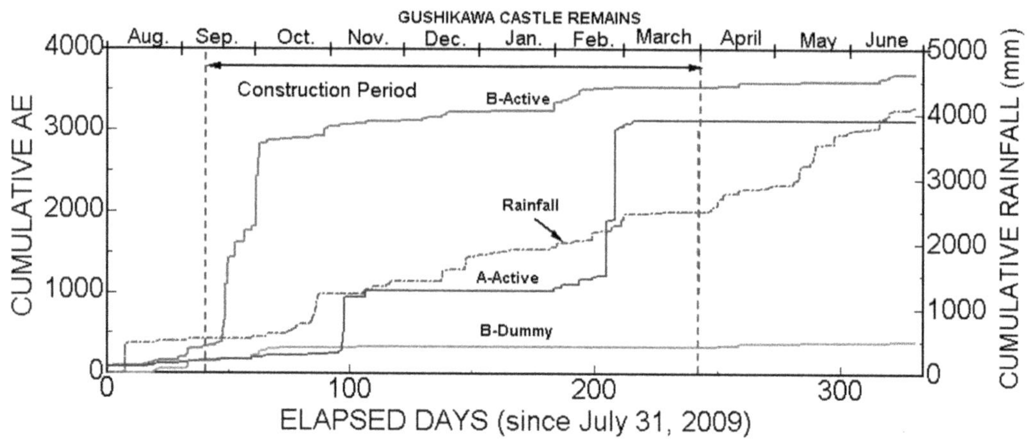

Figure 8.24 Acoustic emission responses of caves during August 2009–June 2010. (From Aydan et al., 2013.)

events drastically decreased following the completion of construction. This trend continued for a considerable period after construction, verifying that the backfilling was effective.

8.6 Areal Spreading of Backfilling Material and Its Modeling

As mentioned previously, backfilling material can be either granular or slurry type, which solidifies over time. A brief outline of the areal spreading of backfill materials and its modeling is described herein.

8.6.1 Granular Backfilling Material

Granular backfilling material used in backfilling abandoned mines or sinkholes can consist of gravelly or sandy soil or crushed rock. Under gravity, they tend to form a triangular and/or conical shape, and the inclination of the sides of the triangular body is equal to the repose angle of the backfilled soil. Depending on the shape and angularity of the granular particles, the repose angle generally ranges between 30° and 45°. Unless vibration is imposed, the lateral spreading is governed by the repose angle of the backfill material. As shown in Figure 4.25, granular material laterally spreads in proportion to the intensity of shaking. Figures 2.25 and 2.26 illustrate that the granular backfill is displaced during earthquakes, causing the granular material to spread into the surrounding space, leading to the subsidence of the backfilled shafts and the formation of sinkholes. This situation can be interpreted as equivalent to the dynamic repose angle. Figure 8.25 shows an example of backfilling with granular material in a sinkhole accessible from an abandoned underground lignite room and pillar mine in Mitake.

8.6.2 Slurry Type Backfilling Material

Slurry-type backfilling materials are generally commercial by-products combined with a cementitious paste as a binding agent. Initially, they behave like a fluid and solidify over time

Figure 8.25 A view of granular backfill material in an abandoned lignite mine.

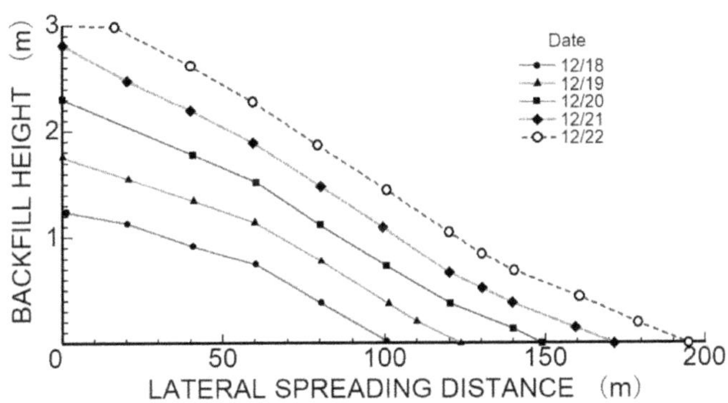

Figure 8.26 Lateral spreading of slurry-type backfilling material as a function of time.

(Aydan, 1994b, 1995b). This type of backfilling material spreads laterally over a large distance. Figure 8.26 shows an example of the lateral spreading of slurry-type backfill material as a function of time. The lateral spread radius is generally greater than 200 m for a backfilling height of 3 m. The average gradient (β) of the hardened state of slurry backfill material ranges between 1% and 2%. The laterally spreading slurry-type backfilling material has a parabolic shape, which can be approximated by the following relation for a given time step.

$$h(x) = h_o\sqrt{1-\frac{x}{L}} \text{ or } h(x) = \sqrt{h_o - \beta \cdot x}$$

where h_o, x, and L are the backfill height, distance from the backfilling location, and maximum spreading distance (or radius). The major issue is always how to estimate the backfill height and spreading distance as a function of time. The following analytical treatment of the problem would provide the necessary estimations. It should be noted that the shape starts to deviate from the parabolic shape as the backfill material hardens, as seen in Figure 8.26.

8.6.2.1 *Fundamental equations for lateral spreading of slurry type backfilling material*

The mass conservation law is given in the following form:

$$\frac{\partial \rho}{\partial t} = -\frac{\partial q_i}{\partial x_i} \pm g \tag{8.1}$$

where ρ, q_i, and g are density, mass flow vector, and mass generation/depletion, respectively. $\partial() / \partial x_i$ is the directional derivative operator. The mass flow vector can be related to density and velocity vector in the following form:

$$q_i = \rho x_i \tag{8.2}$$

The laterally spreading backfilling slurry should also obey the equation of motion given in the following form:

$$\frac{d^2 u_i}{dt^2} = \frac{\partial \sigma_{ij}}{\partial x_j} + b_i \tag{8.3}$$

Slurry-type backfilling material can be considered a visco-elastic or elasto-visco-plastic material. If it is modeled as a visco-elastic material, the constitutive equation takes the following form, known as the Kelvin–Voigt material:

$$\sigma_{ij} = D_{ijkl}\varepsilon_{kl} + C_{ijkl}\dot{\varepsilon}_{kl} \tag{8.4}$$

where

$$\varepsilon_{kl} = \frac{1}{2}\left(\frac{\partial u_k}{\partial x_l} + \frac{\partial u_l}{\partial x_k}\right); \quad \dot{\varepsilon}_{kl} = \frac{\partial \varepsilon_{kl}}{\partial t} \tag{8.5}$$

Navier–Stokes type constitutive relations, which are given below, may also be considered a special form of Eq. (8.4) with the following replacements:

$$\sigma_{ij} = -p\delta_{ij} + \tau_{ij} \text{ or } \sigma_{ij} = -p\delta_{ij} + C_{ijkl}\dot{\varepsilon}_{kl} \tag{8.6}$$

As slurry-type backfilling material is self-hardening, the equations above should be incremental and consider the variation of parameters of the constitutive relations with time.

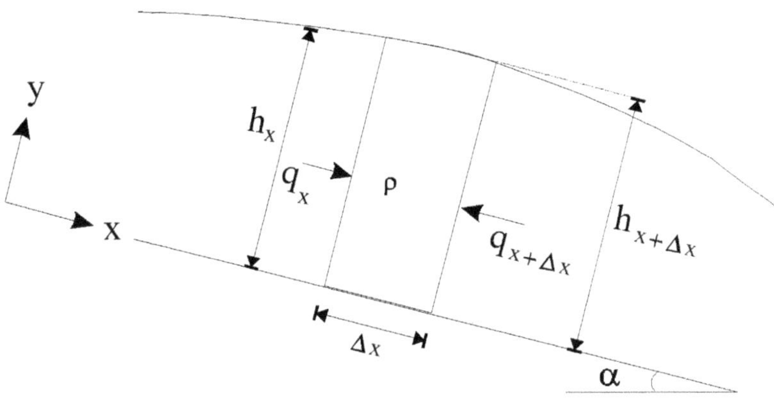

Figure 8.27 Infinitely small element for mass conservation law in 2D.

8.6.2.2 *Simplified form of governing equations*

The governing equations can be simplified for two-dimensional (2D) or two-dimensional axisymmetric conditions (2DA).

8.6.2.2.1 TWO-DIMENSION (2D)

The mass conservation law in 2D (Figure 8.27) can be written as follows:

$$\frac{\partial(\rho h)}{\partial t} = -\frac{\partial q}{\partial x} \tag{8.7}$$

If density (ρ) is constant and the flow vector ($q = \rho\bar{v}h$), the mass conservation law takes the following form:

$$\frac{\partial h}{\partial t} = -\frac{\partial(\bar{v}h)}{\partial x} \tag{8.8}$$

As for the equation of motion (Figure 8.28), it can be written as follows:
x-direction

$$\frac{\partial \tau}{\partial y} = \frac{\partial p}{\partial x} - \rho g \sin\alpha \tag{8.9}$$

y-direction

$$\frac{\partial p}{\partial y} = \rho g \cos\alpha \tag{8.10}$$

where τ, p, ρ, and g are shear stress, pressure, density, and gravitational constant, respectively.

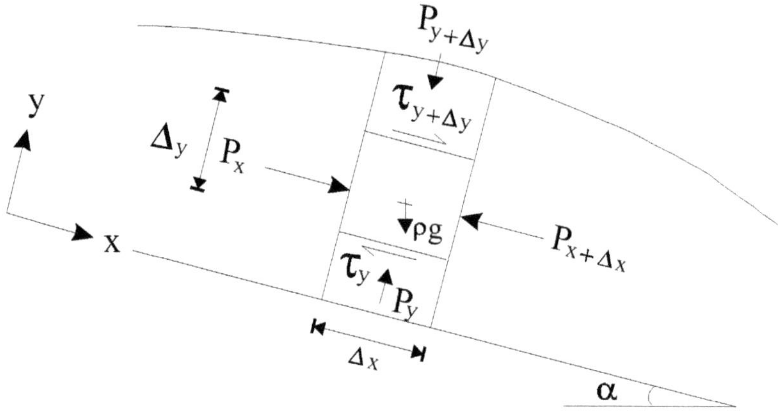

Figure 8.28 Infinitely small element for the equation of motion in 2D.

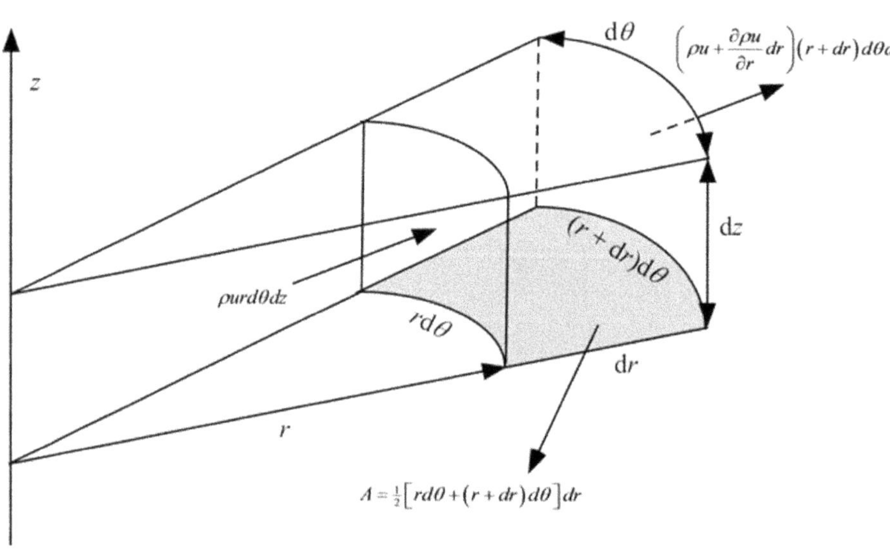

Figure 8.29 Infinitely small element for mass conservation law in 2DA.

8.6.2.2.2 AXISYMMETRIC (PSEUDO 3D) SITUATION

The mass conservation law for a 2D-axisymmetric situation (Figure 8.29) takes the following form:

$$\frac{\partial(\rho h)}{\partial t}=-\frac{1}{r}\frac{\partial}{\partial r}(rq) \tag{8.11}$$

where r is the distance from the central axis.

Similarly, the equation of motion for the axisymmetric situation can be written as follows:

$$\frac{d\tau}{dy}=\frac{dp}{dr}-\rho g\sin\alpha \tag{8.12}$$

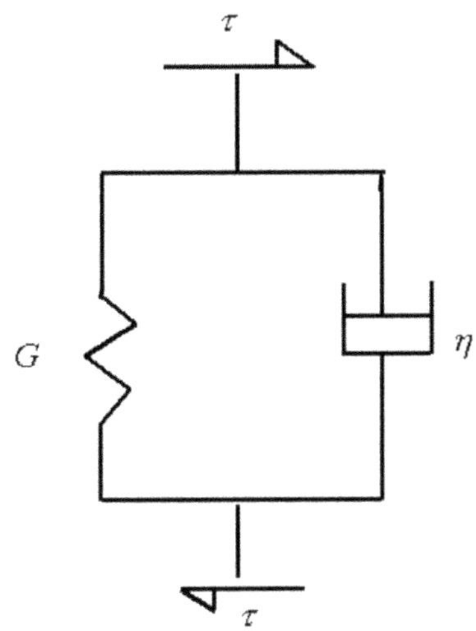

Figure 8.30 Kelvin–Voigt law.

$$\frac{dp}{dy} = \rho g \cos\alpha \tag{8.13}$$

8.6.2.2.3 CONSTITUTIVE MODEL

If volumetric strain is negligible, the Kelvin–Voigt law can be written as follows (Figure 8.30):

$$\tau = G\gamma + \eta\dot{\gamma} \tag{8.14}$$

where

$$\gamma = \frac{\partial u}{\partial y}, \dot{\gamma} = \frac{\partial \dot{u}}{\partial y}, v = \dot{u} = \frac{\partial u}{\partial t} \tag{8.15}$$

8.6.2.2.4 COMBINING FUNDAMENTAL EQUATIONS

Integrating Eq. (8.13) with respect to y, together with the boundary condition $p = 0$ at $y = h$, yields

$$p = \rho g \cos\alpha\left(h(x) - y\right) \tag{8.16}$$

Then, its derivative with respect to x can be written as follows:

$$\frac{\partial p}{\partial x} = \rho g \cos\alpha \frac{\partial h}{\partial x} \tag{8.17}$$

Inserting (8.17) into (8.12) yields

$$\frac{\partial \tau}{\partial y} = \rho g \left(\cos\alpha \frac{\partial h}{\partial x} - \sin\alpha \right) \tag{8.18}$$

Integrating (8.18) with respect to y, together with $\tau = 0$ at $y = h$ gives

$$\tau = \rho g \cos\alpha \left(\tan\alpha - \frac{\partial h}{\partial x} \right)(h - y) \tag{8.19}$$

Using Eq. (8.19) in Eq. (8.14) and solving the resulting differential equation yields the average velocity of laterally spreading soil at a given cross section at x as follows:

$$\bar{v} = \frac{e^{-\frac{G}{\eta}t}}{\eta} \left[\rho g \cos\alpha \left(\tan\alpha - \frac{\partial h}{\partial x} \right) \frac{h^2}{3} \right] \tag{8.20}$$

Inserting (8.20) into (8.13) gives

$$e^{\frac{G}{\eta}t} \frac{\partial h}{\partial t} = -\frac{\partial}{\partial x} \left(\frac{\rho g \sin\alpha h^3}{3\eta} \right) + \frac{\partial}{\partial x} \left(\frac{\rho g \cos\alpha h^3}{3\eta} \frac{\partial h}{\partial x} \right) \tag{8.21}$$

If the inclination α is very small, the effect of the first term on the RHS is negligible. Thus, we have

$$e^{\frac{G}{\eta}t} \frac{\partial h}{\partial t} = \frac{\partial}{\partial x} \left(\frac{\rho g \cos\alpha h^3}{3\eta} \frac{\partial h}{\partial x} \right) \tag{8.22}$$

Let us denote the following

$$A = \frac{\rho g \cos\alpha h^3}{3\eta} \tag{8.23}$$

Thus, the equation above may be rewritten as follows:

$$e^{\frac{G}{\eta}t} \frac{\partial h}{\partial t} = \frac{\partial}{\partial x} \left(A \frac{\partial h}{\partial x} \right) \tag{8.24}$$

It should be noted that this partial differential equation is geometrically non-linear. In this study, A will be treated as a constant during a small-time increment in order to linearize the partial differential Eq. (8.24).

8.6.2.2.5 HYBRID FINITE ELEMENT FORMULATION

Hybrid FEM utilizes closed-form solutions for the y-direction (Eq. 9.20 and 8.21), while the FEM solution is implemented for the x-direction. The specific formulation is as follows:

8.6.2.2.5.1 Weak Form

Taking a variation on δh, the integral form of Eq. (8.24) may be written as

$$\int \delta h e^{\frac{G}{\eta}t} \frac{\partial h}{\partial t} dx = \int \delta h \frac{\partial}{\partial x}\left(A\frac{\partial h}{\partial x} \right) dx \tag{8.25}$$

Integrating this equation by parts gives

$$\int \delta h e^{\frac{G}{\eta}t} \frac{\partial h}{\partial t} dx + \int \frac{\partial \delta h}{\partial x} A \frac{\partial h}{\partial x} dx = \delta h \left(A\frac{\partial h}{\partial x} \right)\Bigg|_{x=a}^{x=b} \tag{8.26}$$

Equation (8.26) is called the weak form of Eq. (8.24).

8.6.2.2.5.2 Discretization

Equation (8.26), which holds for the whole domain, must also hold for a typical element if the domain is discretized into n elements. Let us assume that $h(x)$ can be interpolated element-wise by the following function:

$$h = [N]\{H\}; \frac{\partial h}{\partial x} = \frac{\partial}{\partial x}[N]\{H\} = [B]\{H\} \tag{8.27}$$

where

$$[N] = [N_i, N_j];\ N_i = \frac{x_j - x}{L},\ N_j = \frac{x - x_i}{L},$$

$$[B] = [B_i, B_j];\ B_i = -\frac{1}{L},\ B_j = \frac{1}{L},\ L = x_j - x_i$$

Using the above relations in Eq. (8.27) yields

$$[M]_e\{\dot{H}\}_e + [K]_e\{H\}_e = \{F\}_e \tag{8.28}$$

where

$$[M]_e = e^{\frac{G}{\eta}t} \int_{x_{il}}^{x_j} [N][N] dx,\ [K]_e = A\int_{x_{il}}^{x_j} [B][B] dx,\ \{F\}_e = [\bar{N}]^T \left(A\frac{\partial h}{\partial x} \right)\Bigg|_{x_i}^{x_j}$$

Using Taylor expansion of $\{H\}$ with respect to time step t_n, Eq. (8.28) results in the following simultaneous equation system:

$$[M^*]\{H\}_{n+1} = \{F\}_{n+1} \tag{8.29}$$

where

$$\left[M^*\right]=\frac{1}{\Delta t}\left[M\right];\left\{F\right\}_{n+1}=\left\{F\right\}_n+\left(\frac{1}{\Delta t}\left[M\right]-\left[K\right]\right)\left\{H\right\}_n$$

The boundary conditions are implemented in the following forms with $n_a = -1$ and $n_b = 1$:

At $x=a$,

$$v_a = \frac{\partial h}{\partial t} n_a = -\frac{\partial h}{\partial t} \approx -\frac{h_{n+1}^{x=a} - h_n^{x=a}}{\Delta t} \Rightarrow h_{n+1}^{x=a} = h_n^{x=a} - v_a \Delta t$$

At $x = b$,

$$v_b = \frac{\partial h}{\partial t} n_b = \frac{\partial h}{\partial t} \approx \frac{h_{n+1}^{x=b} - h_n^{x=b}}{\Delta t} \Rightarrow h_{n+1}^{x=b} = h_n^{x=b} + v_b \Delta t$$

As observed from the relations given above, the boundary conditions are updated at each time step.

Since the 2D-axisymmetric formulation can be done in the same manner, the specific formulations will not be provided.

Treatment of large deformation can be handled using the updated-Lagrangian method.

8.6.2.3 Specific examples

The presented formulation is utilized to simulate the spreading behavior of slurry-type backfill material in a channel that is 5,400 mm long, 300 mm high, and 340 mm wide, as shown in Figure 8.31. The backfill material was supplied at one end of the channel at a constant flow rate,

Figure 8.31 Views of experimental setup.

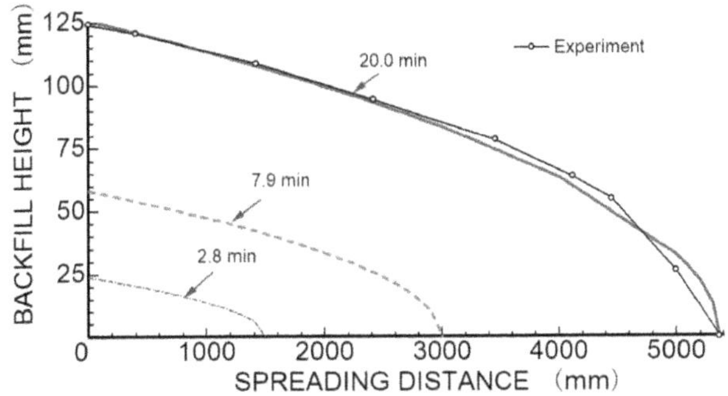

Figure 8.32 Comparison of computed profiles with experimental profiles.

and the flow of the backfill material was monitored at nine points. Figure 8.32 shows the simulated profile of the backfill material at different time steps. As noted from the figure, the profiles have a parabolic shape. The computed profiles are compared with the measured profiles, and the computed profiles closely match the measured profiles.

Chapter 9

Monitoring of Abandoned Room and Pillar Mines

Any physical phenomenon in nature involves several directly measurable or indirectly measurable parameters. These parameters include displacement, velocity, acceleration, strain, force, stress, temperature, water pressure, concentration, magnetic field, and electrical potential. Other indirect parameters may include acoustic emissions (AEs), micro-tremors, electrical potential, and gravitational changes. These parameters can be measured directly by attaching sensors or remotely using lasers, optical lights, gravimetry, and magnetic sensors. The topic is broad and involves interdisciplinary knowledge and background. Therefore, the monitoring of abandoned room and pillar mines will be limited to those techniques employed by the author and his research group. Specifically, the monitoring aspects will cover the following items (Aydan, 2008):

1 Monitoring of ground surface deformation and subsidence
2 Ground vibrations induced by earthquakes or other dynamic events
3 Multi-parameter measurement system for real-time long-term stability monitoring
4 Micro-tremor measurements

The state of the art, along with specific applications, will be presented. Some materials also include sections written by the author for the final report of the Mine Closure Commission of ISRM, chaired by Dr. C. Didier (Didier et al., 2008, 2009).

9.1 Ground Movement Monitoring

Monitoring of ground movements resulting from subsidence in areas worked by long-wall mining or caving methods, as well as sinkholes and subsidence due to creep degradation of surrounding rock in areas of room and pillar mining, may, in some contexts, be required for environmental safety and urban development. The monitoring of ground movements can be accomplished using both direct and indirect techniques. Direct techniques may involve inclinometers, extensometers (mechanical or fiberoptic), geodetic measurements, or GPS techniques. Indirect methods may include aerial photogrammetric techniques or interferometric synthetic aperture radar (InSAR) methods. The most important aspect of monitoring is the long-term reliability and repeatability of measurements, as these measurements must be carried out over extended periods, i.e., years. When ground movements are accelerated, it may be necessary to shorten the measurement intervals to time units such as hours or minutes, and a combination of several techniques may also be required. A brief outline of the major techniques is presented in the following paragraphs.

DOI: 10.1201/9781003517214-9

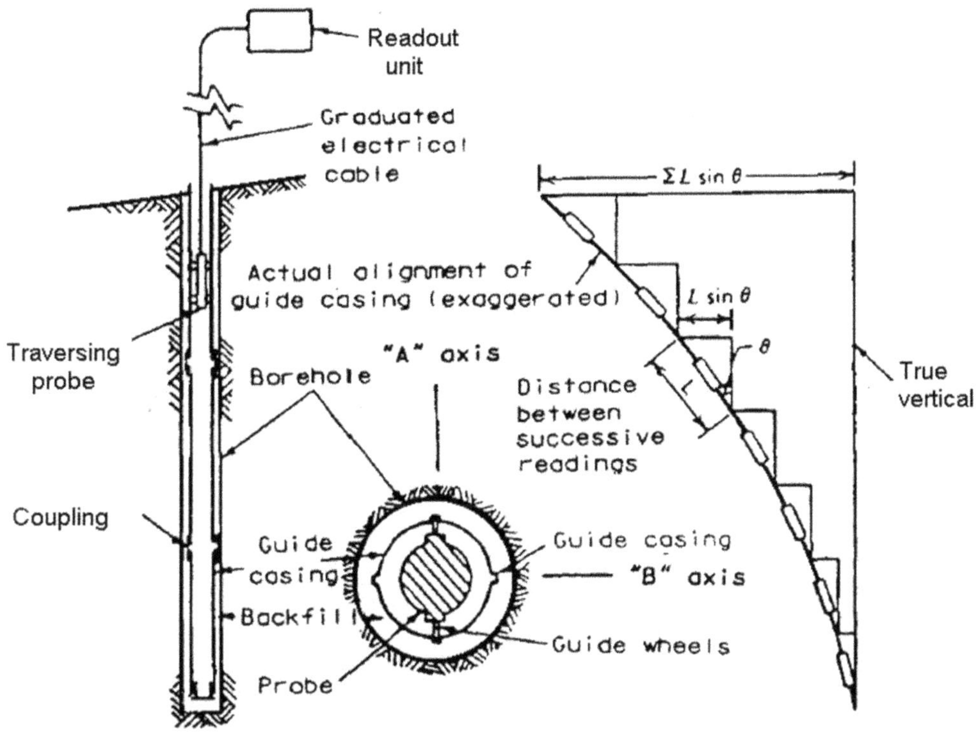

Figure 9.1 Illustration of ground movement measurement by an inclinometer. (From ISRM, 1977.)

9.1.1 *Direct Measuring Techniques*

9.1.1.1 *Inclinometers*

An inclinometer primarily measures the deviation of inclination of each segment of the casing (Figure 9.1). Ground movements are converted to displacements based on the measured deviations and the location of each casing segment. The device consists of an inclinometer casing, a traversing probe, and a readout unit. The casing is permanently installed in a borehole. Important features of the casing include its diameter, coupling mechanism, groove dimensions, straightness, and strength.

The traversing inclinometer probe is the standard device for surveying the casing. Recently, some traversing probes have been equipped with fiberoptic sensors. The traversing probe obtains a complete profile by being drawn from the bottom to the top of the casing. The first survey establishes the initial profile of the casing, while subsequent surveys reveal changes in the profile if movement has occurred.

9.1.1.2 *Extensometers*

An extensometer monitors changes in the distance between one or more anchors fixed to the ground and a reference head at the borehole collar (Figure 9.2). Components of an extensometer

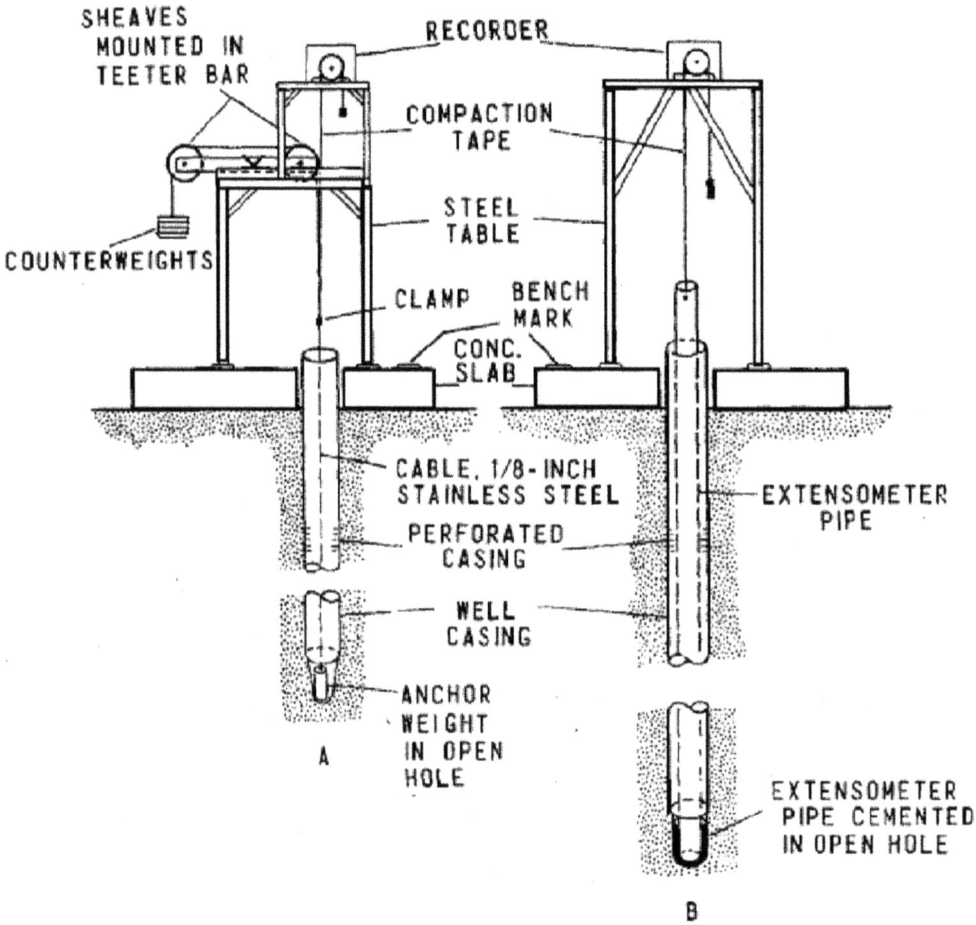

Figure 9.2 Illustration of ground movement measurement by extensometers. (a) Cable type. (b) Pipe-type (From ISRM, 1977.)

include anchors, rods or wires with protective tubing, and a reference head. The anchors, with wires or rods attached, are installed in a borehole. Borehole extensometers have anchored benchmarks. A borehole is drilled to a depth where the strata are stable and is then lined with a steel casing with slip joints to prevent crumpling as subsidence occurs. An inner pipe rests on a concrete plug at the bottom of the borehole and extends to the top, transferring the stable elevation below to the surface. The wires or rods span the distance between the anchors and the reference head, which is installed at the borehole collar. Measurements are obtained at the reference head using a sensor or a micrometer, which measures the distance between the top (near) end of the anchor and a reference surface. A change in distance from the initial measurement indicates that movement has occurred. Movement may be referenced to a borehole anchor installed in stable ground or to the reference head, which can be surveyed. In recent years, fiber optic extensometers have been developed and applied in the practical monitoring of ground movements.

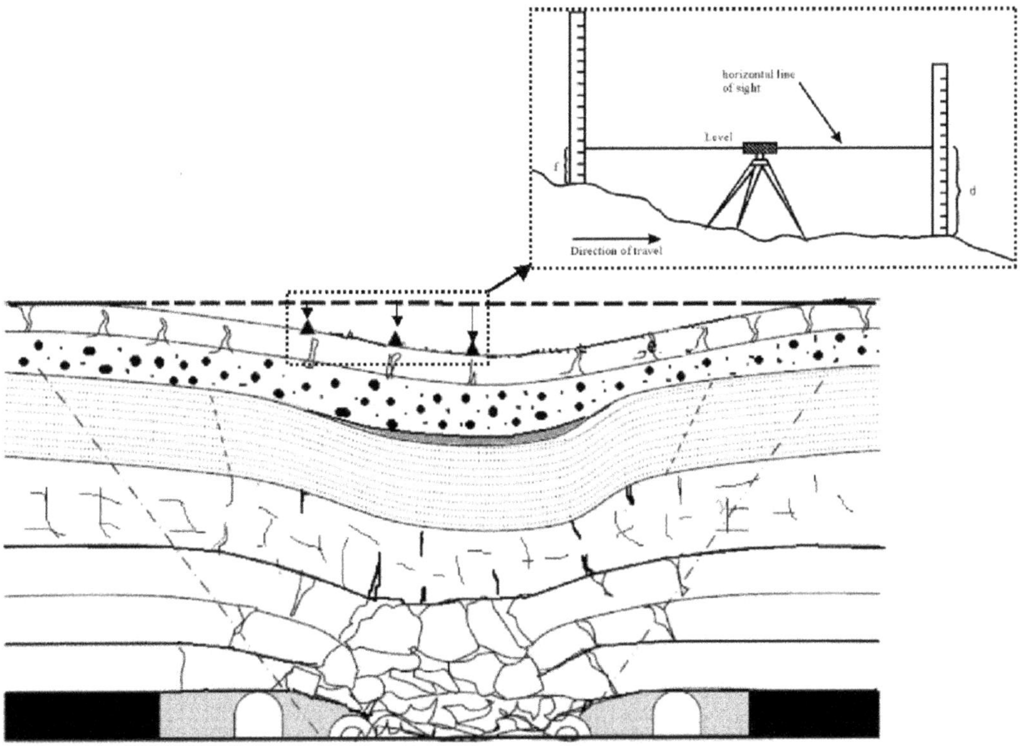

Figure 9.3 An illustration of ground movement measurement by leveling technique. (From Aydan, 2008 (Section 6.3) in Didier et al., 2008.)

9.1.1.3 *Optical leveling technique*

Ground movements may be systematically measured using optical leveling devices (Figure 9.3). For this purpose, a network of leveling points is established, and measurements are performed at specific time intervals. The repeatability of the measurements obtained from different surveys, different loops, and the stability of the monuments with respect to their local environment are the main criteria for this technique. The leveling technique can provide very precise height differences with an accuracy of up to a few millimeters. This technique is relatively flexible for implementation in areas that typically have dense housings, buildings, and/or vegetation. The benchmarks can also be easily located. The data processing and analysis of the measured results are also not complicated.

Although it yields very accurate height differences, the leveling technique is relatively slow and time-consuming in its execution, especially when precise levelling procedures are being implemented. Its operation also depends on time, weather, and environmental conditions along the levelling routes. It should also be noted that the monitored points should generally be associated with a specific benchmark located in a stable zone outside the subsiding area. When the subsiding area is large, connecting to stable benchmarks becomes another limiting constraint for the implementation of the levelling technique.

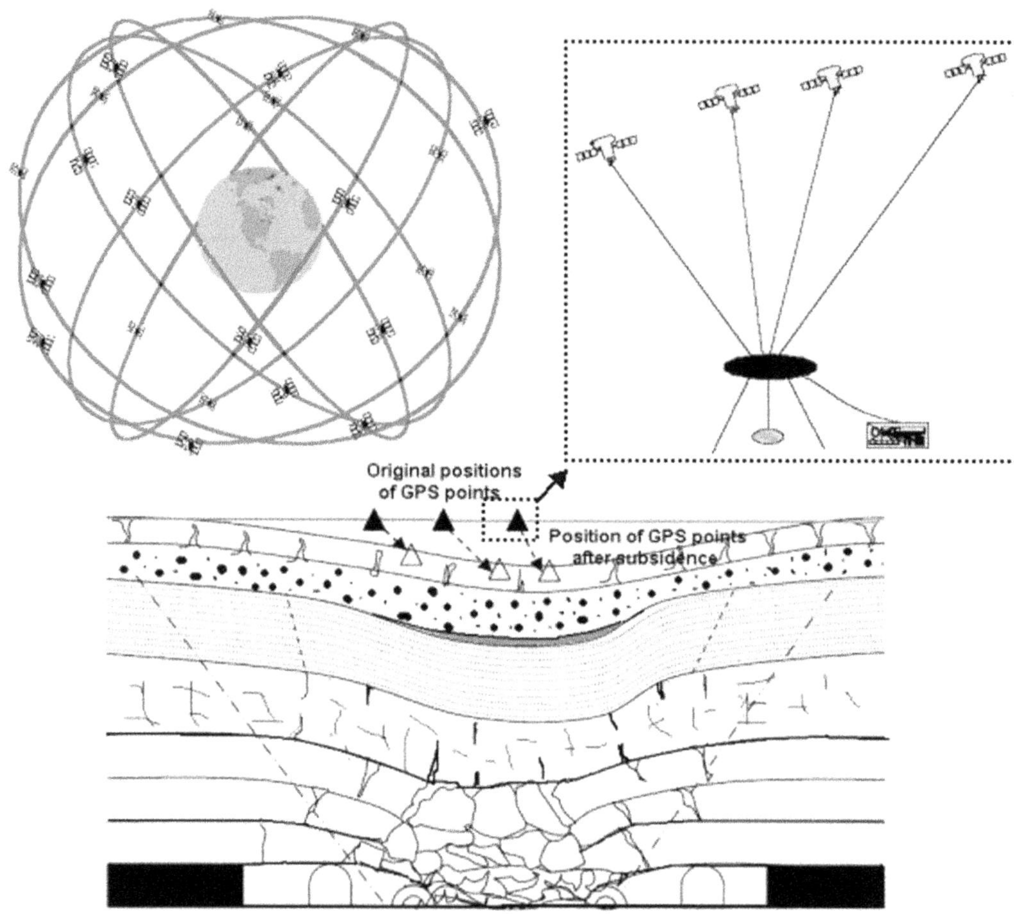

Figure 9.4 An illustration of ground movement measurement by GPS technique. (From Aydan, 2008 (Section 6.3) in Didier et al., 2008.)

9.1.1.4 GPS leveling technique

The GPS technique has also been applied to subsidence measurements in recent years (Figure 9.4). First, a network of GPS stations is established. The GPS surveys are carried out at these stations using dual-frequency geodetic-type GPS receivers. There are several advantages to using the GPS technique, such as

- GPS provides the three-dimensional displacement vector with two horizontal and one vertical component, allowing for three-dimensional subsidence measurements;
- GPS provides the displacement vectors in a unique coordinate reference system, making it effective for monitoring subsidence over a relatively large area;
- GPS can yield displacement vectors with a precision level of several millimeters, which is relatively consistent in both temporal and spatial domains, enabling the detection of even relatively small subsidence signals;

- GPS can be utilized continuously, day and night, independent of weather conditions, allowing for flexible field operations.

However, surveys using the GPS technique may show slightly worse standard deviations due to signal obstruction by trees and/or buildings around the station. Therefore, the precision level obtained in surveys with signal obstructions and/or multi-path effects may be in the order of 1–3 cm, whereas surveys with good signals generally achieve precision levels of several millimeters. This issue may arise from the destruction or alteration of observation monuments in urbanized areas. Furthermore, expertise in GPS data acquisition and precise data processing is required for accurate detection of ground subsidence.

9.1.2 *Indirect Measuring Techniques*

9.1.2.1 *Aerial photogrammetry*

It is ideal to use aerial photographs taken at certain time intervals with scales such as 1:5,000 or 1:10,000 for determining ground deformations. The area of interest must be covered with photographs that are continuous and have an overlapping ratio of at least 60% (Figure 9.5). Data on the internal elements and inner factors of the camera used in aerial photography are required to enable geometric analysis. Points used in the national triangulation network and water level points serve as permanent signals.

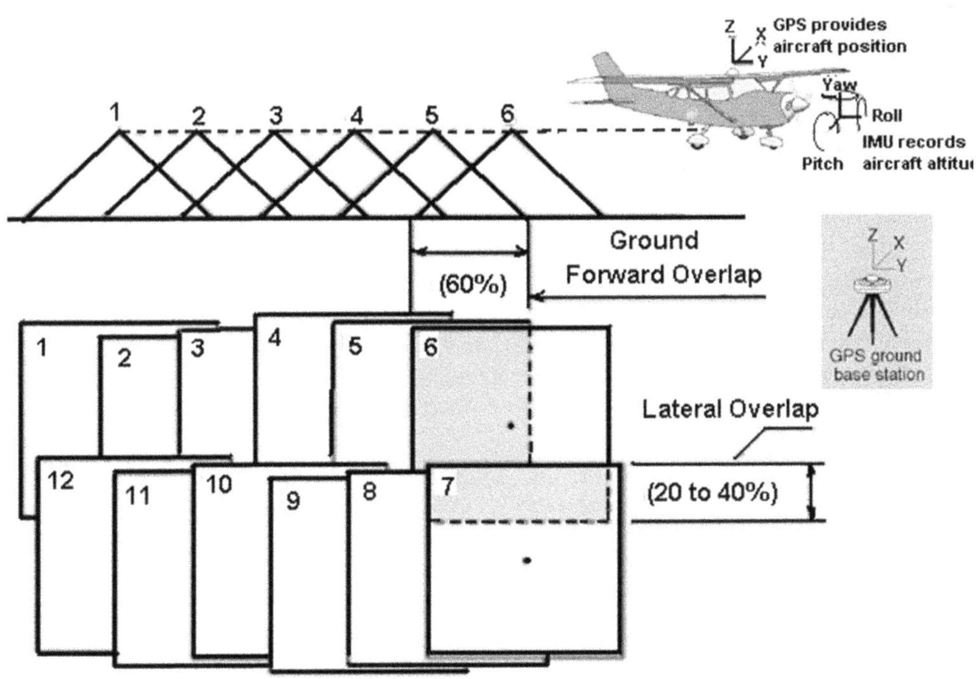

Figure 9.5 An illustration of the fundamental concept of aerial photogrammetry technique. (From Aydan, 2008 (Section 6.3) in Didier et al., 2008.)

The selection of control points is a very important factor for determining their spatial coordinates. It is essential and ideal to determine the control points independently in the field. In addition, it is preferable to determine the coordinates using GPS. Control points are generally marked by white plates on the ground, making them easily distinguishable in aerial photographs. They are assumed to be stationary points. Measurement points, which are located on the ground surface in the area of interest, are marked on photographs taken at certain intervals. It is required to establish measurement points on the ground surface that can be identified in aerial photographs.

In aerial photogrammetry, the variation in relative precision depends on the scale of the photographs and their inner factors, such as the precision in determining the coordinate system of the photographs. The precision of measurements obtained through aerial photogrammetry, for example, in the "Bundle" method, is represented by the standard deviation value of residuals. The vertical relative precision is generally six times the horizontal relative precision. Generally, horizontal absolute precision is smaller than absolute horizontal relative precision.

9.1.2.2 InSAR technique

Although it is a relatively new technique, InSAR has great potential for long-term monitoring of mine subsidence problems. Similar to aerial photogrammetry, SAR images of the same area taken at specific time intervals are necessary (Figure 9.6). Several methods can be applied to monitor mining subsidence using the InSAR technique. Several InSAR images of the same area are used to generate an interferogram, which displays the surface topography as well as the subsidence that has occurred during the period.

A digital elevation model of the area is then used to remove the topography from the interferogram and to generate a "differential interferogram." This differential interferogram shows the changes in surface height that occurred during the period due to mining subsidence. Each fringe in the image represents a height change of 28 mm. From the differential interferogram, points affected by the same surface subsidence are extracted and contoured as lines of "iso-subsidence," which are overlaid onto a satellite image. This combination gives valuable information for assessing the actual and potential damage caused by subsidence.

The InSAR technique yields subsidence information on a regional scale. However, its accuracy is restricted to several centimeters. The use of InSAR technique for studying subsidence phenomena is expected to be increase with the availability of more radar satellites in space (e.g., ERS, Radarsat, Envisat, and ALOS) and more InSAR data processing packages (e.g., Atlantis, Gamma, Vexcel, Roi-Pac, and Doris).

In order to use InSAR technique for studying ground subsidence, multi-temporal radar images of the area are needed, together with InSAR processing software and hardware, and expertise to process the images. Time frames for studying land subsidence will also be influenced by the timing of radar satellites' passes over the studied area. In terms of data processing, relatively dense vegetation, rapid environmental changes, and dynamic atmospheric conditions can also limit the potential of InSAR for accurately detecting land subsidence.

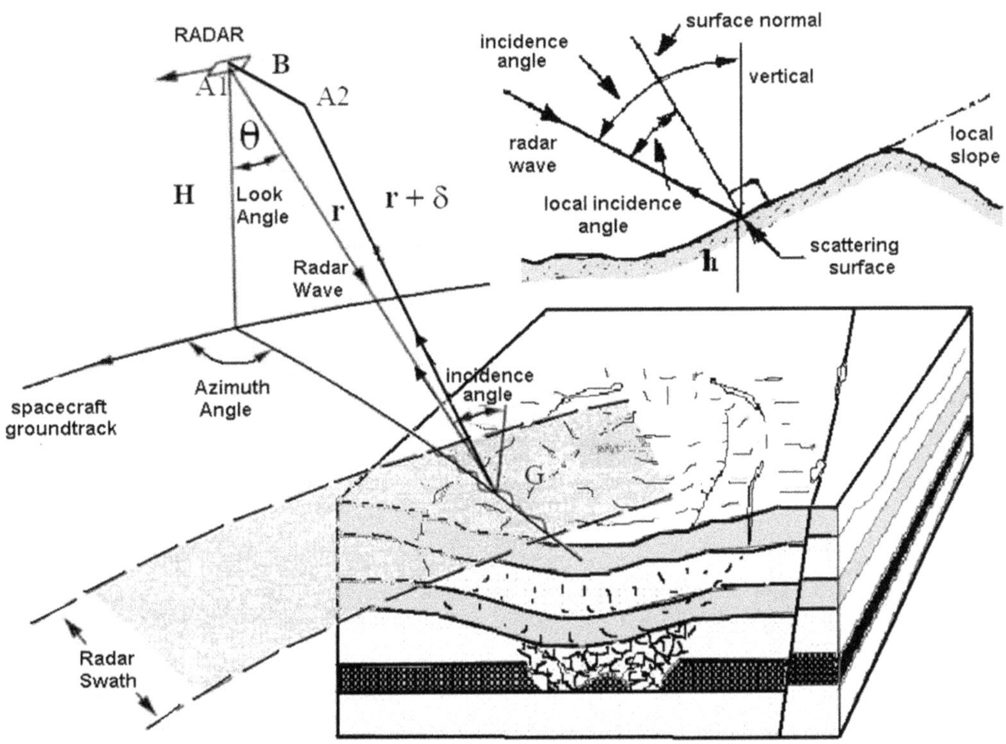

Figure 9.6 An illustration of ground movement measurement by InSAR technique. (From Aydan, 2008 (Section 6.3) in Didier et al., 2008.)

9.2 Monitoring Ground Vibrations Induced by Mine-Induced Seismicity and Earthquakes

9.2.1 *Monitoring Mine-Induced Seismicity to Identify Potential Instability Sources*

Degradation and creep phenomena in the surrounding rocks of abandoned mines may result in the failure of roofs and/or pillars, sooner or later following mine closure. As described earlier, these failures may sometimes lead to catastrophic events at the ground surface, such as large and sudden collapses or subsidence troughs, which may affect or destroy structures and cause casualties. The seismo-acoustic monitoring principle involves monitoring the occurrence, location, and magnitude of such failures and their progression over time. Collected data can then be used to warn authorities responsible for public safety, if necessary. These systems fundamentally consist of sensors (accelerometers, geophones (velocity meters)), A/D digitizers, and data acquisition and processing units. Figure 9.7 shows an example of such a system designed by NIOSH (Swanson et al., 2002). Sensors can be fixed into boreholes, competent rock surfaces, or steel waveguides fixed into the rocks.

Seismic signals are either transmitted via cable to an A/D converter attached to a data acquisition unit or digitized near the sensor, with data transmitted to a data acquisition unit digitally.

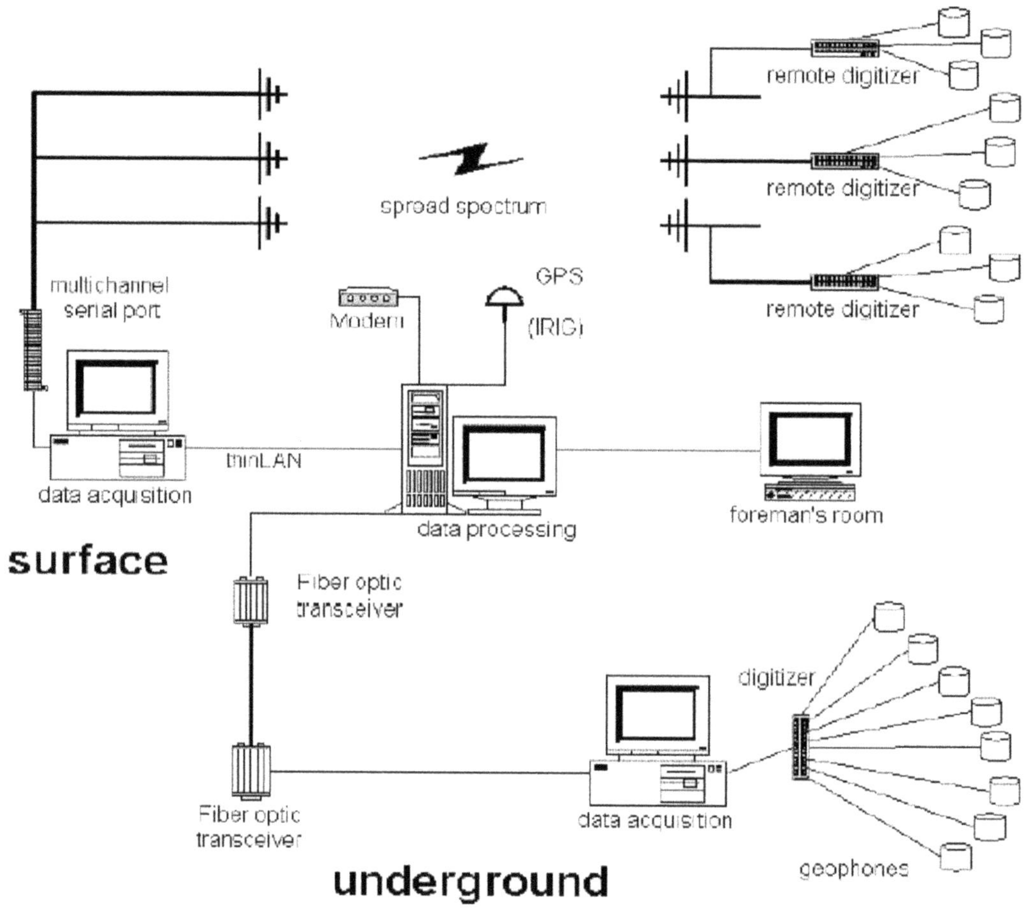

Figure 9.7 An illustration of induced seismicity by abandoned mines. (From Aydan, 2008 (Section 6.3) in Didier et al., 2008 based on Swanson et al., 2002.)

In each case, the data acquisition systems collect waveform files either continuously or in a triggered event capture mode. The data received from the data acquisition unit are processed by the data processing unit to determine the location and magnitude of seismic events within the failing rock body. Some advanced systems also identify the source characteristics of the seismic events. The rate of seismic energy release may be used to warn local authorities and residents of the collapse risk in likely locations of the mined area.

9.2.2 *Monitoring of Abandoned Mines Located in Areas Sensitive to Earthquakes*

In countries with high seismicity, earthquakes may lead to ground amplifications above abandoned mines or cause the collapse of overburden. Such events have been reported in Japan and the USA. For example, the recent M6.2 event in 2003 near Yamoto town during the Miyagi-ken Hokubu Earthquake and the M9.0 2011 Great East Japan Earthquake caused severe damage to abandoned lignite mines, resulting in sinkholes and the discharge of muddy underground water to the ground surface (Aydan & Kawamoto, 2004, Aydan & Tano 2012a,b).

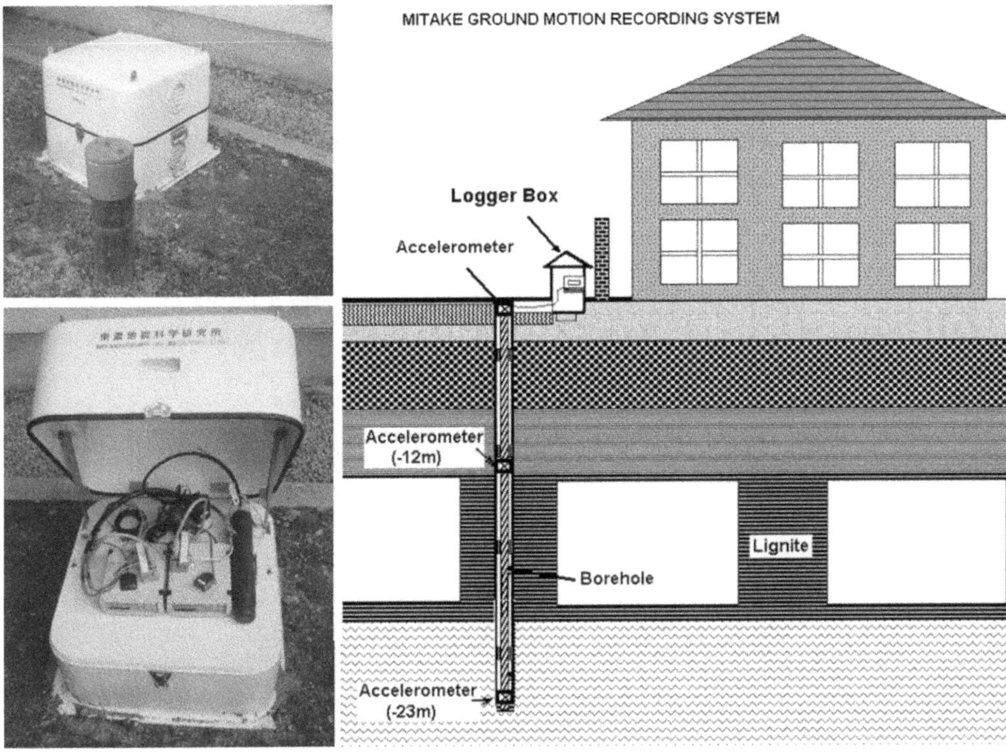

Figure 9.8 Strong motion monitoring system for an abandoned lignite mine at Mitake Town. (From Aydan, 2008 (Section 6.3) in Didier et al., 2008.)

Severe ground amplification problems may also arise due to the existence of abandoned mines that have been exploited using room and pillar methods, as shown by recent physical and numerical models presented in Chapters 4–6.

Urban areas with abandoned room and pillar lignite mines (i.e., Mitake in Gifu Prefecture, Yamoto in Miyagi Prefecture of Japan) are particularly concerned about the risk of disturbances above disused mining works. The monitoring system for this purpose is fundamentally similar to those used for monitoring mine-induced seismicity. However, sensors are installed at several depths to monitor amplification characteristics as well as surface ground movement. At least three or four sensors (one located 10–20 m below the extracted seam, two just below and/or above the seam, and one at the ground surface) are installed, and trigger-type monitoring is conducted. Such a system has been recently installed and operated in Mitake town in Gifu Prefecture, Central Japan (Figure 9.8).

9.3 Multi-Parameter Measurement System for Real-Time Long-Term Stability Monitoring of Abandoned Room and Pillar Mines

As explained in Chapters 2 and 3, the mechanical properties of soft sedimentary rocks are influenced by variations in water content. The compressive strength and elastic modulus of soft rocks generally decrease with an increase in water content or vice versa. The degradation of rock due to cycles of wetting and drying causes flaking near the surface, leading to rock falling under the effect of gravity. This process repeats endlessly, subsequently reducing the supporting area of the pillars over the long term. Therefore, it is crucial to monitor environmental parameters such as temperature, humidity, and air pressure within the abandoned mines, as well as water content and temperature variations in the surrounding rocks in association with the degradation process.

It is also known that when rocks subjected to high stresses begin to fail, the stored mechanical energy in the rock tends to transform into different forms of energy according to the law of energy conservation. Some of these transformations involve variations in electrical potential, magnetic fields, heat release, and kinetic energy in the surrounding rock mass. Recent experimental studies on various types of rocks by Aydan and his co-workers (Aydan et al., 2001, 2002, 2003), including some from abandoned mines, indicated distinct variations in the multi-parameters mentioned above during deformation and fracturing processes. Therefore, these parameters may be used to assess the short-term stability of abandoned mines.

The real-time multi-parameter measurement system involves monitoring electric potential (EP) variations, AEs, temperature, and water content of rocks (RT), as well as temperature, humidity, and air pressure within the abandoned lignite mines above the groundwater table and outside (Aydan et al., 2005b) (Figure 9.9). The rocks around abandoned mines above the groundwater level are much more prone to degradation due to environmental variations over time (Aydan et al., 2005a). Temperature, humidity, and air pressure sensors are installed at

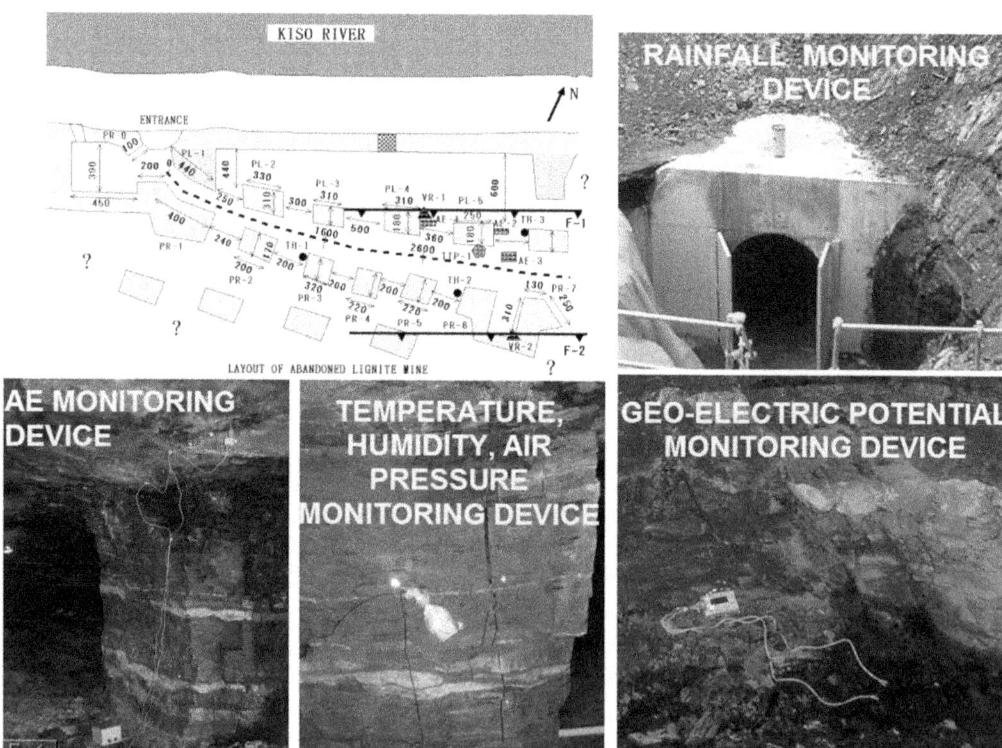

Figure 9.9 Application of multi-parameter system for monitoring degradation and long-term response of surrounding rock mass at an abandoned lignite mine. (From Aydan et al., 2005b.)

specific intervals from the mine entrance to monitor their spatial distribution and variations over time. Temperature and water content sensors are also installed at several depths in selected pillars and monitored over time.

Geo-electric potential monitoring devices and electrodes are set up within the abandoned mine. It is generally desirable to install the devices near geological discontinuities such as faults and fracture zones, which are much more sensitive to stress changes within the surrounding rock mass. When the rock mass is damp or saturated, the electrical resistance of the ground is on the order of kΩ. Therefore, it is sufficient to use devices with impedance on the order of MΩ; otherwise, devices with impedance on the order of GΩ would be necessary. Therefore, the measured EP variations by the devices are directly related to those of the surrounding rock mass. The amplitude and orientation of geo-electric potential variations may be used to infer the likely location and magnitude of sources of instability.

The AE system limited only to counting AE events (Aydan et al., 2005b, Tano et al., 2005), has proven useful for long-term monitoring of rock fracturing in abandoned lignite mines. Pulse signals, which correspond to AE waves exceeding a threshold, are discriminated through a pulsar and recorded onto a pulse counter (logger) as AE rate counts. This limited specification of rate counting reduces system costs, allowing the use of two AE systems as one set. One of the AE systems is called the active unit, while the other is referred to as the dummy unit. The active unit is directly attached to the rock mass, while the dummy AE sensor is not in contact with it. If signals are counted on both systems, the count from the active unit is neglected in the measured data. This active-dummy counting system increases the reliability of AE monitoring and checks noise conditions in the field.

The author and his group implemented a multi-parameter measurement system in two abandoned mines in Mitake Town in March 2004. Measurements of mine deformation, pH values of underground water, seismic wave velocity, and some index tests, such as Schmidt hammer rebound value and needle penetration tests, are conducted intermittently. Furthermore, certain areas within the mine are photographed to observe the degradation process.

Temperature and humidity sensors are installed at specific intervals from the mine entrance, as shown in Figure 9.9. The pillar denoted as PL5 (or LP5) (see Figures 9.9 and 9.10) has two

Figure 9.10 Instrumented pillar PL5 (or LP5). (From Aydan et al., 2005b.)

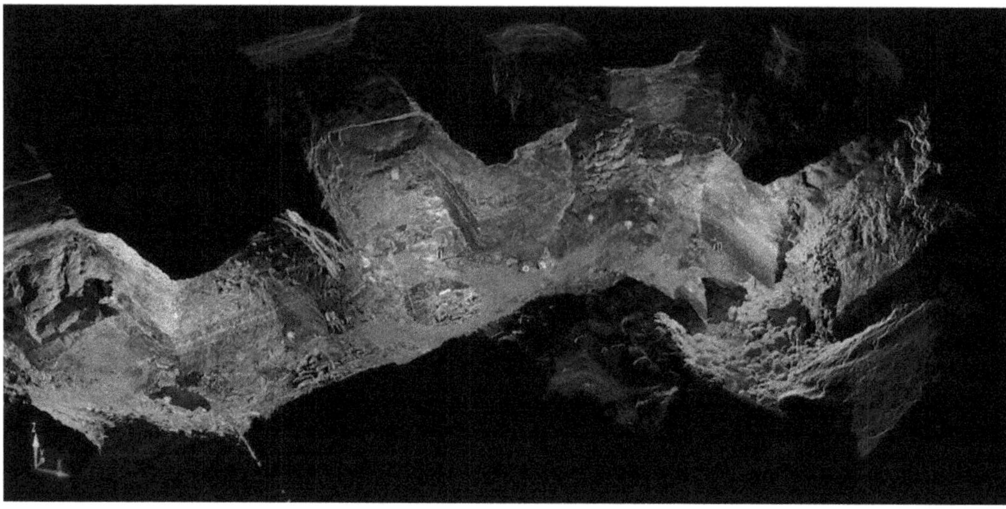

Figure 9.11 A laser scanning example of an abandoned mine in Mitake.

temperature sensors, one of which is embedded at a depth of 10 cm from the pillar surface, while the other is fixed on the pillar surface. Initially, two geo-electric potential monitoring devices are set up, as shown in Figures 9.9 and 9.10. These two devices are positioned on normal faults. An additional geo-electric potential monitoring device was installed in the pillar denoted PL5. The electrical resistance of the ground is on the order of KΩ, while the impedance of the devices is on the order of MΩ. Therefore, the measured EP variations by the devices are directly related to those of the surrounding rock mass.

Furthermore, in certain areas of the abandoned mine, the geometry was measured using laser scanners (Figure 9.11). As the sensitivity of current laser scanners is on the order of several centimeters, they are still not mature enough to monitor changes in abandoned mines. Nevertheless, they may become useful tools in the future if the accuracy of laser scanners improves. Particularly, borehole laser scanners could also be beneficial for monitoring geometric changes in abandoned mines as well as other underground structures over time.

9.3.1 *Environmental Monitoring*

The environmental measurement items include the temperature and humidity of the mine and the pH of groundwater. Temperature and humidity variations in the abandoned mine between March 10, 2004, and May 14, 2005, are shown in Figure 9.12. The overall temperature fluctuation over one year is about 12°C, and the daily temperature variation ranges between 2° and 4°. The humidity variation ranges between 60% and 100%. These results indicate that the soft sedimentary rocks are subjected to cycles of wetting and drying. This observation clearly indicates that rocks would degrade if they are prone to variations in water content.

Figure 9.13 shows the temperature variations on the surface and inside the pillar denoted PL5. The general trend of temperatures at the pillar surface and 10 cm inside the pillar surface is quite similar. However, the fluctuations in temperature at the surface are much larger than those inside the pillar. Furthermore, the yearly variation in temperature is somewhat asymmetric. This implies that thermal loading in rock would be very high near the pillar surface. As a result, the rock

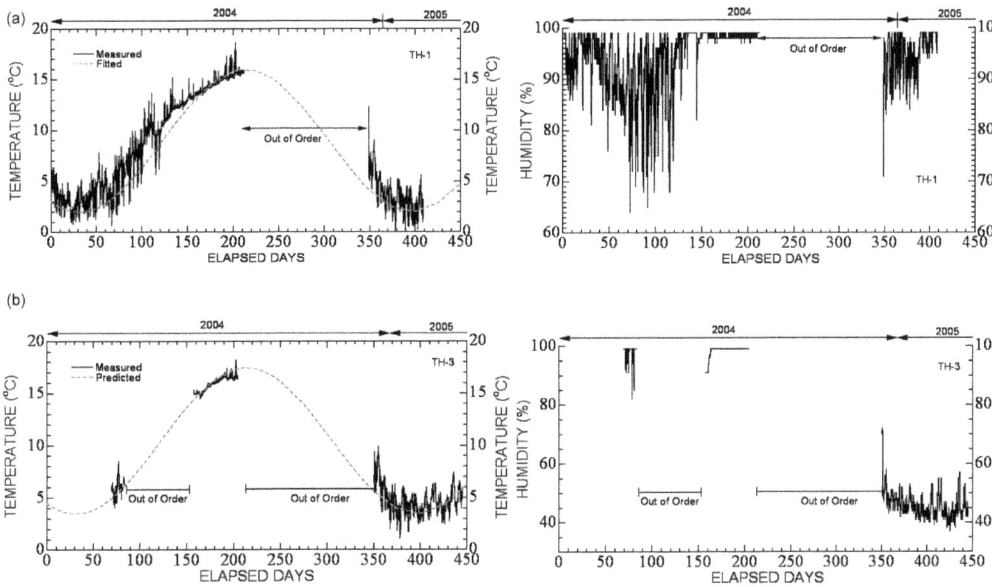

Figure 9.12 Temperature and humidity variations in the mine. (a) Temperature and humidity variations near the entrance of the mine. (b) Temperature and humidity variations inside the mine (40 m away from the entrance). (From Aydan et al., 2005b.)

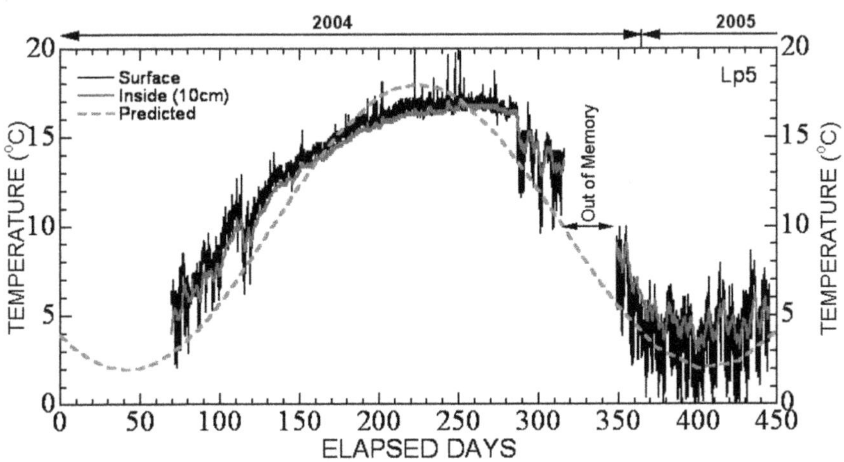

Figure 9.13 Variations of temperature on the surface and inside pillar denoted PL5. (From Aydan et al., 2005b.)

would be prone to degradation due to the daily and yearly cyclic thermal loading, which may cause AE and EP variations shown in Figure 9.14 and flaking of rocks as seen in Figure 9.15.

The pH measurements of groundwater in abandoned mines were carried out at four different locations in the mine. The pH values are almost the same at all measurement locations, generally varying between 6.4 and 8.2. The pH values decrease during the rainy summer season and increase during the dry winter season.

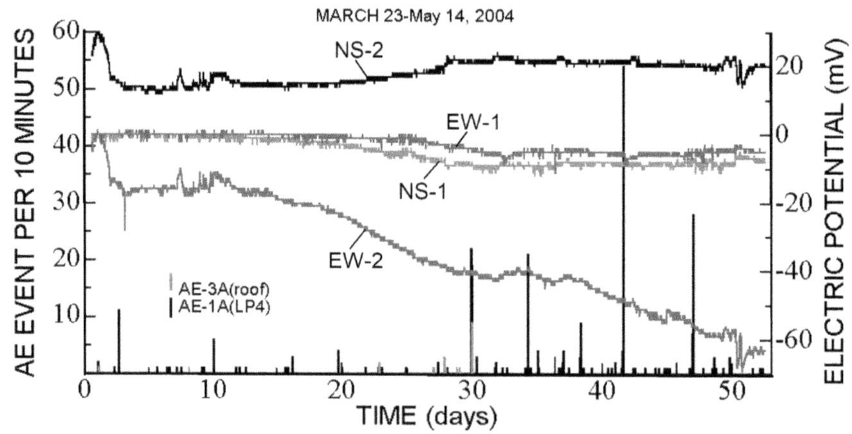

Figure 9.14 Variations of geo-electric potential and acoustic emission (AE) counts. (From Aydan et al., 2005b.)

Figure 9.15 Some fracturing, crack opening in pillars, and flaking of roof layers. (From Aydan et al., 2005b.)

9.3.2 Earthquakes during the Monitoring Period

One of the purposes was to monitor the behavior of the abandoned mine during earthquakes. Two earthquakes relevant to the site occurred during the monitoring period. The first earthquake, officially named the Tokaido-oki Earthquake, had a magnitude of 7.3 and occurred on September 5, 2004. Although the epicenter of the earthquake was 236 km away from the mine, the maximum ground acceleration caused by this earthquake was more than 20 gal, and the shaking period was greater than 20 seconds (Figure 6.52a). On January 9, 2005, an earthquake named the Komaki Earthquake, with a magnitude of 4.7, took place 25 km away from the mine, and the maximum ground acceleration was more than 50 gal at a nearby strong-motion station (Figure 6.51) (K-Net, NIED). These two earthquakes also provided unique data on the response of the abandoned mine during earthquakes. A far-field earthquake occurred in the Chuetsu-oki region (epicentral distance of 253 km) in Niigata Prefecture on July 16, 2007. Although the magnitude of the earthquake was 6.8 and it induced ground accelerations of less than 5 gals, this earthquake significantly influenced one of the monitored abandoned mines, as discussed below.

9.3.3 *Instability Problems during Monitoring*

In addition to instrumental monitoring, visual inspections and photographs of the abandoned mine were carried out periodically. As data downloads are conducted at two-month intervals, visual inspections and photographs of certain locations were taken each time data was downloaded. The first significant collapse was observed when the authors visited the mine on May 14, 2004, following a visit on March 23, 2004. Since the collapse was not observed during the March 23 visit, it must have occurred during the period between these two visits. The collapsed area (3 m × 2 m) was bounded by pillars denoted PL3, PL4, PR3, and PR4, and a 25 cm thick sandstone roof layer had fallen down, as seen in Figure 9.16a.

There were no significant collapses within the area of instrumentation until the earthquake on September 5, 2004. When the mine was visited on September 17, some notable roof collapses were observed, as seen in Figure 9.16b. However, there was no significant collapse induced by the Komaki Earthquake on January 9, 2005. In addition to these major collapses in the mine, the rock mass in the roof and pillars continues to degrade. The degradation of rock mass occurs in the form of flaking rock pieces from the roof and fall of pillars due to cyclic saturation and drying. Furthermore, some cracks in the pillars continue to open, as seen in Figure 9.15.

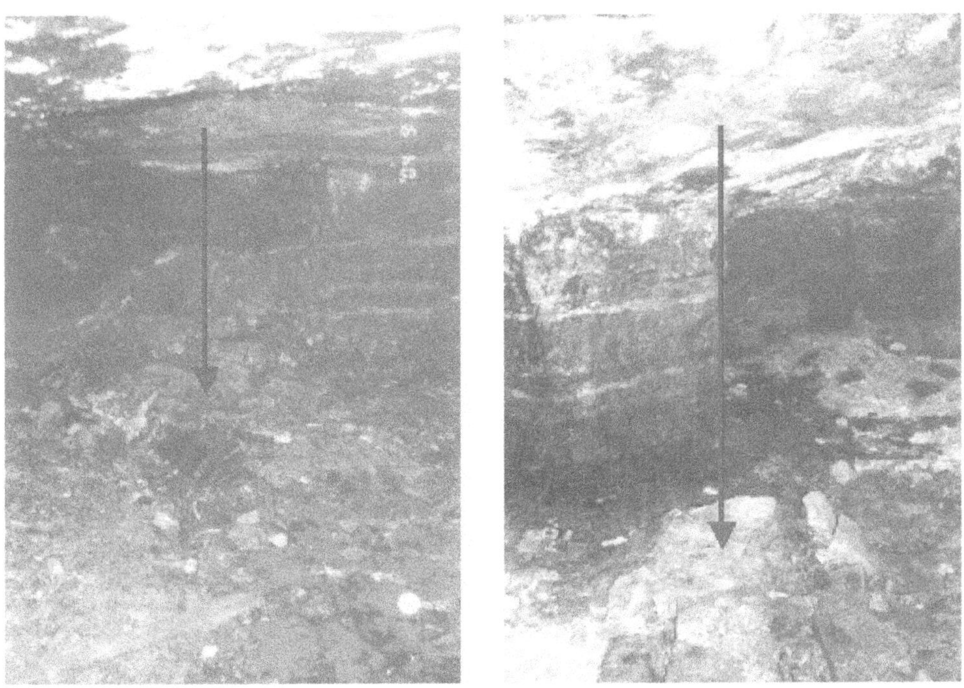

(a) (b)

Figure 9.16 Some views of instability in the abandoned lignite mine. (a) Roof collapse (March 23–May 17). (b) Roof collapse due to Tokaido-oki earthquake. (From Geniş & Aydan, 2008.)

Figure 9.17 Ruptured pillar PL2 in the abandoned lignite mine (A-site). (From Aydan &Tano, 2012b; Aydan et al., 2015.)

Figure 9.18 Fallen roof in the abandoned lignite mine (A-site). (From Aydan & Tano, 2012b; Aydan et al., 2015.)

Another major stability problem was observed in the abandoned lignite mine in the Fushimi region next to the Kiso River. The Chuetsu Earthquake occurred on July 16, 2007, and the abandoned mine was visited on August 17, 2007. The primary instability issue was the shear rupture of pillar PL2, located near the entrance (Figure 9.17). Additionally, there were significant roof failures. A major roof failure (4 × 4 m, 35 cm thick) occurred between pillars PR2 and PR3, which crushed a toolbox that was kept in the mine (Figure 9.18). Monitoring of this abandoned mine was temporarily suspended and restarted in early January 2010.

9.4 Multiparameter Measurements Associated with Instability Problems

Three major events associated with mine stability took place during the monitoring period from March 23, 2004 to July 2, 2011, and the monitoring results are discussed separately in the following subsections.

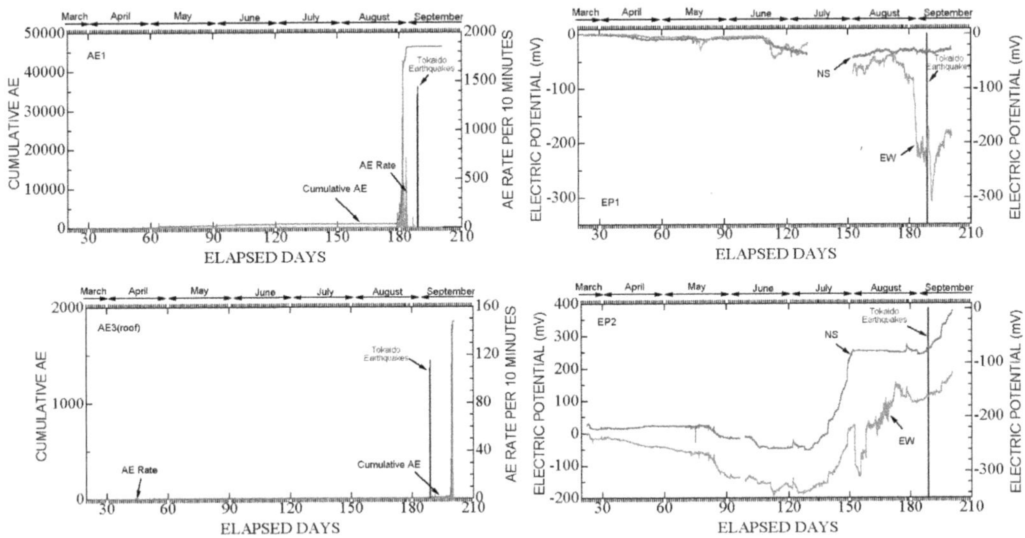

Figure 9.19 Variations of geo-electric potential and acoustic emission (AE) counts. (From Aydan et al., 2005b.)

9.4.1 *Period between March 23, 2004 and May 17, 2004*

Geo-electric potential variations and AE counts per 10 minutes are shown in Figure 9.19 for the period from March 23 to May 17. As seen in the figure, there is almost a one-to-one correspondence between geo-electric variations and AE counts. However, the fluctuations in geo-electric potential begin much earlier than the AE events, as observed in laboratory tests (Aydan et al., 2002, 2003, 2005) on samples from this mine and other sites. The collapse area is very close to the AE device set on pillar PL4. More than 50 AE counts were measured on May 2, 2004. However, high AE activity began to occur after April 20, 2004. Geo-electric potential variations started to occur about 10 days before the high AE activity. The geo-electric potential variation vectors indicated that the instability was located to the west of the devices. The actual event was observed approximately 5–10 m west of the geo-electric devices.

9.4.2 *Period before and after Tokaido-oki Earthquake of September 5, 2004*

Both geo-electric potential variations and AE counts per 10 minutes are shown in Figure 9.20. AEs were observed at the device numbered AE1 (PL4) before the earthquake. This pillar has a secondary normal fault to the principal one, and a crack on the opposite side of the pillar continues to open. These AEs should be directly related to the fracturing of this pillar. The EP1 device is also next to this pillar, and very large geo-electrical variations occurred during the AE activity. The AE3 device, located between pillars LP5 and PR5, indicated high AE activity just after the earthquake, with an AE count of more than 400 on September 12. The EP2 geo-electric device is about 5 m south of AE3, and the geo-electric potential started to increase after the earthquake. Major collapses were observed near pillars PL5 (or LP5) and PL7, which are about 10 m east of the AE device. The geo-electric potential vector of EP2 indicated that the instability locations were directed to the NE, while the location of the EP2 device was to the east. Particularly, it is

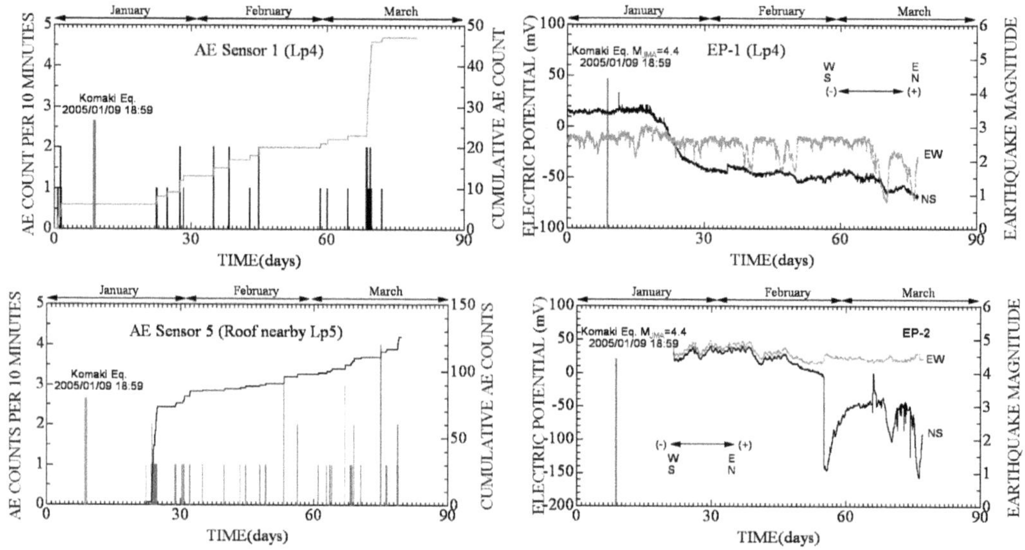

Figure 9.20 Variations of geo-electric potential and acoustic emission (AE) counts. (From Aydan et al., 2005b.)

of great interest that the geo-electric potential vectors can be used to locate the spot of instability. Furthermore, it seems that geo-electric potential variations are also influenced by the crustal geo-electric potential variations associated with earthquakes. Since the Tokaido-oki Earthquake was to the SW of the site, the variations in geo-electric potentials indicate the location of the source of geo-electric potential signals. It should also be noted that AE counts are generally associated with events whose sources are within a vicinity of 10 m. Therefore, a dense network of AE devices could be useful for locating the exact locations of the instability spots. The combination of these measurements, together with geo-electric potential measurements, can distinguish between near-field and far-field events.

9.4.3 Period before and after the Komaki Earthquake of January 9, 2005

Geo-electric potential variations and AE counts per 10 minutes are shown in Figure 9.20 in association with the 2005 Komaki Earthquake. Some AEs were observed at the device numbered AE1 (PL4) before and after the earthquake. As previously mentioned, this pillar has a secondary normal fault in addition to the principal one, and a crack on the opposite side of the pillar (its picture is shown in Figure 9.15a) continues to open. These AEs should be directly related to the fracturing of this pillar. The EP1 device is also next to this pillar, and very large geo-electrical variations occur during the AE activity. Furthermore, the geo-electrical potential vector indicates that the events have been occurring to the southwest of the device. A new AE device numbered AE5 was installed in December 2004 between pillars PL4 and PL5. The AE events measured by AE1 and AE5 occur almost simultaneously. During these events, the geo-electric potential measuring device EP1, which is close to AE1 and AE5, shows fluctuations in the measured responses. Both EP1 and EP2 geo-electric potential measuring devices indicate that some instability problems to the south or SW of their location may be taking place. Since the observations are limited to a certain area due to concerns about roof collapses and pillar failures, we could not confirm the implications of these geo-electric potential measurements through visual inspections. However, the effects of the 2005 Komaki Earthquake could not be observed in light of the measurements shown in Figure 9.19.

9.4.4 Multi-Parameter Observations on Pillar PL5(LP5) during March 21, July 02, 2005

Pillars resemble samples under uniaxial compression tests in the laboratory. A geo-electric potential measuring device was recently installed in the pillar denoted LP5, in addition to temperature and AE measurements. Compared with the initial state of this pillar, vertical cracks in this pillar have opened further. The geo-electrical potentials were measured in vertical (UD) and horizontal (NS) directions. Figure 9.10 shows the instrumented pillar LP5. Figure 9.21 shows the variations of geo-electric potential, temperature, and AE events during March 21 and July 02, 2005. As noted from the figure, some AE events have been taking place, indicating that crack opening and propagation have been occurring. Particularly, the AE events have been increasing since the beginning of June, which roughly corresponds to the rainy season in Japan. Simultaneously, geo-electric potential variations become larger, which may indicate a decrease in the overall electric resistivity of the pillar. This decrease allows the geo-electrical potentials associated with crack development to become more amplified. The temperature variations do not seem to be associated with the AE and geo-electric potential variations.

9.4.5 Multi-Parameter Observations on Pillar PL5(LP5) during March–August, 2007

The multi-parameter responses during March–August 2007 are shown in Figure 9.22. Specifically, AE and E responses are plotted in the figure. As noted from the figure, EP responses are quite sensitive to the earthquakes that occurred in 2007: the March 2 Noto Earthquake (Mw 6.7), the April 15 Kameyama Earthquake (Mw 5.5), and the July 16 Chuetsu Earthquake (Mj 6.8). Notably, EP starts to decrease following the Chuetsu-oki Earthquake, while AE events increase. It is interesting to note that the variation of EP in the NS direction and the maximum value of AE events occur almost simultaneously. Compared to the EW component, the NS

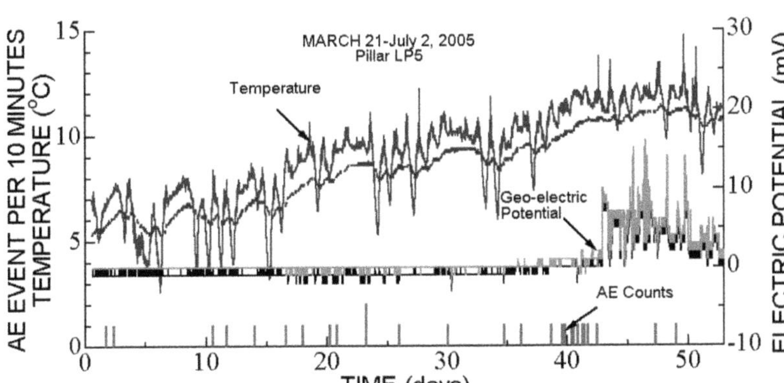

Figure 9.21 Multi-parameter response of Pillar LP5 during March 21 and July 02, 2005. (From Aydan et al., 2005b.)

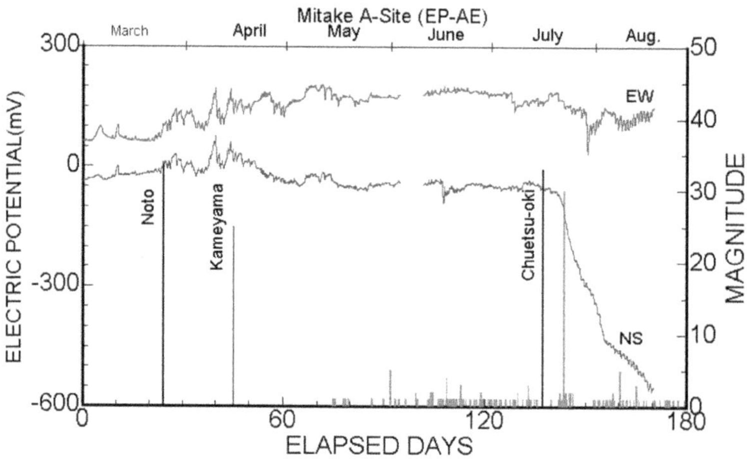

Figure 9.22 EP and AE responses at the monitoring stations in an abandoned mine (A-site). (From Aydan et al., 2015.)

Figure 9.23 Views of slope failures near the abandoned mine. (From Aydan et al., 2015.)

component shows the maximum variation, and the epicenter of the earthquake is located 253 km north of the Mitake abandoned mine. These measurements also confirm that instability issues were directly related to the 2007 Chuetsu-oki Earthquake.

9.4.6 EP Variations during July 15, 2010 in Relation to Large-Scale Slope Failure

A large-scale slope failure occurred at the abandoned mine in the Fushimi region next to the Kiso River, as shown in Figure 9.23, following heavy rainfall on July 15, 2010. Figures 9.24 and 9.25 show the EP variations. As noted, very large EP variations occurred on the day of the slope failure. Furthermore, when the EP variations are projected onto a horizontal plane, one can easily notice the direction of the signals indicating the location of the slope failure. This example demonstrates that multi-parameter monitoring can be useful for identifying nearby large instability issues.

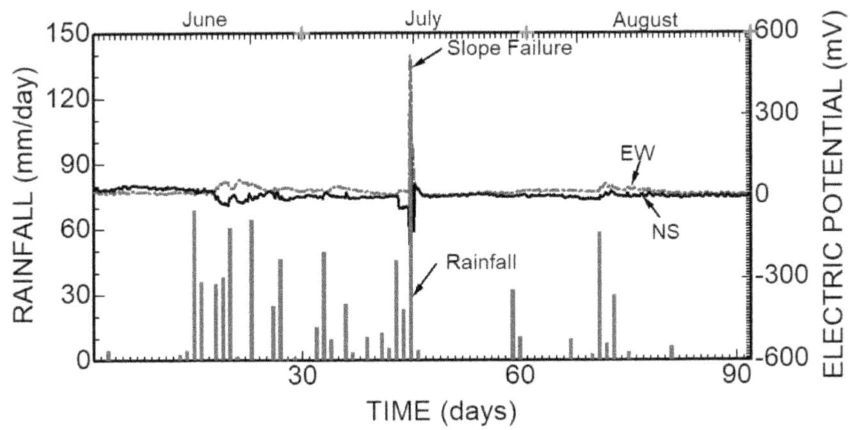

Figure 9.24 EP variations from June to August 2010. (From Aydan et al., 2015.)

Figure 9.25 Projected EP variations on the horizontal plane from June to August 2010. (From Aydan et al., 2015.)

9.5 Strong Motion Monitoring for Abandoned Mines

A strong motion monitoring system is essential for evaluating the stability of abandoned mines. Nevertheless, the high cost involved in monitoring prevents its widespread implementation. The author and his group have attempted to address this issue by developing a low-cost monitoring system (Aydan et al., 2016b). The system consists of triaxial accelerometers with a power supply system and is named QV3-OAM-SYN (Figure 9.26).

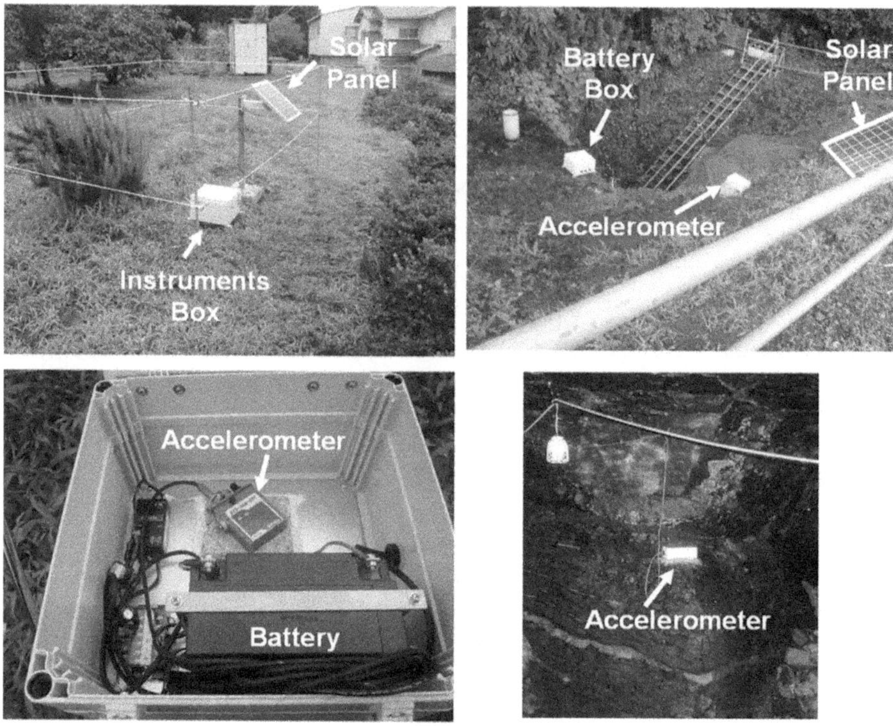

Figure 9.26 Views of the components of the strong motion measurement system. (From Aydan et al., 2016b.)

The accelerometers are set in trigger mode and record and store acceleration data in their internal memory. Another major issue in practice is the power source. If electricity is available in the environment, a backup battery is necessary in case of power loss during short periods. If there is no electricity, solar panels can be used for this purpose, implemented together with battery backup systems (Figure 9.26). In the worst case, charged batteries may be used, which can last about two months. This system was utilized in the Mitake abandoned mines, where one accelerometer measured strong motions at the ground surface, while another was installed underground, with their power supply sourced from a common solar panel, as seen in Figure 9.26. Recently, a new accelerometer named OAMIT_AI was developed to measure accelerations and inclination.

Herein, some applications of the system are briefly explained. Soon after the installation of the system in Mitake town, a sinkhole occurred approximately 120 m away from the strong motion station. Figure 9.27 shows a view of the sinkhole, and the acceleration records induced by the collapse of the abandoned mine cavity are shown in Figure 9.28. As noted from the acceleration records, several roof falls can be distinguished, lasting about 20 seconds. It is interesting to note that the maximum acceleration was about 500 gals.

Figure 9.29 shows an acceleration record taken on December 1, 2015, at 16:00:58. Although the source of the event cannot be related to a nearby earthquake, the accelerations were about 27 gals.

Figure 9.27 A view of the sinkhole that occurred on July 20, 2015. (From Aydan et al., 2016b.)

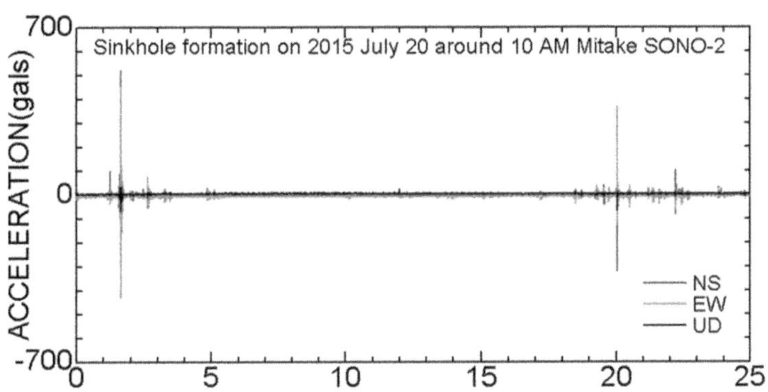

Figure 9.28 Acceleration records induced by the sinkhole. (From Aydan et al., 2016b.)

9.6 Micro-Tremor Measurement

Micro-tremor systems are often utilized to determine the vibration characteristics of the ground as well as those of structures. Regarding the abandoned mines, one application is to check the vibration characteristics of abandoned mine areas in relation to the evaluation

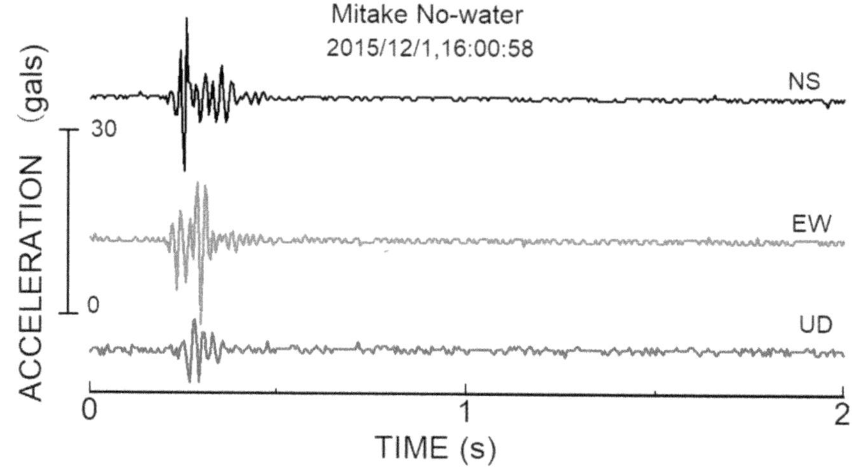

Figure 9.29 Acceleration records over an abandoned mine area.

Figure 9.30 Views of the micro-tremor devices utilized in abandoned mines in Mitake. (From Aydan et al., 2016b.)

of vibration response of structures on the ground surface (Figure 9.30). Another purpose could be the evaluation of the effect of backfilling through the difference in the vibrational characteristics of the ground.

Figure 9.31 shows the H/V spectra of micro-tremor measurements over the abandoned mines with and without groundwater. As noted from the figure, the H/V spectra differ for the abandoned mines with and without groundwater.

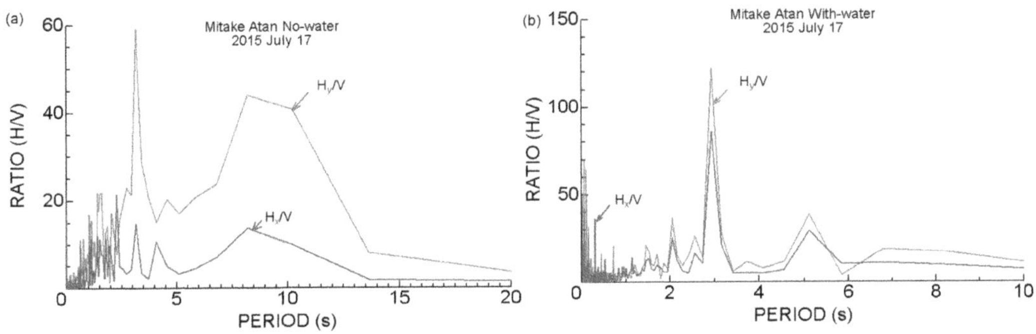

Figure 9.31 H/V spectra of micro-tremors of ground surface over the abandoned mines (a) without and (b) with groundwater. (From Aydan et al., 2016b.)

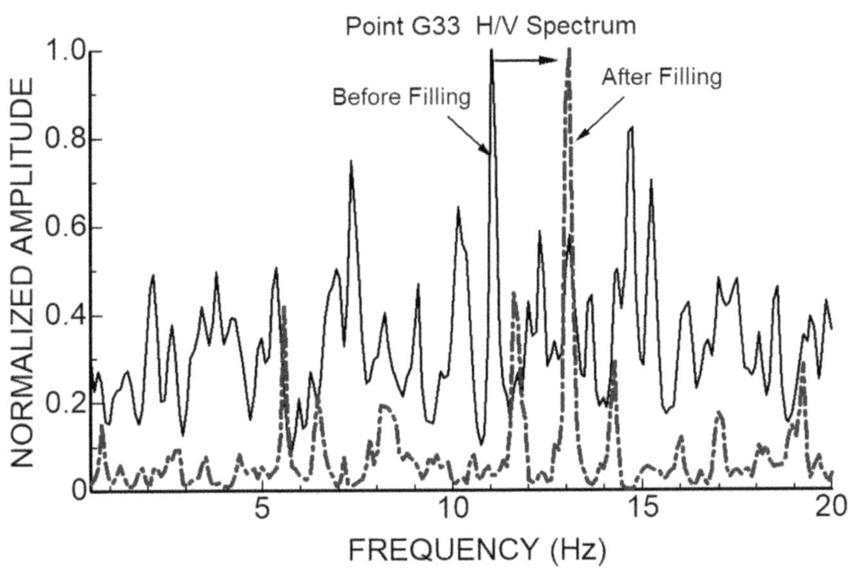

Figure 9.32 H/V spectra of micro-tremors of ground surface over the abandoned mines before and after backfilling.

To assess the effect of backfilling of abandoned lignite mines in Mitake Town, micro-tremor measurements were implemented in the same area before and after backfilling. Figure 9.32 shows an example of measurements in Mitake Town. As explained in Chapters 4–6, the stiffness of the ground increases after backfilling. As a result of this action, the H/V spectra imply that the dominant natural frequency increases due to backfilling.

References

Ağan, C., Yeşilnacar, M.I., Geniş, M., Kulaksız, S., Ulusay, R. and Aydan, Ö. (2013). A geoengineering assessment of Bazda antique underground quarries in Şanlıurfa, Turkey. *The 2013 ISRM EUROCK International Symposium, Wroclaw*, 93–98.

Amadei, B. and Savage, W.Z. (1985). Gravitational stresses in regularly jointed rock masses. *International Symposium on Fundamentals of Rock Joints*, Bjorkliden, 463–473.

Amadei, B. and Stephanson, O. (1997). *Rock Stress and Its Measurement.* Chapman & Hall, 490p.

Anderson, E.M. (1951). *Dynamics of Faulting*. Oliver & Boyd, Edinburgh.

Anderson, L.D. and Hart, R.S. (1976). An earth model based on free oscillations and body waves. *Journal of Geophysical Research*, 81(8), 1461–1475.

Angellier, J. (1984). Tectonic analysis of fault slip data sets. *Journal of Geophysical Research*, 89(B7), 5834–5848.

Aydan, Ö. (1989). The stabilization of rock engineering structures by rockbolts. Doctorate Thesis, Nagoya University, 204p.

Aydan, Ö (1993). A consideration on the stress state of the earth due to the gravitational pull. *The 7th Annual Symposium on Computational Mechanics, Tokyo*, 243–252.

Aydan, Ö. (1994a). A model study on the caving mechanism of longwalls in heavily jointed media. *The 3rd International Symposium on Mine Planning and Equipment Selection*, Istanbul, October, 765–770.

Aydan, Ö. (1994b). The dynamic shear response of an infinitely long visco-elastic layer under gravitational loading. *Soil Dynamics and Earthquake Engineering*, 13, 181–186.

Aydan, Ö. (1995a). The stress state of the earth and the earth's crust due to the gravitational pull. *The 35th US Rock Mechanics Symposium*, Lake Tahoe, 237–243.

Aydan, Ö. (1995b). Mechanical and numerical modelling of lateral spreading of liquefied soil. *International Conferences on Earthquake Geotechnical Engineering*, IS-Tokyo, 881–886.

Aydan, Ö. (2000a). A stress inference method based on structural geological features for the full-stress components in the earth's crust. *Yerbilimleri*, 22, 223–236.

Aydan, Ö. (2000b). A stress inference method based on GPS measurements for the directions and rate of stresses in the earth' crust and their variation with time. *Yerbilimleri*, 22, 21–32.

Aydan, Ö. (2003). An experimental study on the dynamic responses of geomaterials during fracturing. *Journal of School of Marine Science and Technology, Tokai University*, 1(2), 1–7.

Aydan, Ö. (2004). Damage to abandoned lignite mines induced by 2003 Miyagi-Hokubu earthquakes and some considerations on its possible causes. *Journal of School of Marine Science and Technology*, 2(1), 1–17.

Aydan, Ö. (2008). Monitoring. *Mine Closure and Post-Mining Management – International State of Art*, C. Didier (ed.), 111–123. International Commission on Mine-Closure, ISRM.

Aydan, Ö. (2012). The inference of physico-mechanical properties of soft rocks and the evaluation of the effect of water content and weathering on their mechanical properties from needle penetration tests. *ARMA 12-639, 46th US Rock Mechanics/Geomechanics Symposium*, Chicago, Paper No. 639, 10p (on CD).

Aydan, Ö. (2013). In-situ stress inference from damage around blasted holes. *Journal of Geo-system Engineering*, 16(1), 83–91.

Aydan, Ö. (2015). Large rock slope failures induced by recent earthquakes. *Rock Mechanics and Rock Engineering*, Special Issue on Deep-seated Landslides, 49, 2503–2524.

Aydan (2016). *Time dependency in Rock Mechanics and Rock Engineering*. CRC-Press.

Aydan, Ö. (2017). *Rock Dynamics*. CRC Press, Taylor and Francis Group, 462p, ISRM Book Series No. 3, ISBN 9781138032286.

Aydan, Ö. (2018). *Rock Reinforcement and Rock Support*. CRC Press, Taylor and Francis Group, 486p.

Aydan, Ö. (2020a). A revisiting of photo-elasticity technique and its renewed potential use in Rock Mechanics and Rock Engineering. *47th Japan Rock Mechanics Symposium*, JSCE, Tokyo, 299–304.

Aydan, Ö. (2020b). *Rock Mechanics and Rock Engineering: Applications*, Vol. 2. CRC Press, Taylor and Francis Group, 383p.

Aydan, Ö. (2024). Crustal stress states associated with the surface ruptures of the 2023 Pazarcık and Ekinözü earthquakes. *Yerbilimleri*, 45 (1), 52–65.

Aydan, Ö. (2021). *Continuum and Computational Mechanics for Geomechanical Engineers*. ISRM Book Series, No. 7, CRC Press, Taylor and Francis Group, 324p.

Aydan, Ö. and Geniş, M. (2008). Assessment of dynamic stability of an abandoned room and pillar underground lignite mine, 23–44. *Turkish Journal of Rock Mechanics Bulletin*, Turkish National Rock Mechanics Group, ISRM, No. 16, 23–44 (in Turkish).

Aydan, Ö. and Geniş, M. (2015). A numerical study on the response and stability of abandoned lignite mines in relation to the excavation of a large underground opening below. IACMAG 2015, *Computer Methods and Recent Advances in Geomechanics*, Kyoto, 1031–1036.

Aydan, Ö. and Ito, T. (2015). The effect of the depth and groundwater on the formation of sinkholes or ground subsidence associated with abandoned room and pillar lignite mines under static and dynamic conditions. *Proceedings of the International Association of Hydrological Sciences*, 372, 281–284.

Aydan, Ö. and Kawamoto, T. (1992). The stability of slopes and underground openings against flexural toppling and their stabilisation. *Rock Mechanics and Rock Engineering*, 25(3), 143–165.

Aydan, Ö. and Kawamoto, T. (1994). A consideration on the stress state of the earth and earth-crust due to the gravitational pull (in Japanese). *The 9th Domestic Rock Mechanics Symposium, Tokyo*, 635–640.

Aydan, Ö. and Kawamoto, T. (2004). The damage to abandoned lignite mines caused by the 2003 Miyagi-Hokubu earthquake and some considerations on its causes. *3rd Asian Rock Mechanics Symposium*, Kyoto, 525–530.

Aydan, Ö. and Kim, Y. (2002). The inference of crustal stresses and possible earthquake faulting mechanism in Shizuoka Prefecture from the striations of faults. *Journal of the School of Marine Science and Technology, Tokai University*, No.54, 21–35.

Aydan, Ö. and Nawrocki, P. (1998). Rate-dependent deformability and strength characteristics of rocks. *International Symposium on the Geotechnics of Hard Soils-Soft Rocks*, Naples, 1, 403–411.

Aydan, Ö. and Paşamehmetoğlu, G. (1994). In-situ stress measurements and lateral stress coefficients in various parts of the earth-crust (in Turkish). *Kaya Mekaniği Bülteni, TSRM, Ankara*, 10, 1–17.

Aydan, Ö. and Tano, H. (2012a). The damage to abandoned quarries and mines by the M9.0 east Japan Mega Earthquake on March 11, 2011. *The 41st Rock Mechanics Symposium of Japan*, Paper No. 410011, Tokyo, 129–134.

Aydan, Ö. and Tano, H. (2012b). The damage to abandoned mines and quarries by the Great East Japan Earthquake on March 11, 2011. *Proceedings of the International Symposium on Engineering Lessons Learned from the 2011 Great East Japan Earthquake*, March 1–4, 2012, Tokyo, Japan, 981–992.

Aydan, Ö. and Tano, H. (2012c). The observations on abandoned mines and quarries by the Great East Japan Earthquake on March 11, 2011 and their implications. *Journal of Japan Association on Earthquake Engineering*, 12(4), 229–248.

Aydan, Ö. and Tokashiki, N. (2011). A comparative study on the applicability of analytical stability assessment methods with numerical methods for shallow natural underground openings. *The 13th International Conference of the International Association for Computer Methods and Advances in Geomechanics*, Melbourne, Australia, 964–969.

Aydan, Ö. and Tokashiki, N. (2013). An experimental study on the supporting effect of backfilling on abandoned room and pillar mines and quarries and karstic caves. *The 47th US Rock Mechanics/Geomechanics Symposium*, San Francisco, ARMA 13-379, 10p.

Aydan, Ö., Akagi, T., Ito, T. and Kawamoto, T. (1992). Deformation behaviour of tunnels in squeezing rocks and its prediction. *Journal of Japan Society of Civil Engineers, Geotechnical Division,* 448(III-19), 73–82.

Aydan, Ö., Akagi, T. and Kawamoto, T. (1996). The squeezing potential of rock around tunnels: Theory and prediction with examples taken from Japan. *Rock Mechanics and Rock Engineering*, 29(3), 125–143.

Aydan, Ö., Minato, T. and Fukue, M. (2001). An experimental study on the electrical potential of geomaterials during deformation and its implications in geomechanics. *38th US Rock Mechanics Symposium*, Washington, 2, 1199–1206.

Aydan, Ö, Kumsar, H. and Ulusay, R. (2002). How to infer the possible mechanism and characteristics of earthquakes from the striations and ground surface traces of existing faults. *JSCE, Earthquake and Structural Engineering Division*, 19(2), 199–208.

Aydan, Ö., Tokashiki, N., Ito, T., Akagi, T., Ulusay, R. and Bilgin, H.A. (2003). An experimental study on the electrical potential of non-piezoelectric geomaterials during fracturing and sliding. *9th ISRM Congress*, South Africa, 73–78.

Aydan, Ö., Daido, M. and Ogura, Y. (2004a). Vibration characteristics and dynamic responses of ground in abandoned lignite mines. *Journal of Japan Cavity Filling Association, JUTEN*, No. 45, 11–17.

Aydan, Ö., Daido, M., Ogura, Y. and Kawamoto, T. (2004b). A research on dynamic characteristics of areas with abandoned mines. *15th Central Japan Geotechnical Symposium (Geotechnical Environmental Problems)* (in Japanese), Nagoya, 19–24.

Aydan, Ö., Sakamoto, A., Yamada, N., Sugiura, K. and Kawamoto, T. (2005a). The characteristics of soft rocks and their effects on the long-term stability of abandoned room and pillar lignite mines. *Post Mining* 2005, Nancy.

Aydan, Ö., Sakamoto, A., Yamada, N., Sugiura, K. and Kawamoto, T. (2005b). A real time monitoring system for the assessment of stability and performance of abandoned room and pillar lignite mines. *Post Mining* 2005, Nancy.

Aydan, Ö., Daido, M., Tano, H. and Ito, T. (2005c). The instability modes of abandoned lignite mines and the assessment of their long-term stability. *35th Japan Rock Mechanics Symposium*, Tokyo, 57–62.

Aydan, Ö., Daido, M., Tano, H., Tokashiki, N. and Ohkubo, K. (2005d). A real-time multi-parameter monitoring system for assessing the stability of tunnels during excavation. *ITA Conference*, Istanbul, 1253–1259.

Aydan, Ö., Daido, M., Ito, T., Tano, H. and Kawamoto, T. (2006). Instability modes of abandoned lignite mines and the assessment of their stability in long-term and during earthquakes. *3rd Asian Rock Mechanics Symposium*, Singapore, on CD.

Aydan, Ö., Tano, H. and Geniş, M. (2008). Assessment of long-term stability of an abandoned room and pillar underground lignite mine. *Turkish Journal of Rock Mechanics Bulletin*, Turkish National Rock Mechanics Group, ISRM, No. 16, 1–22 (in Turkish).

Aydan, Ö., Ohta, Y., Geniş, M., Tokashiki, N. and Ohkubo, K. (2010). Response and stability of underground structures in rock mass during earthquakes. *Rock Mechanics and Rock Engineering*, 43(6), 857–875.

Aydan, Ö., Ohta, Y., Daido, M., Kumsar, H., Genis, M., Tokashiki, N., Ito, T. and Amini, M. (2011). Chapter 15: Earthquakes as a rock dynamic problem and their effects on rock engineering structures. *Advances in Rock Dynamics and Applications*, Y. Zhou and J. Zhao (eds.), 341–422. CRC Press, Taylor and Francis Group.

Aydan, Ö., Geniş, M., Sugiura, K. and Sakamoto, A. (2012a). Characteristics and amplification of ground motions above abandoned mines. *International Symposium on Earthquake Engineering, JAEE*, 1, 75–84.

Aydan, Ö., Tokashiki, N. and Geniş, M. (2012b). Some considerations on yield (failure) criteria in rock mechanics ARMA 12-640. *46th US Rock Mechanics/Geomechanics Symposium*, Chicago, Paper No. 640, 10p (on CD).

Aydan, Ö., Sugiura, K. and Sakamoto, A. (2013a). Dynamic characteristics of ground above abandoned mines. *13th Japan Rock Mechanics Symposium, Okinawa,* 519–524.

Aydan, Ö, Sato, T., Hikima, R. and Tanno, T. (2013b). Inference of in-situ stress by blasthole damage method (BDM) at mizunami URL and its comparison with other direct and indirect methods. *RS2013, 6th International Symposium on In-Situ Rock Stress*, Sendai, 359–368.

Aydan, Ö., Tokashiki, N. and Tano, H. (2013c). An experimental study on the supporting effect of backfilling on abandoned room and pillar mines, quarries and karstic caves, and its in-situ verification. *47th US Rock Mechanics/Geomechanics Symposium*, ARMA 13-379, 10p.

Aydan, Ö., Geniş, M., Tokashiki, N. and Tano, H. (2013d). *Some Considerations on the Appropriate Dimension in the Numerical Analysis of Geoengineering Structures*. APCOM & ISCM, Singapore, 6p.

Aydan, Ö., Ulusay, R. and Tokashiki, N. (2014a). A new rock mass quality rating system: Rock mass quality rating (RMQR) and its application to the estimation of geomechanical characteristics of rock masses. *Rock Mechanics and Rock Engineering*, 47(4), 1255–1276.

Aydan, Ö., Ito, T., Özbay, U., Kwasniewski, M., Shariar, K., Okuno, T., Özgenoğlu, A., Malan, D.F. and Okada, T. (2014b). ISRM suggested methods for determining the creep characteristics of rock, *Rock Mechanics and Rock Engineering*, 47(1), 275–290.

Aydan, Ö., Sato, A. and Yagi, M. (2014c). The inference of geo-mechanical properties of soft rocks and their degradation from needle penetration tests. *Rock Mechanics and Rock Engineering*, 47, 1867–1890.

Aydan, Ö., Sugiura, K., Sakamoto, A., Yamada, N. and Tano, H. 2015. Observations on creep and degradation phenomena on abandoned lignite mines in Mitake Town and the results of multiparameter monitoring system for a decade. *43th Japan Rock Mechanics Symposium*, Tokyo, 19–24 (in Japanese).

Aydan, Ö., Ito, T. and Rassouli, F. (eds.). (2016a). Chapter 11: Tests on creep characteristics of rocks. *Rock Mechanics and Engineering*, 2, 333–364. CRC Press.

Aydan, Ö., Sakamoto, A. Sugiura, K. and Wada, K. (2016b). Dynamic monitoring for the evaluation of the effect of back-filling of abandoned lignite mines and its applications. *44th Japan Rock Mechanics Symposium*, Tokyo, 86–91 (in Japanese).

Aydan, Ö., Sugiura, K. and Sakamoto, A. (2018). The quality of backfilling of abandoned lignite mines on ground settlement under static and dynamic conditions. *The Fourth Australasian Ground Control in Mining Conference (AusRock)*, Paper No. 57, Sydney, 134–141.

Aydan, Ö., Tomiyama, J., Horiuchi, K., Aasim, B., Karimi, A.K. and Nasiry, N.Z. (2021). A comparative study on stress distributions in physical models using photo-elasticity and FEM. *Challenges and Innovations in Geomechanics*, M. Barla, A. Di Donna and D. Sterpi (eds.), 3–11. IACMAG.

Barton, N. (1995). The influence of joint properties in modelling jointed rock masses. *Proceedings of the 8th International Congress on Rock Mechanics*, Tokyo, Japan, September 1995, 1023–1032.

Barton, N., Lien, R. and Lunde, J. (1974). Engineering classification of rock masses for the design of tunnel support. *Rock Mechanics*, 6, 189–236.

Belem, T. and Benzaazoua, M. (2008). Design and application of underground mine paste backfill technology. *Geotechnical and Geological Engineering*, 26(2), 147–174.

Bell, F.G. and Genske, D.D. (2001). The influence of subsidence attributable to coal mining on the environment, development and restoration: Some examples from Western Europe and South Africa. *Environmental & Engineering Geoscience*, 7, 81–99.

Beck, B.F. and Herring, J.G. (2001). Geotechnical and environmental applications of karst geology and hydrology. *Proceedings of the International Conference on 8th Multidisciplinary Conference on Sinkholes and the Engineering and Environmental Impacts of Karsts*, Louisville Kentucky, Balkema.

Bieniawski, Z. (1984). The design process in rock engineering. *Rock Mechanics and Rock Engineering*, 17, 183–190.

Brady, B.H.G. and Brown, E.T. (1999). *Mining*-Induced Surface Subsidence. *Rock Mechanics*. Springer, Dordrecht.

Brace, W.F. and Kohlstedt, D.L. (1980). Limits on lithospheric stress imposed by laboratory experiments. *Journal of Geophysical Research*, 85, 6248–6252.

Brewster, D. (1815). Experiments on the depolarization of light as exhibited by various mineral, animal and vegetable bodies with a reference of the phenomena to the general principle of polarization. *Philosophical Transactions*, 2, 29–53.

Brown, E.T. and Hoek, E. (1978). Trends in relationships between measured in-situ stresses and depth. *International Journal of Rock Mechanics and Mining Sciences*, 15(4), 211–215.

Byerlee, J. (1978). Friction of rocks. *Pure and Applied Geophysics*, 116, 615–626.

Didier, C., Merwe, V.D., Betournay, M., Mainz, Aydan, Ö., M., Kotybra, A., Song, W.K. and Josien, J.P. (2008). Mine Closure and Post-Mining Management, ISRM, International State of the Art International Commission on Mine Closure. Technical Report, 165p. DOI: 10.13140/2.1.3267.8407.

Didier, C., Merwe, V.D., Betournay, M., Mainz, Aydan, Ö., M., Kotybra, A., Song, W.K. and Josien, J.P. (2009). Presentation of the ISRM mine closure state of the art report. *ISRM-Sponsored International Symposium on Rock Mechanics: "Rock Characterisation, Modelling and Engineering Design Methods"* (SINOROCK 2009), May 2009, Hong Kong, China.

Doherty, J.P., Hasan, A., Suazo, G.H. and Fourie, A. (2015). Investigation of some controllable factors that impact the stress state in cemented paste backfill. *Canadian Geotechnical Journal*, 52, 1901–1912.

Edmonton, J.M. and Paterson, M.S. (1972). Volume changes during the deformation of rocks at high pressures. *International Journal of Rock Mechanics and Mining Sciences*, 9, 161–182.

Eringen, A.C. (1980). *Mechanics of Continua, R.E.* Krieger Pub. Co., New York.

Fowler, C.M.R. (1990). *The Solid Earth – An Introduction to Global Geophysics*, Cambridge University Press, Cambridge.

Fresnel, A. (1820). Resume d'une Memoire sur la reflexion da la lumiere. *Annales de Chimie et de Physique*, XV, 379–386.

Frocht, M.M. (1965). Photoelasticity. John Wiley and Sons, London.

Fujii, Y., Ishijima, Y., Ichihara, Y., Kiyama, K., Kumakura, S., Takada, M., Sugawara, T., Narita, T., Kodama, J., Sawada, M. and Nakata, E. 2011. Mechanical properties of abandoned and closed roadways in the Kushiro Coal Mine, Japan. *International Journal of Rock Mechanics and Mining Sciences*, 48(4), 585–596.

Geniş, M. and Aydan, Ö. (2006). Assessment of dynamic stability of an abandoned room and pillar underground lignite mine. *8th Regional Rock Mechanics Symposium*, Istanbul, 29–38.

Geniş, M. and Aydan, Ö. (2008). Assessment of dynamic response and stability of an abandoned room and pillar underground lignite mine. *The 12th International Conference of International Association for Computer Methods and Advances in Geomechanics (IACMAG)*, Goa, India, 3899–3906.

Geniş, M. and Aydan, Ö. (2013). A numerical study on the ground amplifications in areas above abandoned room and pillar mines and longwall old mines Assessment of dynamic response and stability of an abandoned room and pillar underground lignite mine. *EUROCK*, 2013, Wroclaw, Poland, CRC Press, 733–737.

Geniş, M. and Aydan, Ö. (2022). Dynamic analyses of abandoned mines in Mitake (Japan) during earthquakes. *Environmental Geotechnics*, 9(3), 159–170.

Geniş, M., Tokashiki, N. and Aydan, Ö. (2009). The stability assessment of karstic caves beneath Gushikawa Castle remains (Japan). *EUROCK2009 ISRM Regional Symposium – Rock Engineering in Difficult Ground Conditions-Soft Rocks and Karst*, Vrjklan (ed.), Dubrovnik, Croatia, 449–454.

Geographical Information Authority of Japan (GSI). (2011). Special web site for East Japan mega earthquake. https://www.gsi.go.jp/

Ghabraie, B. (2016). Multi-seam mining-induced ground surface subsidence, characteristics and prediction. PhD thesis, RMT University. https://researchbank.rmit.edu.au/eserv/rmit:161862/Ghabraie.pdf.

Grice, A.G. (1998). Underground mining with backfill. *Proceedings of the 2nd Annual Summit—Mine Tailings Disposal Systems*, Brisbane, Australia, 234–239.

Griggs, D. (1939). Creep of rocks. *The Journal of Geology* 47, No.3, pp 225–251.

Hast, N. (1969). The state of stress in the upper part of the Earth's crust. *Tectonophysics*, 8, 169–211.

Hawkins, A.B. and McDonald, C. (1994). Hazards associated with old stone mine workings: A case history from Combe Down, Bath. In *Risk and Reliability in Ground Engineering*, B.O. Skipp (ed.), 215–226. Thomas Telford.

Herget, G. (1986). Changes of ground stresses with depth in the Canadian shield. *International Symposium on Rock Stress and Rock Stress Measurements*, Stockholm, 61–68.

Hiramatsu, Y., Okamura, H. and Sugawara, K. (1979). Surface subsidence and horizontal displacament caused by mining induced coal seams. *Proceedings of the 4th Congress International Society for Rock Mechanics*, Balkema, Vol. 1, 665–670.

Hirth, G. and Tullis, J. (1994). The brittle-plastic transition in experimentally deformed quartz aggregates. *Journal Geophysical Research*, 99, 11731–11747.

Hoek, E. (2007). *Practical Rock Engineering*. Rocscience, Toronto, Canada.

Hoek, E. and Brown, E.T. (1980). Empirical strength criterion for rock mass. *Journal of the Geotechnical Engineering Division*, 106, 1013–1035.

Hoek, E. and Brown, E. (1997). Practical estimates of rock mass strength. *International Journal of Rock Mechanics and Mining Sciences*, 34, 1165–1186.

Hubbert, M.K. (1951). Mechanical basis for certain familiar geological structures. *Bulletin of the Geological Society of America*, 62, 355–372.

Ikeda, K. (1970). A classification of rock conditions for tunnelling. *1st International Congress of Engineering* Geology, IAEG, Paris, 1258–1265.

Itasca. (2005). FLAC3D-Fast Lagrangian Analysis of Continua-User Manual (Dynamic Option) *(Version 2.1)*. Itasca Consulting Group Inc., Minneapolis.

Ito, T., Akagi, T., Aydan, Ö., Ulusay, R. and Seiki, T. (2016). Time-dependent properties of tuffs of Cappadocia, Turkey. *EUROCK2016*, Ürgüp, 229–234.

Jaeger, J.C. and Cook, N.G.W. (1979). *Fundamentals of Rock Mechanics*. Chapman and Hall Ltd, London.

Jeffreys, H. and Bullen, K.E. (1940). *Seismological Tables*. British Association for the Advancement of Science, London.

Kaiser, J. (1953). Erkenntnisse und Folgerungen aus der Messung von Gerauschen bei Zugbeanspruchung von metallischen Werksto#en, *Archiv fur das Eisenhuttenwessen*, 24, 43–45.

Kanagawa, T., Hayashi, M. and Kitahara, Y. (1981). Acoustic emission and overcoring methods. *Proceedings of the International Symposium on Weak Rock*, Tokyo, 1205–1210.

Kaptan, E. (1990). Findings related to the history of mining in Turkey. *Bulletin of the Mineral Research and Exploration*, 111, 75–84.

Karfakis, M.G. and Topuz, E. (1991). Post mining subsidence abatements in Wyoming abandoned coal mines. *Mining Science and Technology*, 12. 215–231.

von Karman, T. (1911). Testigkeitsversuche unter allseitigem Druck. *Zeitschrift des Vereines Deutscher Ingenieure*, 55, 1749–1757.

Kastner, H.. (1962). *Statik des Tunnel-und Stollenbaues*. Springer, Berlin Heidelberg, New York.

Kawamoto, T. 2005. Damage to the ground caused by caving and investigation of underground openings (in Japanese). *Butsuri-Tansa*, 58(6), 589–597.

Kawamoto, T., Aydan, Ö. and Tsuchiyama, S. (1991). A consideration on the local instability of large underground openings. *International Conferences*, GEOMECHANICS'91, Hradec, 33–41.

Kawamoto, T., Aydan, Ö., Okura, K. and Fujii, K. (2004). Reconnaissance report on the sinkholes in the abandoned lignite mine area caused by the 2003 Miyagi-Hokubu earthquake (in Japanese). Japan Filling Association (Nihon Juten Kyokai), 36p.

KiK-net. (2011). https://www.kyoshin.bosai.go.jp/kyoshin/

Kirsch, G. (1898), Die Theorie der Elastizitat und die Bedurfnisse der Festigkeitslehre. *Zantralblatt Verlin Deutscher Ingenieure*, 42, 797–807.

K-NET. (2011). https://www.kyoshin.bosai.go.jp/kyoshin/

K- Net NIED. (2004, 2005, 2011). Strong motion records of 2004 Tokaido-oki, 2005 Komaki and Great East Japan earthquakes. https://www.kyoshin.bosai.go.jp/kyoshin/

Koketsu, K. (2012). Seismological and geodetic aspects of the 2011 Tohoku earthquake and Great East Japan Disaster. *Proceedings of the International Symposium on Engineering Lessons Learned from the 2011 Great East Japan Earthquake*, Tokyo, 1–8.

Lanaro, F. and Matsui, H. (2008). BEM-DDM modelling of rock damage and its implications on rock laboratory strength and in-situ stresses. Japan Atomic Energy Agency, 2007-093, 237.

Matsui, H. and Hirano, T. (2010). Preliminary validation of conceptual model developed during the subsurface-based investigation phase in the Mizunami URL Project. *Proceedings of the 39th Rock Mechanics Symposium*, JSCE, Tokyo, 48–53.

Maury, V. (1969). *Mechanique des Milieux Stratifies, Experiences et Calcuis*, Dunod, Paris, Vol. 9, 128p.

Nadai, A.L. (1950). Theory of Flow and Fracture of Solids, Vol. II, McGaw-Hill, New York, Toronto, London, 572p.

Nishida, N., Esaki, T., Aoki, K. and Kameda, K. (1984). Evaluation and prediction of subsidence in old working areas and practical and preventive measures against mining damage to new structures. *Proceedings of Third International Symposium on Land Subsidence, Venice*, A.I. Johnson, L. Carbognin and L. Ubertini (eds.), 717–725. *International Association of Hydraulic Sciences*.

Ohta, Y., Aydan, Ö., Sakamoto, A., Sugiura, K. and Yamada, N. (2005). Failure modes of abandoned lignite mines induced by earthquakes and the evaluation of their stability (in Japanese). *Proceedings of the 35th Rock Mechanics Symposium, JSCE*, Tokyo, 57–62.

Poulsen, B.A. and Shen, B. (2013). Subsidence risk assessment of decommissioned bord-and-pillar collieries. *International Journal of Rock Mechanics and Mining*. 60, 312–320.

Rocscience. (1998). *Phase2' User Guide*. RocScience Inc., Toronto, Ontario, Canada.

Sakamoto, A. (2007). A study on ground failures and subsidence due to existing underground cavities and counter-measures. Doctorate Thesis, Waseda University.

Sakamoto, A., Yamada, N., Iwaki, K. and Kawamoto, T. (2005b). Applicability of re-cycling materials to cavity filling materials. *Zairyo, Japan Society of Material Science*, 15(11), 1123–1128.

Sakamoto, A., Yamada, N., Sugiura, K., Aydan, Ö., Tano, H. and Hamada, M. (2006). Failure modes and responses of abandoned lignite mines induced by earthquakes and the evaluation of their stability. *4th Asian Rock Mechanics Symposium*, Singapore, Paper No. A0493 (on CD).

Sakamoto, A., Yamada, N., Sugiura, K. and Kawamoto, T. (2005a). Some examples of the cavity filling along transportation routes above abandoned room and pillar lignite mines in Tokai Region. *Post-Mining* 2005, Nancy, 10p, GISOS.

Salustowicz, A. (1968). Einige Bemerkungen über den Spannungszustand im Inneren der Erdkugel. *Felsmekanik*, Suppl. 7, 93–98.

Shimada, M. (1993). Two types of brittle fracture of silicate rocks and scale effect on rock strength: their implications in the earth crust. *Scale Effects in Rock Masses*, Lisbon, 55–62.

Sibson, R. (1974). Frictional constraints on thrust, wrench and normal faults. *Nature*, 249, 542–544.

Singh, D.P. (1975). A study of the creep of rocks. *International Journal of Rock Mechanics and Mining*, 12(9), 271–276.

Stephanson, O, Sarkka, P. and Myrvang, A. (1986). State of stress in Fennoscandia. *International Symposium on Rock Stress and Rock Stress Measurements*, Stockholm, 21–32.

Stone D (2014). The evolution of paste for backfill. *Proceedings of the 11th International Symposium on Mining with Backfill: Mine Fill 2014*, Australia, 31–38.

Sugawara, K. and Obara, Y. (1993). Measuring rock stress. *Comprehensive Rock Engineering*, Pergamon Press, Chapter 21, Vol. 3, 533–552.

Sugito, M., Furumoto, Y. and Sugiyama, T (2000). Strong motion prediction on rock surface by superposed evolutionary spectra. *12th World Conference on Earthquake Engineering, 2111/4/A*, CD-ROM, Christchurch.

Sugiura, K. and Yamada, N. (2003). On the completion of the cavity filling works of Tokai By-Pass Expressway (in Japanese). *Juten, Journal of Japan Grouting (Juten) Association*. 44, 8–17.

Sugiura, K., Ishiai, N. and Wada, S. (2005). Cavity filling in urbanized area: Changes and state of construction environment in cavity filling works (in Japanese). *Juten, Journal of Japan Grouting (Juten) Association*. 47, 8–13.

Sugiura, K., Sakamoto, A., Tano, H. and Aydan, Ö. (2019). An integrated system for the cave-filling of an abandoned lignite mine below Kyowa Secondary School in Mitake, Japan against an anticipated mega-earthquake. *Proceedings of 2019 Rock Dynamics Summit in Okinawa*, 428–433.

Swanson, P., Kenner, B. and Todd Krahenbuhl, T. (2002). Seismic event data acquisition and processing: distribution and coordination across PC-based networks. *Application of Computers and Operations Research in the Mineral Industry: Proceedings of the 30th International Symposium (APCOM 2002)*, Littleton, 637–647.

Tano, H., Abe, T. and Aydan, Ö. (2005). The development of an in-situ AE monitoring system and its application to rock engineering with particular emphasis on tunneling. *ITA Conference*, Istanbul, 1245–1252.

Tokashiki, N., Akagi, T., Ito, T. and Aydan, Ö. (2004). An experimental study on the short and long-term behavior of pillars of Ryukyu limestone. *3rd Asian Rock Mechanics Symposium*, Kyoto, Vol. 2, 885–890.

Tokashiki, N., Aydan, Ö, Akagi, T. and Ito, T. (2005). An experimental study on the short and long-term behaviour of pillars. *The 34th* Japan Rock Mechanics Symposium, Tokyo, Paper No. 50, 459–464 (in Japanese),

Ulusay, R. (2014). The ISRM Suggested Methods for Rock Characterization, Testing and Monitoring: 2007–2014, Springer, 293p.

Ulusay, R. and Hudson, J.A. (ed.) (2007). *The complete ISRM suggested methods for rock characterization, testing and monitoring: 1974–2006*. ISRM Turkish National Group, Ankara, 628p.

Walton, G. and Cobb, A.E. (1984). Mining subsidence. *Ground Movements and Their Effects on Structures*, P.B. Attewell and R.K. Taylor (eds.), Surrey University Press, Chapman and Hall, 216–241.

Watanabe, H., Tano, H., Ulusay, R., Yüzer, E., Erdoğan, M. and Aydan, Ö. (1999). Initial stress state in Cappadocia. *Japan-Korea Rock Engineering* Symposium, Fukuoka, Japan.

Whittaker, B.N. and Reddish, D.J. (1989). *Subsidence: Occurrence, Prediction and Control*. Elsevier Science Publishing Company Inc., New York.

Zoback, M.D., Mooss, D., Mastin, L. and Anderson, R. (1985). Well bore breakouts and in-situ stress. *Journal of Geophysical Research*, 90, 5523–5530.

Index

For Product Safety Concerns and Information please contact our EU
representative GPSR@taylorandfrancis.com
Taylor & Francis Verlag GmbH, Kaufingerstraße 24, 80331 München, Germany

www.ingramcontent.com/pod-product-compliance
Ingram Content Group UK Ltd.
Pitfield, Milton Keynes, MK11 3LW, UK
UKHW052212190926
475702UK00006B/984

* 9 7 8 1 0 3 2 8 5 0 7 3 3 *